AF807239

Transcription Factors for Biotic Stress Tolerance in Plants

Shabir Hussain Wani
Vennampally Nataraj • Gyanendra Pratap Singh
Editors

Transcription Factors for Biotic Stress Tolerance in Plants

Springer

Editors
Shabir Hussain Wani
MRCFC Khudwani
Sher e Kashmir University of Agricultural
Sciences and Technology
Srinagar, Jammu and Kashmir, India

Gyanendra Pratap Singh
ICAR-Indian Institute of Wheat
and Barley Research
Karnal, India

Vennampally Nataraj
Division of Crop Improvement
ICAR-Indian Institute of Soybean Research
Indore, India

ISBN 978-3-031-12989-6 ISBN 978-3-031-12990-2 (eBook)
https://doi.org/10.1007/978-3-031-12990-2

© The Editor(s) (if applicable) and The Author(s), under exclusive license to Springer Nature Switzerland AG 2022
This work is subject to copyright. All rights are solely and exclusively licensed by the Publisher, whether the whole or part of the material is concerned, specifically the rights of translation, reprinting, reuse of illustrations, recitation, broadcasting, reproduction on microfilms or in any other physical way, and transmission or information storage and retrieval, electronic adaptation, computer software, or by similar or dissimilar methodology now known or hereafter developed.
The use of general descriptive names, registered names, trademarks, service marks, etc. in this publication does not imply, even in the absence of a specific statement, that such names are exempt from the relevant protective laws and regulations and therefore free for general use.
The publisher, the authors, and the editors are safe to assume that the advice and information in this book are believed to be true and accurate at the date of publication. Neither the publisher nor the authors or the editors give a warranty, expressed or implied, with respect to the material contained herein or for any errors or omissions that may have been made. The publisher remains neutral with regard to jurisdictional claims in published maps and institutional affiliations.

This Springer imprint is published by the registered company Springer Nature Switzerland AG
The registered company address is: Gewerbestrasse 11, 6330 Cham, Switzerland

Preface

With the erratic changes in climate, crop plants are facing many forms of biotic stresses. Employing genetic resistance in their management is the most economical, effective, and eco-friendly approach. However, limited genetic variation in the gene pool is hindering the rapid progress in the field of plant genetic resistance. Moreover, major resistance genes are knocked-down due to continuous evolution of novel virulent races/biotypes. Therefore, the concept of durable resistance is ever lasting since ages in management of biotic stresses. Under natural conditions, plants face different biotic and abiotic stresses simultaneously. Therefore, broad spectrum resistance and resistance against multiple stress forms can be of prime focus to combat economic yield losses. When plants are under stress, among several gene families, regulatory genes play a vital role in signal transduction in modulating the expression of genes underpinning several defense pathways, and targeting regulatory proteins (viz, transcription factors (TFs)) can be the alternative. Transcription factors directly regulate the downstream R genes and are excellent candidates for disease resistance breeding. Till date, numerous transcription factors have been identified and characterized structurally and functionally. Of them, TF families, such as WRKY, NAC, Whirly, Apetala2 (AP2), and ethylene responsive elements (ERF), are found to be associated with transcriptional reprogramming of plant defense response. These TFs are responsive to the pathogen's PAMPs/DAMPs – host's PRR protein interactions, and specifically bind to the *cis*-elements of defense genes and regulate their expression. With this background, realizing the importance of TFs in resistance breeding, current book has been proposed.

This book provides an authoritative review account of different aspects and progress in the field that have been made in the recent past. Book includes chapters prepared by specialists and subject experts on different aspects of gene editing techniques, role of synthetic promoters and microbial bio-agents as elicitors in plant defense regulation, and role of TFs in disease resistance. The first chapter introduces various genome editing techniques, whereas six chapters deal with the role of TFs in biotic stresses in crops like wheat, sugarcane, maize, pearl millet, tomato, and potato. Three chapters are exclusively about the transcription factors associated with defense response against fungal biotrophs, necrotrophs, and viruses. One chapter is exclusively about the synthetic promoters in regulating disease gene expression and one chapter about the role of microbial bio-agents as elicitors in plant defense regulation

The book provides state-of-the-art information on the potential of TFs in supplementing and complimenting the conventional methods of crop improvement against biotic stresses. We earnestly feel that this book will be highly useful for students, research scholars, and scientists working in the in the area of crop improvement and biotechnology at universities, research institutes, R&Ds of agricultural MNCs for conducting research, and various funding agencies for planning future strategies.

We are highly grateful to all learned contributors, each of whom has attempted to update scientific information of their respective area and expertise and has kindly spared valuable time and knowledge.

We apologize wholeheartedly for any mistakes, omissions, or failure to acknowledge fully.

We would like to thank our families (Sheikh Shazia and Muhammad Saad Wani (wife and son of SHW), Keerthi and Adhvay Rishi (wife and son of NV)) for their continuous support and encouragement throughout the completion of this book.

We highly appreciate the all-round cooperation and support of Springer International Publishing AG, Cham for their careful and speedy publication of this book.

Srinagar, India — Shabir Hussain Wani
Indore, India — Vennampally Nataraj
Karnal, India — Gyanendra Pratap Singh

Contents

1 **Targeted Genome-Editing Techniques in Plant Defense Regulation** . 1
Vineeta Dixit and Priti Upadhyay

2 **Synthetic Promoters in Regulating Disease Gene Expression** 33
Badrinath Khadanga, Tsheten Sherpa, Jeky Chanwala, and Nrisingha Dey

3 **Transcription Factors Associated with Defense Response Against Fungal Biotrophs**. 51
Thirunarayanan Permal, Tulasi Korra, Sushree Suparna Manopatra, K. Manikandan, and Jyostrani Prathan

4 **Transcription Factors Associated with Defense Response Against Fungal Necrotrophs** 61
L. Mathan, Namo Dubey, Swati Verma, and Kunal Singh

5 **Role of Plant Transcription Factors in Virus Stress** 79
Bipasha Bhattacharjee and Vipin Hallan

6 **Role of Microbial Bioagents as Elicitors in Plant Defense Regulation** 103
Mukesh Meena, Garima Yadav, Priyankaraj Sonigra, Adhishree Nagda, Tushar Mehta, Andleeb Zehra, and Prashant Swapnil

7 **Transcriptional Factors' Response Under Biotic Stress in Wheat**. . . 129
Fahad Masoud Wattoo, Rashid Mehmood Rana, and Sajid Fiaz

8 **Potential Transcription Factors for Biotic Stress Tolerance in Sugarcane** 143
Moutoshi Chakraborty, Saurab Kishore Munshi, Tofazzal Islam, and Muhammad J. A. Shiddiky

9 The Role of Transcription Factors in Response to Biotic Stresses in Maize 175
Jyoti Prakash Sahoo, Deepali Dash, Ankit Moharana, Manaswini Mahapatra, Amit Kumar Sahoo, and Kailash Chandra Samal

10 The Role of Transcription Factors in Response to Biotic Stresses in Pearl Millet 195
Jeky Chanwala, Deepak Kumar Jha, I. Sriram Sandeep, and Nrisingha Dey

11 The Role of Transcription Factors in Response to Biotic Stresses in Tomato 213
Lopamudra Jena, Subhasmita Sahu, Pranaya Pradhan, Kailash Chandra Samal, Jyoti Prakash Sahoo, Laxmipreeya Behera, and Siddhartha Kumar Palai

12 The Role of Transcription Factors in Response to Biotic Stresses in Potato (*Solanum tuberosum* L.) 235
Namo Dubey and Kunal Singh

Chapter 1
Targeted Genome-Editing Techniques in Plant Defense Regulation

Vineeta Dixit and Priti Upadhyay

Abstract Domestication of crop plants coexisted with human civilisation. With the progress in the scientific arena, the skill to modify the plant characteristic sharpened, and new tools and techniques are searched and invented almost every decade to meet the nutraceutical, economical or agronomical needs. Improper selection method was successfully replaced by conventional breeding of distant crop population. While conventional breeding techniques depend on ambiguous needs of rigorous selection after successful crossing between likely close species, advanced genetic engineering methods that have the ability to modify the genome need stable integration of foreign desired genes, whereas recently evolved targeted genome editing entails breaking particular sequences with sequence specificity in the target DNA and incorporating modifications during the repair process. At the moment, targeted genome-editing technologies provide the most modern biotechnological approaches for accurate, effective and precise site-specific genome change in an organism. In a range of plant species, genome-editing technologies have been used to improve certain features in order to increase agricultural yield and build resilience and adaptive capacity and disease proliferation. This chapter discusses the current uses of genome editing in plants, with an emphasis on its prospective applications for defensive management against diverse stressful conditions, resilient growth and hence enhanced end-use. The future potential for merging this breakthrough technique with traditional and next-generation breeding strategies, as well as novel breakthroughs that are broadening the possibilities of genome-edited crops, is also discussed.

Keywords CRISPR · Meganuclease · Stress · TALENS · Zinc-finger nuclease

Vineeta Dixit and Priti Upadhyay contributed equally with all other contributors.

V. Dixit
Department of Botany, Guru Ghasidas University, Bilaspur, India

P. Upadhyay (✉)
Department of Botany, University of Delhi, New Delhi, India

© The Author(s), under exclusive license to Springer Nature Switzerland AG 2022
S. H. Wani et al. (eds.), *Transcription Factors for Biotic Stress Tolerance in Plants*, https://doi.org/10.1007/978-3-031-12990-2_1

1.1 Introduction

Domestication of crop plants coexisted with human civilisation. With the progress in the scientific arena, the skill to modify the plant characteristic sharpened, and various tools and techniques are searched and invented almost every decade to meet the nutraceutical, economical or agronomical needs. Improper selection method was successfully replaced by conventional breeding of distant crop population. While conventional breeding techniques depend on ambiguous needs of rigorous selection after successful crossing between likely related/distant species, advanced genetic engineering methods have the ability to modify the genome with stable integration of foreign desired genes. Narrow genetic base of plant species was broadened using mutagenetic tools, and later other modern techniques were used to create target-specific variations. Recombinases, transposons and TILLING technologies, in addition to chemical mutagens, were utilised in functional genomics and reverse genetic investigations. A special objective of molecular and plant biologists was/is induced variation at target locus. In the last few decades, considerable improvement has been observed in the field of targeted genome modifications. Diverse fields of genetics and life science including human genetics, clinical genetics, gene therapy, precision medicine, synthetic biology, drug development, plant biology and agricultural research have utilised them and produced the desired set of traits. Gen/Ed (gene/genome editing) tools at present are the most advanced and preferred applications that facilitate specific and efficient site-specific amendments in a chosen genome/organism. Gene editing utilising locus-specific nucleases enables for rapid and accurate reverse genetics, genome remodelling and targeted transgene insertion (Bortesi and Fischer 2015). Genome-edited GMO tagged crops are subjected to a variety of biosafety issues, and differences in regulatory legislation between countries provide significant impediments to the quick adoption of new GM features (Prado et al. 2014), limiting the benefits of GM traits to a small number of commercial crops. Targeted Gen/Ed produces sequence-specific nicks in the target DNA, and specific edits are incorporated during repair, and thus products of Gen/Ed can be designed for non-GMO tag (genetically modified organism). These approaches produce modifications that are only a few nucleotides long and mimic spontaneous mutation in the crop, implying that they potentially pose fewer risks than GMO crops (Voytas and Gao 2014). Thus, incorporating genome editing by Gen/Ed into contemporary breeding programmes would allow for expedited and accurate crop improvement, ensuring that future food demand is met and food security is assured. Plant breeding can employ a gene–/genome-editing system to make point mutations that mimic natural SNPs, integrate foreign genes, adjust gene function, gene pyramiding and knockout and inhibit or activate gene expression, as well as epigenetic editing (Kamburova et al. 2017). With advances in sequencing technology, genomic information on an increasing number of plant species is becoming available, enabling genome-editing tools for precise gene editing in a wide range of crops and opening up new avenues for modern agriculture.

Gene editing (Gen/Ed) are broadly based on either DNA-guided editing and RNA-guided editing mechanism. The core technologies now most commonly used to facilitate DNA-guided genome editing are (1) meganucleases or homing endonucleases, (2) TALENs (transcription activator-like effector nucleases) and (3) ZFNs (zinc-finger nucleases). CRISPR (clustered regularly interspaced short palindromic repeats) and CRISPR-associated protein such as CRISPR/Cas 9 are solely based on RNA-guided editing mechanism. All the aforementioned Gen/Ed tools have the potential to catalyse the formation of double-strand breaks (DSBs) at the target DNA sequence, which activates cellular DNA repair mechanisms and enables the incorporation of site-specific genetic alterations (Rouet et al. 1994; Choulika et al. 1995). DNA repair can be achieved either through homologous recombination (HR) or non-homologous end joining (NHEJ). The artificial template provided by DSB-stimulated gene targeting is an exogenous template for a natural repair mechanism. The HR approach uses a homologous donor DNA segment as a template, and homologous recombination is employed to repair the DSB. This process might be used to perform precise gene changes or gene insertions. DSBs stimulate both mutagenesis and gene replacement locally in most organisms, including higher plants, even though the generation of breaks in both DNA strands induces recombination at specific genomic loci. In most organisms, including higher plants, NHEJ is the most common DSB repair process, whereas targeted integration by HR is significantly less common than random integration (Puchta 2005). In non-homologous end joining, broken ends are commonly joined erroneously, generating random indels (insertions or deletions) and substitutions at the break site. Thus, NHEJ is expected to cause frameshift mutations in the majority of cases and, if it happens in a gene's coding domain, can essentially result in a gene knockout. If overhangs are generated in the DSB, NHEJ can manage the targeted introduction of a DNA template with compatible overhangs efficiently (Cristea et al. 2013; Maresca et al. 2013). Other strategies, including the use of negative selection markers outside the homology region of the insertion cassette to avoid random integration events, or overexpressing proteins engaged in HR, can result in modest improvements in gene targeting efficiency (reviewed in Puchta and Fauser 2013). The design and cloning of targeted nucleases have become easier as a result of freely available software tools and knowledge, expanding the capacity of medium-funded laboratories. In addition to ZFNs, TALENs and CRISPR, other designed nucleases like homing endonucleases or meganucleases have been employed for targeted Gen/Ed (Roth et al. 2012), although their application is limited in contrast to the aforementioned nucleases. In this chapter, we first go through the many genome-editing techniques that are utilised for precise editing in plants, as well as their strengths and limitations. The possible uses of each technology for defensive regulation and resilient development in various plant species are then discussed.

1.2 Homing Endonucleases or Meganucleases

Homing endonucleases (HEs) or meganucleases are found in microbes that are enzymes that generate double-strand breaks at specified genomic invasion locations to mobilise their own reading frames (Fig. 1.1) and thus splitting DNA at particular sequence. HEs are molecular scissor proteins that display economies of size with an attribute to recognise long DNA sequences (typically 14–40 base pairs) (Belfort and Roberts 1997); hence, these are sequence-specific endonucleases (SSN) (Arbuthnot 2015). HEs may break double-stranded DNA at particular identified base pairs and have a broad range of precision at individual nucleotide sites having significant effect of host constraints on the targeted gene's coding sequence. These proteins' action creates recombination interactions that are very much site specific and it may produce DNA mutation having different mechanisms like insertion, deletion, etc. Researchers have been working on these proteins for over 15 years, and they have solved the crystal structure of various homing endonuclease families. Since the mechanism of creating variations by applying these enzymes is known and also that these cleave and create novel DNA targets, engineered homing endonuclease proteins are currently being employed in a number of biotech and medicinal applications to induce targeted genomic alterations.

Unlike restriction enzymes, which protect microbes from invasive DNA, HEs let genetic components to move around freely within an organism. HEs get their name from the process, which is known as "homing", a self-splicing mechanism where intervening sequence of group I or group II introns or inteins is precisely replicated into host gene receiver alleles that lack such a sequence (Belfort and Perlman 1995; Belfort and Roberts 1997; Chevalier and Stoddard 2001; Dujon 1989).

Homing endonuclease's (HE's) presence has been documented in all three biological kingdoms. Studies on budding yeast in the 1970s provided the first evidence of the presence of HEs (Belfort and Robert 1997). In another study in yeast, the transmission of the genetic marker omega (ω), that was reported as a group I intron of large ribosomal RNA, among yeast strains was proven (Chevalier and Stoddard 2001). The production of double-strand breaks (DSBs) at specific spots was used to

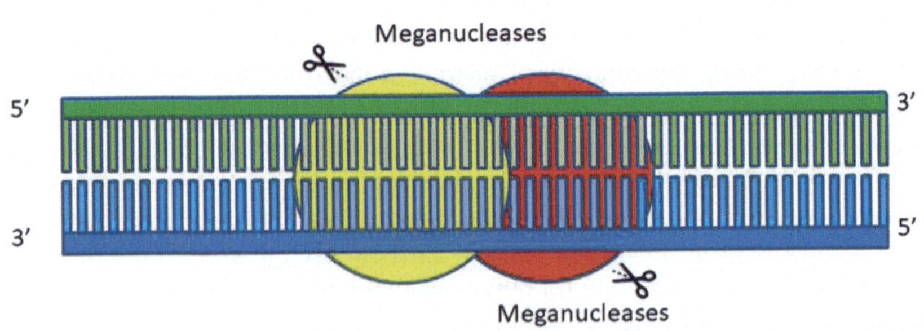

Fig. 1.1 Schematic representation to show mechanism of meganuclease gene-editing system. A meganuclease *has a homodimer structure*. Meganucleases are highly specific and easy to deliver to cells but difficult to redesign for new targets

transfer the genetic material, and the study discovered that the endonuclease responsible for the split was encoded by own DNA sequences of the group I intron. The first of numerous HEs to be characterised was I-SceI. At the target, cleft or break was followed by homology-directed repair which resulted in the intron sequence being introduced into the "intronless" target. In the target sites, there is some tolerance for sequence variation, which is considered to be crucial for homo endonucleases to accept a variation in the frequency of an existing gene variation in the population of a host organism (genetic drift). Degeneracy is tolerated at places that coincide with the wobble positions of protein-coding regions, which is an interesting coincidence.

It has wide application in targeted gene editing as it has an attribute of sequence specificity. The efficiency and success of sequence insertion mediated by homologous recombination employing homing endonucleases in maize were investigated by induction of a targeted DNA double-strand break at the desired integration location, and numerous significant numbers of carefully designed events were discovered in maize DNA where integration happened in extremely correct way with improved and optimised protocol with I-SceI gene for expression. This improved procedure worked for both *Agrobacterium* and particle bombardment DNA delivery methods, but the results indicated that targeted double-strand break-induced homologous recombination is an effective way to ensure precise changes in the maize genome and that targeted genome alteration of agronomic crops is possible (D'Halluin et al. 2008).

A transgene integrated with intrins was inserted at the exact locus using meganuclease in the model plant *Arabidopsis* (*A. thaliana*) to achieve an independent (not affected by transformation methods) and effective targeted insertion that established the development of premeditated endonucleases with site specificity. It was considered that such targeted insertion may boost the establishment of gene targeting (GT) techniques in a variety of species. Research in this emerging field of modifying gene is growing day by day, and a patent has been submitted in the United Kingdom for an engineering technique of I-CreI homing endonuclease variants capable of cleaving mutant I-CreI sites with variations in positions 8–10. An I-CreI homing endonuclease variation obtained by this strategy resulted in phytophthora-resistant potato with enhanced yield, as demonstrated by experiments in potato (Hogler and Timo 2012).

Meganuclease mutants are easily accessible and may be successfully used in plants for precise genetic alteration. Meganucleases are smaller (40 kD) than ZFNs and TALENs, which enables them to be used in vectors with smaller coding sequences specially that belongs to viruses (Iqbal et al. 2020). However, due to several restrictions, such as DNA binding and cleavage domains overlap (Stoddard 2011) that cause compromised catalytic activity of meganuclease, lack of the modular DNA-binding domain design and sometimes issue of sequence degeneracy for meganuclease, their use in genome editing/engineering is not as widespread as ZFNs or TALENs (Argast et al. 1998).

1.2.1 Zinc-Finger Nucleases

Plant phenotypes are the outcome of a complex array of biochemical, physiological and developmental processes culminating in physical appearance. All these activities are essentially governed by nucleotide base sequences found in nuclear, plastid and mitochondrial genomes, which supply both configurational and regulatory instructions to the live cell and, as a result, the growing organism. However, while the nucleotide sequences found in live creatures are similar, they differ from one another owing to changes within and recombinations among these sequences. The phenotypic variety observed across organisms is based on variations in their sequence and structure (Petolino 2015). Plant breeders can use naturally occurring and/or produced sequence changes and recombinations after analysing the sequence information. Plant breeders can use naturally occurring and/or produced sequence changes and recombinations after analysing the sequence information. They can adjust or alter the nucleotide sequence to suit their needs and change the phenotype. As a result, significant progress may be made in terms of improving the quality and performance of crops for agricultural and industrial purposes.

Sequences on DNA can be altered by using molecular scissors, and there are many present in living system. ZFNs (zinc-finger nucleases) are a type of DNA-binding protein that permits for customised genome editing by causing double-strand breaks in DNA at user-specified places. (Fig. 1.2). Individual ZFNs' DNA-binding domains generally include three to six zinc-finger repeats, each of which can identify between 9 and 18 bps (Ramirez et al. 2008). At present, most of the engineered ZFs arrays that are available are based on three individual zinc-finger domain that can recognise a nine base pair target location with high affinity (Christy and Nathan 1989). Other approaches that can build zinc-finger (ZF) arrays comprising six or more individual zinc fingers are combination of one-finger and two-finger modules (Shukla et al. 2009). A following research employed modular assembly to make zinc-finger nucleases with both three-finger and four-finger arrays, finding that the four-finger arrays had a substantially greater success rate (Kim et al. 2009).

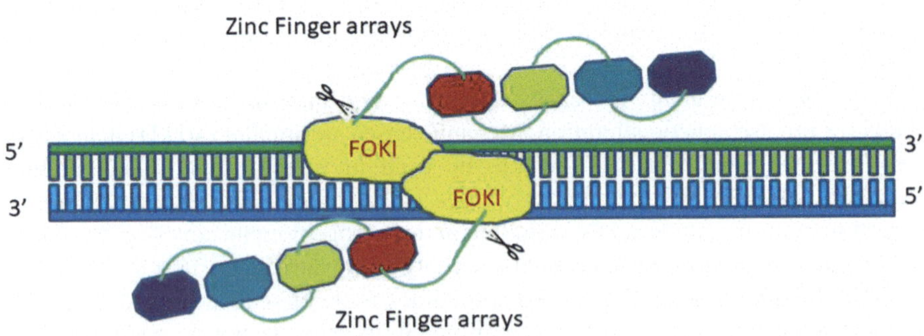

Fig. 1.2 Schematic representation to show mechanism of *zinc-finger nuclease (ZFN)* gene-editing system. *Zinc finger nuclease (ZFN) is composed of two monomers, and hexagon represent a zinc finger DNA-binding domain.* Each zinc finger typically recognises 3 bp

To construct ZF arrays capable of targeting specified sequences, a variety of selection approaches have been applied. Initially, phage display was used for identifying the proteins that bind a target DNA sequence from a huge pool of partly randomised ZF arrays, but recent research has focused on yeast one-hybrid systems, bacterial one-hybrid and two-hybrid systems and mammalian cells to select the constructed ZF arrays that are capable of targeting specified sequences (Chandrasegaran and Carroll 2016). The inventors of "oligomerised pool engineering (OPEN)", a promising new strategy for selecting innovative zinc-finger arrays, have named it after a bacterial two-hybrid system (Maeder et al. 2008). This technique combines pools of individually selected ZFs, each of which was preferred to bind a certain triplet, and then employs a second round of selection to generate three-finger arrays competent of binding a nine base pair sequence. This technique was developed by the Zinc Finger Consortium as an alternative to commercially available zinc-finger arrays.

Plant and animal genomes can both benefit from zinc-finger nucleases. In a study with *Arabidopsis*, researchers identified an effective technique for targeted mutagenesis of two genes (ADH1 and TT4) by controlling the production of zinc-finger nucleases that cause a double-strand breaks at specific target loci in DNA. The mutations produced were typically insertions of base pairs or deletions of base pairs, and the size of these varied from 1 bp to 142 bp. These mutations were found to be localised near the zinc-finger nuclease cleavage site and most probably resulted from non-homologous end-inaccuracy joining's in repairing chromosomal breaks. For about 70 percent of primary transgenics expressing the ADH1 ZFNs and around 33% of primary transgenics expressing the TT4, mutations created through use of ZFNs were passed down to the following generation. The findings revealed the applicability of ZFNs for obtaining the mutants in any target gene in *Arabidopsis* and it would have independent mutant phenotype (Zhang et al. 2010).

Zinc-finger nucleases enzymes was applied to create double-strand breaks at specific loci in acetohydroxyacid synthase (SuRA and SuRB) genes in tobacco. Some specific mutations in this gene are responsible for resistance to imidazolinone and sulfonylurea herbicides. Through this study, it was observed that utilisation of zinc-finger nuclease enzymes in tobacco was an efficient method for directed DNA sequence modifications (Townsend et al. 2009). The high rate of mutants with focused gene editing suggested that making precise sequence alterations in endogenous plant genes may be efficient. Curtin et al. (2011) employed ZFN genome engineering to target mutagenesis of nine endogenous genes and a transgene in soybean (*Glycine max*). Under an oestrogen-inducible promoter, cloning was done for specific zinc-finger nuclease targeting DICER-LIKE (DCL) genes and other genes that are involved in RNA silencing. The effectiveness of zinc-finger nuclease-induced mutagenesis at each marked locus was investigated using a hairy-root transformation technique. Transgenic roots demonstrated somatic mutations in genes DCL4a and DCL4b that were introduced through whole-plant transformation into soybean and generated independent mutation events to get mutants for seven out of nine targeted genes. The ZFN-induced mutation was efficiently heritable transmitted in the subsequent generation with the dcl4b mutation. The findings of this study indicated that mutagenesis based on zinc finger nuclease can be an efficient method for making mutations in duplicate genes.

Custom-made ZFNs have been engineered to split DNA at specific sequences and are proven to be an effective tool in targeted gene manipulations. Also, they have the unusual property of blocking dominant mutations in heterozygous people. It causes breaks in both the strands of DNA (DSBs) in the mutant allele, which are restored by non-homologous end-joining in the absence of a homologous template (NHEJ). Durai et al. (2005) explored the gene targeting utilising the zinc-finger nuclease for plant and mammalian genome and found that there is great potential of ZFNs for "directed mutagenesis" and targeted "gene editing", that makes it more applicable for ZFN-based gene therapy for human therapeutics in future. It is possible to entirely erase whole vast portions of genomic sequence using numerous pairs of ZFNs in an experiment to inhibit the mutation (Paschon et al. 2019).

ZFNs are synthetic restriction enzymes that have been utilised in *Arabidopsis* to cause mutagenesis at particular sequence or homologous recombination at the repair location, and the result showed that no gene-targeted plants were produced at the end of the experiment. The study also demonstrated that in *Agrobacterium* T-DNA constructs, ZFNs improved creation of mutation at specific location and gene targeting by fully eliminating that occurrence (de Pater et al. 2009). ZFNs can also be utilised to redraft an allele's alignment or pattern by calling a machinery of recombination, i.e. homologous in nature to repair a double-strand splits or break (DSB) using the provided DNA fragment as a template. In an individual homozygous for the concerned allele, the technique of gene targeting using ZFN's efficacy would be reduced because the undamaged copy of allele can be used as a template for repair rather than the given fragment. ZFNs have also been used in genome/gene therapy, with the effectiveness of this method relying on the precise and proper insertion of genes under therapy into an appropriate and specific chromosomal site within the human genome without causing cell damage, cancer-causing alterations or an immune response. Vectors for this technique that are plasmid based can be created easily and quickly.

Off-target cleavage and immunogenicity are two possible issues with ZFNs. When zinc finger domains lack specificity and selectivity for their particular DNA location, off-target cleavage occurs, which can lead to genomic changes that aren't wanted. This causes chromosomal rearrangements, encourages random donor DNA integration and may even be lethal to the cells (Durai et al. 2005). When multiple foreign proteins are injected into the human body, an immune reaction to the therapeutic drug has been reported. As the protein must only be produced transiently, this raises the issue of immunogenicity (Durai et al. 2005).

Despite these two drawbacks mentioned above, ZFNs' capacity to accurately change the living organism's genomes offers a variety of effective applications in fundamental and applied research such as in the field of agriculture and human health. Improved ways of creating zinc-finger domains along with better supply of ZFNs from a commercial provider have made this technology available and assessable for increasing number of researchers, and it is now being utilised in conjunction with CRISPR to enhance plant agronomic features. Artificial zinc-finger nucleases (AZFNs) were created based on the ones with the highest DNA-binding affinities for *Geminiviridae* DNA as an example of generating plants with begomovirus resistance. In vitro digestion and transient expression assays revealed that these AZFNs can effectively cleave the target sequence and suppress the reproduction of several begomoviruses (Chen et al. 2014), signifying that this strategy might be beneficial for the aforementioned goal.

1.2.2 Transcription Activator-Like Effector Nucleases (TALENs)

Transcription activator-like effector nucleases, in short TALENs, are another molecular scissor with a structure similar to ZFNs. The building block of TALENs is a highly conserved base sequence that are found to be expressed naturally in *Xanthomonas* proteobacteria as TALEs, i.e. transcription activator-like effectors. These are delivered into recipient cells of plants through a system of type III secretion, where they attach to DNA present in nucleus of cell and modify transcription, allowing harmful bacteria to colonise the cells more easily (Boch and Bonas 2010). TALEs mediate DNA binding by using arrays of highly preserved 33–35 amino acid repeats bordered by extra TALE-derived domains at the amino- and carboxy-terminal ends of the array. TALEs (DNA-binding proteins of 33–35 amino acids) (Fig. 1.3) are found in TALENs, derived from naturally existing plant pathogenic bacteria, and have ability to precisely detect one base pair of DNAs. Transcription activator-like effectors is connected together in the form of chain which may recognise and split a single location within the genome, similar to ZFNs. These nucleases are fusions of the cleavage domain *Fok*I and TALE protein-derived DNA-binding

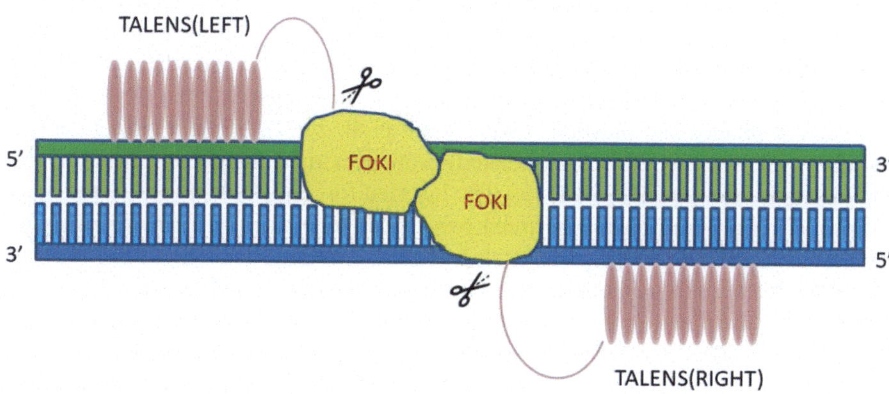

Fig. 1.3 Schematic representation to show mechanism of *transcription activator-like effector nuclease (TALEN)* gene-editing system. *TALEN comprises of two monomers, and light brown rectangles represent the DNA-binding domain.* The two TALEN target sites are typically separated by a 15–20-bp spacer sequence

domains. The structure of TALEs has depicted that it contains multiple amino acid repeat domains and each can distinguish a single base pair as ZFNs. TALENs induce double-strand splits in DNA that activate its damage response pathways and permit custom alterations like zinc-finger nuclease (Gaj et al. 2013). TALENs are comparable to ZFNs in that they can detect a single base rather than a triplet, which provides them more versatility than ZFNs (Gaj et al. 2016). Many effector domains, like as transcriptional activators and site-specific recombinases, have been created that may be joined to TALEN chains for targeted genetic alterations (Li et al. 2020).

One important difficulty with TALENs is their creation, which necessitates the assembling of many, virtually identical repeat sequences, which is a technical hurdle for a researcher (Cermak et al. 2011). Several revolutionary laboratory approaches, such as fast ligation-based automatable solid-phase high-throughput (FLASH) (Reyon et al. 2013), iterative capped assembly (ICA) (Briggs et al. 2012) and commercial DNA synthesis, have emerged as a result of this (Cermak et al. 2011). The ability to change any gene sequence quickly and effectively using TALENs assures a significant influence on research in the field of biosciences including health and agriculture, and it has the potential to boost yield potential as well as tolerance to biotic and abiotic stressors.

1.3 Clustered Regularly Interspaced Short Palindromic Repeats (CRISPR)/Cas9

The term "genome editing" is a collective form of technologies that provides biotechnologists and breeders the ability to modify an organism's DNA by adding or removing genetic material to it. Above we discussed different techniques that include different nucleases that allow the researcher to make a sequence-specific cut in genome and alter it. CRISPR and CRISPR/Cas9 are some other recent approaches that are easier, quicker, more efficient, less expensive and by far the most adaptable and simple to use, and its efficacy and accuracy have revolutionised the area of plant biology. It's a natural defence mechanism in bacteria against external DNA sources like bacteriophages and plasmids (Wiedenheft et al. 2009). When a virus infects a bacterial cell, a Cas (CRISPR-associated) protein extracts a piece of foreign DNA and inserts it into the CRISPR locus. The inserted foreign DNA, now referred to as a "spacer", is accommodated between two repeat sequences in a lengthy array of such repeat-spacer-repeat triplets, each from a distinct invader. The CRISPR array allows the bacteria to "remember" the viruses, even when the cell divides (Fig. 1.4), and thus the information is carried to the daughter cells (Horvath and Barrangou 2010; Sawyer 2013), and if the virus infects again, the bacteria employ Cas9 or a similar enzyme to cleave the virus's DNA apart, thereby rendering it inactive. Based on the structure and function of the Cas protein, the CRISPR/Cas systems may be classified into two classes (class I and class II), which can then be further divided into six types (type I–VI) (Makarova et al. 2015) among which type I, III and IV belongs to the previous class and the rest belongs to class II (Mohanraju et al. 2016).

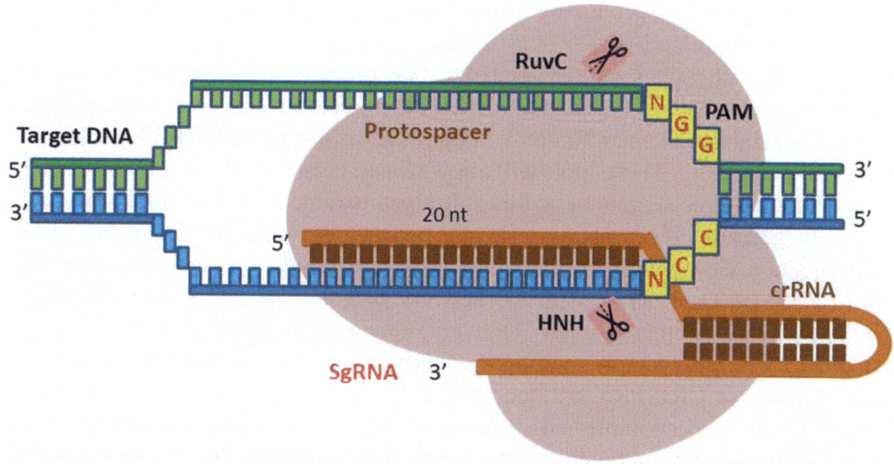

Fig. 1.4 Schematic representation to show mechanism of CRISPR/CAS9 gene-editing system. CRISPR/Cas9 system comprises a Cas9 protein (depicted in skin colour) with two nuclease domains (RuvC and HNH), and a single guide RNA (sgRNA). The sgRNA guides the Cas9 protein to the complementary sequences of the DNA target. The presence of a protospacer-adjacent motif (PAM) in yellow is a prerequisite for DNA cleavage by Cas9

Types I, II and V recognise the specific sequence in DNA and cleave it, whereas type VI has a feature to edit RNA and type III has editing attribute for both DNA and RNA (Terns 2018). Soyars et al. (2018) explored several types of Cas proteins and factors that are adjustable during optimisation of CRISPR/Cas9 systems for plants. There are several additional research and review publications that have covered the CRISPR/Cas9 approach (Wada et al. 2020; Zhu et al. 2020). Nucleases are employed in CRISPR to cleave DNA at particular sequences. Cas9 was the first nuclease found in this system that was tailed by Cpf1, discovered in *Francisella novicida*'*s* as CRISPR/Cpf1 system (Fonfara et al. 2016). CRISPR/C2c2, an RNA-guided CRISPR system with RNA as target rather than DNA, was identified later in *Leptotrichia shahii*, a bacterium. It can cleave/knock down single-stranded RNA targets (Abudayyeh et al. 2016).

Development of CRISPR/Cas9 system has permitted efficient and precise targeted mutagenesis. Because of its precision and effectiveness in altering the genome, CRISPR/Cas9 technology has exploded in popularity. CRISPR mutants were produced to examine the complete cleave gene PMR4 in tomato, which is responsible for powdery mildew pathogen susceptibility (S). For this, a CRISPR/Cas9 construct with four single-guided RNAs (sgRNAs) was applied that targeted PMR4 gene. This enhanced the likelihood of substantial deletions in mutants, as well as mutants with varying numbers of base pairs inversion, which were discovered following PCR-based transformant selection and sequencing. Visual assessment of symptoms and analysis of relative fungal biomass can be considered as the basis for grading these mutants that show a decreased sensitivity towards the pathogen. The efficacy and adaptability of this system as a valuable tool for studying and characterising susceptibility genes by producing a number of mutations were established in the investigation (Santillán Martnez et al. 2020).

Equipped with novel edited gene delivery method, these newly discovered CRISPR/Cas systems in combination with other recent technologies for targeted gene editing thus, in the near future, will increase the use of the CRISPR toolset for plant genome editing. These tools will allow researchers to explore new approaches for specific and precise genome editing. It also guarantees that no transgenes will remain in genome-edited plants once the product is produced. There are many research and review articles available that elaborated and explained the methodology of CRISPR/Cas9 along with the delivery of genes in host genome in detail (Wada et al. 2020; Zhu et al. 2020). Delivery of gene or fragment of DNA is a tough task during CRISPR genome editing. Recently many delivery methods have been experimented for CRISPR/Cas9 genome-editing technique. Construct delivery in plant cells is largely accomplished by three methods: PEG-mediated *Agrobacterium*-mediated transformation, bombardment and biolistic transformation. Various delivery systems, their efficiency and accomplishments were explored in depth in a recent review paper (Sandhya et al. 2020). The paper found that genome editing's high efficiency is dependent on a number of variables. Using *Agrobacterium*-mediated transport of CRISPR/Cas9 components, 100% editing efficiency was reported for the banana plant (Naim et al. 2018). The effectiveness of various delivery strategies is determined by the tissue type and subsequent regeneration into entire plants. The characteristics of the plant species, tissue type, and culture method all influence regeneration problems. Naim et al. (2018) also emphasised the need to develop new methods for delivering CRISPR/Cas9 components, such as nanoparticle-mediated delivery (directly into the meristematic region) and pollen-mediated delivery, which would allow researchers to skip the time-consuming and labour-intensive tissue culture. Through the development of innovative delivery techniques, CRISPR/Cas technologies in agriculture will be boosted, and crops will be transformed. This technology will also overcome ethical and regulatory barriers, as it does not require any vector DNA for editing (Sandhya et al. 2020).

Above all, the CRISPR research community's open access policy might be one of the causes for the technology's recent rise in popularity. Through Addgene (a non-profit repository), the community makes plasmids available to the public, various web tools for gRNA sequences and predicting specificity, viz. http://cbi.hzau.edu.cn/cgi-bin/CRISPR; http://www.genome.arizona.edu/crispr/; http://www.rgenome.net/cas-offinder; and http://www.e-crisp.org/E-CRISP/index.html and also do hosts for forums for discussion groups, e.g. https://groups.google.com/forum/#!forum/crispr.

1.4 Gene-Editing Tools: Comprehensive Strengths and Limitations

In theory, all GenEd methods can cause identical variation in the nuclear genome, but each one differs in terms of mechanism of action, specificity, simplicity and, of course, cost effectiveness.

Following the initial reports suggesting the use of CRISPR/Cas9 technology in plants (Feng et al. 2013; Nekrasov et al. 2013; Jiang et al. 2013; Xie and Yang 2013), a large number of reports based on the CRISPR/Cas9 technology have found their way into PubMed, clearly demonstrating that CRISPR technology has outperformed all other Gen/Ed tools in the plant world. CRISPR/Cas9 has made ripples in the scientific world as a ground-breaking genome editing tool, even winning the Nobel Prize in Chemistry in 2020. Agronomic trait manipulation necessitates coordinated genetic regulation of several genes to manage the complicated metabolic network required for trait expression. As a result, CRISPR/Cas technology with multiplexing capabilities (several target sites may be edited at the same time) has leapfrogged the competition and shown to be extremely useful in both fundamental research and commercial applications. Several research papers have used Golden Gate-related cloning or the Gibson Assembly technique to integrate several sgRNAs into single Cas9/sgRNA expression vectors, with multiple sgRNAs driven by distinct promoters (Engler et al. 2008). A generic methodology for the synthesis of sgRNA from a polycistronic gene was developed by Xie et al. (2015). Improvement in the targeting and multiplexing efficiency of CRISPR/Cas 9 was achieved by Xie et al. (2015) by modulating the molecular intrinsic processing properties of t-RNA.

Ding et al. (2018) used this modified and enhanced tRNA-processing machinery in the CRISPR/Cpf1 system to achieve multiplex editing. Cpf1, unlike Cas9, is a binary nuclease that cleaves target DNA while also processing its own CRISPR RNA (Fonfara et al. 2016; Zetsche et al. 2017). Wang et al. (2017) took advantage of this property by engineering a sequence-specific nuclease CRISPR/Cas 9 (C-ERF922) and targeting multiple sites within the OsERF922 region, demonstrating that multiple sgRNAs can also be used to target a single gene in order to further improve editing rates in crops with minimal transformation or editing efficiency.

The CRISPR/Cas system has significant advantages over other sequence specific nucleases. A table that compares the features of various Gene Editing Tools (Table 1.1) are given and discussed below.

1.4.1 Simplicity (Ease of Designing)

CRISPR plasmid construction is simpler than ZFN and TALENS because target specificity is based on ribonucleotide complex generation rather than protein to DNA recognition. ZFN and TALEN both include DNA-binding domains that are connected to the FokI endonuclease, which needs dimerisation in order to cleave DNA. ZFN design necessitates rigorous protein engineering steps, and context-dependent specificity imposes limitations (Sander et al. 2011). Zinc-fingers construction step is simplified by procuring commercially engineered nucleases which are far superior to those designed individually (Ramirez et al. 2008). Sangamo Biosciences (Richmond, CA) has created a unique platform (CompoZr) for zinc-finger building in partnership with Sigma-Aldrich (St. Louis, Missouri), allowing scientists to bypass zinc-finger assembly and validation altogether. The

Table 1.1 Various gene-editing techniques: a comparative analysis

Attributes	Meganucleases	ZFN	TALENs	CRISPR/Cas9
Region of target loci	14–40 bp	9–18 bp	28–40 bp per TALENS pair	19–22 bp + PAM sequence
Specificity	High	High	High	Moderately high
Designing	Extremely difficult	Complex	Moderately difficult	Easy
DNA recognition mechanism	DNA and protein interaction	DNA and protein interaction	DNA and protein interaction	DNA and RNA interaction
DNA breakage and repair mechanisms	Double-stranded break with endonuclease	Double-stranded break by Fok 1	Double-stranded break by Fok 1	Cas 9-induced single- or double-stranded break
Off targeting	Low	Low to moderate	Low	High
Multiplexing	Difficult	Difficult	Difficult	Easily can multiplexes

development of TALENs has been facilitated by efficient DNA assembly and cloning methods such as Golden Gate (Engler et al. 2008), and unlike ZFN, its design has been improved by one-to-one recognition criteria between protein repeats and nucleotide sequences. Each ZNF recognises 3–6 nucleotide triplets on average, and since the cleavage domain Fok1 needs dimerisation to cleave DNA, every particular locus requires two ZNFs to target specific DNA fragment. TALENs are composed of highly repetitive sequences that can promote homologous recombination in vivo (Holkers et al. 2013), and they are also much easier to construct than ZNFs. Guide RNA-based (gRNA) cleavage, on the other hand, is based on a simple Watson–Crick base pairing with the target DNA sequence; therefore, no complex and difficult protein engineering is necessary for each target, and just 20 nt in the gRNA must be modified to recognise a different target. In addition, just 20 nucleotides in the gRNA sequence must be changed to confer a different target specificity, eliminating the need for cloning. Any number of gRNAs may be produced in vitro using two complementary annealed oligonucleotides (Cho et al. 2013). Vector systems for Cas9 expression are available in a variety of formats. SgRNA is available as a DNA expression vector, an RNA molecule, or a pre-loaded Cas9-RNA combination for delivery to cells. This allows for the creation of large gRNA libraries at a relatively low cost, allowing the CRISPR/Cas9 system to be used for high-throughput functional genomics applications and bringing GEN/Ed within reach of any lab interested in using CRISPR. Conventional TALENs and ZFN cannot cleave DNA containing 5-methylcytosine, but methylated cytosine is indistinguishable from thymidine in the major groove. Unlike ZFNs and TALENs, the CRISPR/Cas9 system in human cells can produce incisions in methylated DNA (Hsu et al. 2013), allowing for genomic modifications that other nucleases cannot (Ding et al. 2013). Although this element of the CRISPR/Cas9 system has not been fully researched in plants, it is reasonable to assume that it should be similarly efficient regardless of the kind of genome targeted, given CRISPR's ability to cleave methylated DNA is an inherent characteristic of the system. In plants, the majority of CpG/CpNpG sites (≥70%), particularly CpG islands in promoters and proximal exons, have been found to be methylated (Vanyushin and Ashapkin 2011). CRISPR/Cas9 technology can therefore be more adaptable for genome editing in plants in general, but it's especially good for monocots with high genomic GC content (Miao et al. 2013).

1.4.2 Efficiency

Other targeted gene editing approaches are outperformed by the CRISPR/Cas system. RNAs encoding the Cas protein and gRNA can be infused directly into cell lines to provide modifications. When using classic homologous recombination procedures to create selected mutant lines, this avoids the time-consuming and labour-intensive transfection and selection steps. The relative efficacy of various nucleases (CRISPR associated) in plants is incomparable since the plant species studied by different scientists differs and each has employed a diverse set of CRISPR/Cas. Although CRISPR is more effective than current Gen/Ed methods, the regeneration aspect of engineered plants must be addressed since it significantly increases the tool's efficiency.

1.4.3 Multiplexing

The ease of multiplexing is CRISPR/key Cas9's practical advantage over ZFNs and TALENs. By injecting numerous gRNAs into several genes at the same time, mutations can be introduced in multiple genes at the same time (Li et al. 2013; Mao et al. 2013), which can be very effective for knocking off redundant genes or parallel pathways. By targeting two widely dispersed cleavage sites on the same chromosome, the same technique can be used to construct massive genomic deletions or inversions (Li et al. 2013; Upadhyay et al. 2013; Zhou et al. 2014). The monomeric Cas9 protein and any number of distinct sequence-specific gRNAs are all that's needed for multiplex editing with the CRISPR/Cas9 system. Multiplex editing with ZFNs or TALENs, on the other hand, necessitates separate dimeric proteins specialised for each target location.

All of the technologies – meganuclease, ZNFs, TALENs, and CRISPR/Cas – provide researchers with new ways to produce mutants more quickly than classic gene targeting approaches, but each come with their own set of restrictions and complications. Some of them are discussed below.

1.4.4 Off-Site Effects

One of the most significant drawbacks of these technologies is that mutations are frequently introduced at non-specific sites. These loci exhibit homology to the target locations that is similar but not identical. These can be difficult to spot, requiring a genome scan for mutations at places that are similar in sequence to the gRNA target sequence. CRISPR/Cas9 systems are more likely to elicit off-target actions than other systems (Zhang et al. 2014), because Cas9 can cut at other unintended sites in the genome in addition to the intended target region. Other systems have a high level of precision, but their construction or delivery are difficult. Actual Cas9 off-target activities are lower in *Arabidopsis*, maize, rice, tomato, and tobacco than in mammals (Nekrasov et al. 2013; Feng et al. 2014; Gao et al. 2015; Woo et al. 2015; Ishizaki 2016; Pan et al. 2016; Peterson et al. 2016; Tang et al. 2016). On target indel frequencies in *Arabidopsis* range from 33 to 92 percent of sequencing reads, but no off-target editing events were found elsewhere in the genome at expected or unexpected locations, corroborating findings from smaller scale studies (Peterson et al. 2016). During pathogen-related gene editing (Nekrasov et al. 2017) and targeted deletion of cis-regulatory regions (Rodríguez-Leal et al. 2017), no off-target mutations were observed in tomato. Backcrossing to a parental line can remove these so-called off-targets in some plant species. When targeting members of closely related gene families, the specificity of gene editing tools is particularly noticeable, especially when recent paralogues are co-located in the genome and unlikely to segregate. Another approach is to create a chimeric fusion between the FokI catalytic domain and a catalytically inactive Cas9 protein (dCas9). Guilinger et al. (2014) and Aouida et al. (2015) employed the inactive dCas9 as a targeting module to bring the FokI domain into close proximity and allow dimerisation, and the production of homodimers with the correct spacer sequence then allows the generation of DSBs. As it requires 40 bp of unique sequence and a unique distance between the two monomers, this greatly improves cutting specificity, limiting off-target actions (Yee 2016).

1.4.5 Mosaicism

As Cas9 nucleases may not always cut the DNA during the one cell stage of embryonic development, genetic mosaicism occurs when an individual species has more than two alleles with a mutant allele in only some of their cells. The CRISPR/Cas9 system may continually target and cleave genes at different phases of embryonic development, resulting in mosaicism of the introduced mutations, which is often documented in animal systems (Mizuno et al. 2014; Oliver et al. 2015; Luo et al. 2016). Small indel mutations in plants may have been missed by present detection methods, resulting in overall mosaicism rates being routinely overestimated or ignored.

1.4.6 Delivery

Despite the fact that CRISPR/Cas is a ground-breaking and unrivalled technology, there are still certain barriers to its widespread use in crop improvement and translational research. One of these obstacles is the efficient delivery of transformation vectors into the appropriate host cells, followed by successful plant regeneration. Transient transformation and steady transformation are two processes in the transformation of plants. Stable transformation is responsible for producing edited plants with heritable mutations, from which the nuclease incorporated transgene can be separated to produce transgene-free plants.

1.4.7 Multiple Alleles

Non-homologous end joining can heal the nuclease cleavage site, resulting in cohorts of mutants with different mutations from the same targeting constructs, necessitating genome sequencing to confirm the type and position of the individual mutation. It's also possible to create mutants with mosaics of numerous mutations, and breeding may be required to separate and isolate a cultivar with single mutations. Phenotyping bottlenecks are also created by the generation of mutants with many variations.

Despite these challenges, ZNFs, TALENs, and, in particular, the CRISPR/Cas systems are powerful new genome-editing tools. These methods are expected to be refined further, and they will be modified in novel ways to generate even more sophisticated plant models.

1.5 Plant Defence and Genome Editing

Interactions between plants and bacteria have piqued scientists' interest for ages. Microbes have been discovered to have either a protagonistic association with plants, in which they form a synergistic interaction with the plants that benefits both of them, or an antagonistic association with their hosts, in which they harm their hosts. Plants may undergo entire genome duplication events to counteract abiotic stress, and functional redundancy in multigene families may also be detected (Khan et al. 2018). One of the key goals of plant researchers is to have a full understanding of the molecular basis of abiotic stresses (such as drought, salinity, and heat) and associated tolerance mechanisms in order to engineer stress tolerance in plants. The antagonistic confrontations between plants and diseases, according to the Red Queen dynamic model, result in ever-changing co-evolutionary cycles (Han 2018). In the absence of an adaptive immune system, plants have evolved innate immune systems (including resistance proteins) to detect and respond to both biotic and abiotic stresses. Plants defend themselves against pathogens using innate immune responses triggered by cell surface-localised pattern recognition receptors (PRRs) at the plasma membrane of cells, and cytoplasmic threat recognition is mediated by Nibblers (NB-LRR receptors), which are cytoplasmic nucleotide-binding domain leucine-rich repeat containing receptors (NLRs), resulting in pattern-triggered immunity (PTI) and effector-triggered immunity (ETI) (Jones and Dangl 2006; Liu et al. 2009; Monaghan and Zipfel 2012). The Zigzag model of plant-pathogen interaction suggests two layers of induced defence, the first of which is known as pathogen associated molecular pattern (PAMP and/or MAMP) and is currently recognised as pattern-triggered immunity (PTI). Effector-triggered immunity is the activation of defence responses that serve to suppress the effects of PTI at the second layer (ETI). If pathogens circumvent PTI due to the lack of R proteins, effector-triggered susceptibility (ETS) emerges. Meanwhile, keeping up with disease development and increased food demand is a huge challenge for conventional breeding, particularly in light of global climate change (Zaidi et al. 2019). Genome-editing methods can generate GMO-free resistant cultivars in a reasonably short period to address market demand for resistant and robust crops in today's agricultural system. Gene editing for disease resistance necessitates precise regulation of these defensive regulatory mechanisms, and several genes are targeted to create specific PTI in concert with specific ETS.

In cotton, a re-engineered meganuclease was created to cleave an endogenous target sequence close to a transgenic insect control locus. Targeted DNA breakage combined with homologous recombination-mediated repair allowed for the precise insertion of extra trait genes (hppd, epsps) in cotton. D'Halluin et al. (2013) found that targeted insertion events occurred in roughly 2% of the independently manipulated embryogenic callus lines.

A study with the fungus *Coniothyrium glycines*, the causative agent of soybean red leaf blotch, used meganuclease and yielded promising results, as sequencing of the *C. glycines* mitochondrial genome revealed a circular 98,533 bp molecule containing 12 mitochondrial genes typically involved in oxidative phosphorylation (atp6, cob, cox1–3, nad1–6, and nad4L), 1 for a ribosomal protein (rps3), 4 for hypothetical proteins, 1 for each of the small and large subunit ribosomal RNAs (rns and rnl), and a set of 30 tRNAs. It also identified 32 introns within 8 protein-coding genes and the rnl, accounting for 54.1 percent of the overall mitochondrial genome. Eighteen introns had potential intronic ORFs with either LAGLIDADG or GIY-YIG homing endonuclease motifs, while another 11 introns had truncated or degenerate endonuclease motifs (Stone et al. 2018).

Using zinc-finger nuclease, several characteristics of *Arabidopsis*, *Nicotiana*, Zea mays, *Petunia*, *Glycine max*, *Brassica napus*, *Oryza sativa*, *Malus domestica*, and *Ficus carica* have been successfully modified (Martinez et al. 2017; Ran et al. 2017). By altering the inositol phosphate profile of growing seeds and interrupting the ZmIPK1 gene with the insertion of PAT gene cassettes, ZFN technology was employed to make herbicide-tolerant maize seed (Shukla et al. 2009). Using ZFN-mediated targeted transgenes, trait stacking to combine several advantageous characters in maize has been effectively done (Ainley et al. 2013). Mitchell et al. (2014)

employed ZFNs to find and target possible safe gene integration locations in rice. The found locations were thought to be reliable locus for gene insertion and trait stacking in the future. Since, creating ZFNs is a complex and technically arduous procedure with limited effectiveness, there are few reports of ZFN application (Table 1.2). To overcome the complicated construction of ZFN, TALENS were successfully employed to integrate and modify genes. TALEs may potentially be created to bind any desired DNA sequence, which can then be fused to a nuclease (TALEN) to cause DNA breaks at any specified position (Miller et al. 2011). In human cell lines and animals, the use of TALENs has been shown to be highly effective (Joung and Sander 2013); however, there have been few examples of TALEN uses in plants (Li et al. 2012; Sun et al. 2016). Thus, TALEN was employed and investigated for targeted genome editing in plants in order to improve their quality and tolerance to biotic and abiotic challenges. TAL effectors from *Xanthomonas oryzae* pv. *oryzae* pathogen virulence is mediated by *Oryzae* (Xoo), activating certain rice disease susceptibility (S) genes, specifically one or more members of the SWEET family of sugar transporter genes. *Xanthomonads* have a Hrp-type III secretion (T3S) mechanism that translocates effector proteins into plant cells, which is important for pathogenicity of most *Xanthomonas* pathovars. The effectors are assumed to conduct a range of actions within the eukaryotic cell to assist bacterial pathogenicity, proliferation, and dispersion. They showed that transcription activator-like effector nucleases (TAL effectors) are transcription factors that are translocated into plant cells by *Xanthomonas* bacteria via their type III secretion system, as well as their structure, activity, and host targets (Boch and Bonas 2010; Makino et al. 2006; Doyle et al. 2012). Due of the significant virulence effect of Xoo, SWEET genes such as SWEET11/xa13, SWEET1/Xa25, and SWEET14/Os11N3 are key TALEs (Li et al. 2012; Streubel et al. 2013).

Zhou et al. (2015) discovered a sucrose transporter gene (OsSWEET13a) and a disease-susceptibility gene for the TALE effector PthXo2 of the Xoo2 strain, implying that the occurrence of cryptic recessive resistance to PthXo2-dependent *X. oryzae* pv. *oryzae* was attributed to the variation in the promoter regions of OsSWEET13 in japonica rice. TALEs AvrXa7, PthXo3, TalC, and Tal5, identified in spatially separated Xoo strains, target OsSWEET14, making it a key TALE target in rice-Xoo interactions. According to the study, stable expression of TALE-nuclease (TALEN) constructs in rice resulted in the generation of an allele library of the OsSWEET14 promoter. Plants edited in AvrXa7 or Tal5 EBEs were resistant to bacterial strains relying on the corresponding TALE (Blanvillain-Baufumé et al. 2017).

Targeting homologs of the Mildew Resistance Locus (MLO) and other loci have enhanced fungal disease resistance in various species. Wang et al. (2014) employed both TALEN and CRISPR/Cas9 systems to target the genes of the Mildew Resistance Locus (MLO) in wheat and successfully knocked out all three MLO homoeoalleles (Mildew Resistance Locus (MLO), TaMLO-A, TaMLO-B, and TaMLO-D). The results of this study showed that TALEN-induced mutation of all three TaMLO homoeologs in the same plant confers heritable broad-spectrum resistance to powdery mildew and demonstrated the feasibility of engineering targeted DNA

Table 1.2 Genome editing in plants or disease resistance in plants through various gene-editing tools

System	Disease resistance	Targeted plants	Plant family	Gene target	Author and year
Endonuclease/ meganuclease	Soybean red leaf blotch	Soybean	Fabaceae	Mitochondrial genes (*atp*6, *cob*, *cox*1–3, *nad*1–6, and *nad*4L)	Stone et al. (2018)
	Pest resistance	Cotton	Malvaceae	*hppd*, *epsps*	D'Halluin et al. (2013)
ZFN	DNA viral disease (beet severe curly top virus)	Arabidopsis	Brassicaceae	Replication region	Sera (2005)
	Herbicide tolerant	Maize	Poaceae	ZmIPK1	Shukla et al. (2009)
	DNA viral disease (begomoviruses)	Tobacco	Solanaceae	Rep	Chen et al. (2014)
TALENS	Bacterial blight (*Xanthomonas oryzae*)	Rice	Poaceae	OsSWEET14/promoter	Li et al. (2012)
	Powdery mildew (*Blumeria graminis*)	Wheat	Poaceae	TaMLO/Cds region	Wang et al. (2014)
CRISPR/Cas9	Fire blight (*Erwinia amylovora*)	Apple	Rosaceae	DIPM 1, DIPM 2 and DIPM 4	Malnoy et al. (2016)
	DNA virus (banana streak virus)	Banana	Musaceae	eBSV sequence	Tripathi et al. (2019)
	DNA virus (wheat dwarf virus)	Barley	Poaceae	MP, CP Rep/Rep A Coding region, LIR, C-terminus	Kis et al. (2019)
	General resistance (*Phytophthora tropicalis*)	Cacao	Malvaceae	*TcNPR3/cds region*	Fister et al. (2018)
	RNA Virus (cassava brown streak virus)	Cassava	Euphorbiaceae	nCBP-1, and nCBP-2/ cds region	Gomez et al. (2019)
	RNA virus (cucumber vein yellowing virus, zucchini yellow mosaic virus and *papaya ring spot mosaic virus-W.*)	Cucumber	Fabaceae	Elf4E/cds region	Chandrasekaran (2016)
	Citrus Canker	Orange (*Citrus sinensis* Osbeck)	Rutaceae	CSLOB1/promoter region	Peng et al. (2017)
	Bacterial blight (*Xanthomonas oryzae*)	Rice	Poaceae	OsSWEET13/Cds region	Zhou et al. (2015)

System	Disease resistance	Targeted plants	Plant family	Gene target	Author and year
	Herbicide resistant	Rice	Poaceae	Acetolactate Synthase 1 region	Sun et al. (2016)
	Bacterial blight (*Xanthomonas oryzae*)	Rice	Poaceae	Os8N3/promoter	Kim et al. (2019)
	Bacterial blight (*Xanthomonas oryzae*)	Rice	Poaceae	Xa13/cds region	Li et al. (2019)
	Bacterial blight (*Xanthomonas oryzae*)	Rice	Poaceae	SWEET11, SWEET 13, SWEET 14/promoter	Oliva et al. (2019)
	Bacterial blight (*Xanthomonas oryzae*)	Rice	Poaceae	OsSWEET11, SWEET 14/promoter	Xu et al. (2019)
	Powdery mildew	Tomato	Solanaceae	SIMlo1/cds rgion	Nekrasov et al. (2017)
	Bacterial Speck (*Xanthomonas spp., Pseudomonas syringae, Phytophthora capsici*)	Tomato	Solanaceae	SIDMR6-1/cds region	Thomazella et al. (2016)
	Bacterial speck (*Pseudomonas syringae*)	Tomato	Solanaceae	SIDMR6-1/cds region	Ortigosa et al. (2019)
	DNA virus (*tomato yellow leaf curl virus*)	Tomato	Ranunculaceae	Coat protein and replicase protein	Tashkandi et al. (2018)
	DNA virus (*tomato yellow leaf curl virus*)	Tobacco	Ranunculaceae	IR, CP, RCRII	Ali et al. (2015)
	Virus resistance (Geminivirus, *beet severe curly top virus* (BSCTV))	Tobacco (*Nicotiana benthamiana*)	Ranunculaceae	SgRNA target sites	Ji et al. (2015)
	Powdery mildew (*Blumeria graminis*)	Wheat	Poaceae	TaMLO/cds region)	Wang et al. (2014)
	Powdery mildew (*Blumeria graminis*)	Wheat	Poaceae	TaEDR1/cds region	Zhang et al. (2014)
Cas9/sgRNA	Citrus canker	Duncan grape fruit (*Citrus paradisi* Macf.)	Rutaceae	CSLOB1/cds region	Jia et al. (2016)
CRISPR/Cas9/sgRNA	Citrus canker	Duncan grape fruit (*Citrus paradisi* Macf.)	Rutaceae	CSLOB1/cds region	Jia et al. (2017)

insertion in bread wheat through nonhomologous end joining of the double-strand breaks caused by TALENs.

CRISPR, one of the most recent gene-editing techniques, is a prominent technology now being used to generate suitable plant material for sustainable food supply (Zaidi et al. 2019). Using CRISPR/Cas-mediated targeted mutagenesis in the coding area of *Oryza sativa* mitogen-activated protein kinase 5, Xie and Yang (2013) attempted to establish enhanced rice varieties by creating resistance against *Burkholderia glumae* (OsMPK5). Despite the workers' lack of certainty about the targeted pathogen's resistance status, experimental assessments of mutation efficiency and off-target effect, as well as genome-wide prediction of specific gRNA seeds, revealed that the CRISPR/Cas9 system is a simple and effective tool for plant functional genomics and agricultural development. Broad-spectrum resistant against *Xanthmonas* (causing bacterial blight) was conferred by CRISPR/Cas-mediated editing of three SWEET gene promoters viz. *SWEET 11*, *13*, and *14* genes in rice varieties by Oliva et al. (2019). Five promoter mutations were introduced into the rice line Kitaake; as well as the elite mega varieties IR64 and Ciherang-Sub1 at the same time in rice cv. Kitaake, Xu et al. (2019) used the CRISPR/Cas9- GenEd tool to disrupt the TALE binding elements (EBEs) of two S genes, OsSWEET11 and OsSWEET14. They found two PthXo2-like TALEs, Tal5LN18 and Tal7PXO61, to be key virulence factors in some Xoo strains and discovered that Xoo encodes at least five different PthXo2-like effectors. CRISPR/Cas9 technology was subsequently utilised to create InDels in the EBE of the OsSWEET13 promoter in MS14K, resulting in a novel germplasm with three modified OsSWEET EBEs and broad-spectrum resistance to all Xoo strains examined. Their findings demonstrated how the loss of effector-triggered susceptibility in plants may be used to disarm TALE-S co-evolved loci and build broad-spectrum resistance. Many other scientists followed a similar method in developing resistance to bacterial blight caused by diverse strains of *Xanthomonas oryzae* in various rice genotypes (Kim et al. 2019).

CRISPR/Cas9 system was employed to engineer disease resistance in tomato by inactivating DMR6 orthologue gene. The regenerated plants showed disease resistance against a wide variety of pathogens including *P. syringae*, *P. capsici* and *Xanthomonas* spp. (Thomazella et al. 2016). The SlMlo1 gene was targeted by utilising CRISPR/Cas9 to produce powdery-resistant Tomelo transgene-free tomato (Nekrasov et al. 2017). Antibiotic resistance to bacterial speck disease CRISPR/Cas9 was used to create specific mutations in the tomato genome (Ortigosa et al. 2019). *Pseudomonas syringae* pv. tomato is the bacteria that causes Speck disease in Tomato (Pto) DC3000. The organism creates coronatine (COR), a metabolite that mimics the bioactive jasmonic acid (JA) hormone and hence increases the opening of stomata, allowing bacteria to infiltrate and thrive in the apoplast. Tashkandi et al. (2018) reported the effective use of CRISPR/Cas9 to create resistance against tomato yellow leaf curl virus (TYLCV). They used Cas9-single guide RNA to target the coat protein (CP) or replicase (Rep) regions in the TYLCV genome, asserting that the CRISPR/Cas9-based immunity remained active through numerous generations in N. and tomato plants. Zhang et al. (2017) improved resistance to powdery mildew by modifying three homologs of the wheat TaEDR1 gene simultaneously.

CRISPR/Cas9 was utilised to alter the promoter sequence of the CsLOB1 canker susceptibility gene in citrus, resulting in resistance to canker (Jia et al. 2017; Peng et al. 2017). The CRISPR-Cas9 system was used to inactivate the eukaryotic translation initiation factor gene elF4E in cucumber, resulting in non-transgenic homozygotic mutant plants that were resistant to cucumber vein yellowing virus, zucchini yellow mosaic virus, and papaya ring spot mosaic virus (Chandrasekaran et al. 2016). The fungus *Erwinia amylovora*, which causes fire blight, is a serious threat to apples and other commercial and ornamental plants. CRISPR/Cas was used to improve resistance *to Erwinia amylovora* by targeting three separate genes: DIPM-1, DIPM-2, and DIPM-4 (Tegtmeier et al. 2020; Alphonse et al. 2021). Resistance against fusarium wilt was developed in banana using gene-editing tool CRISPR (Maxmen 2019).

Genome-editing technologies have advanced significantly and have become one of the most significant genetic tools for implementing disease resistance in plants. Table 1.2 provides comprehensive information on this and other works. Overall, the CRISPR/Cas9 system is a successful tool for treating bacterial, viral, and fungal infections in plants, and it has the potential to be used against additional pathogens and the creation of resilience. The discovery of new S genes in various plant species would pave the road for the long-term evolution of disease resistance using GETs like CRISPR/Cas9. The development of various novel methods for targeted gene editing, as well as the discovery of other CRISPR/Cas systems, implies that the CRISPR toolbox for plant engineering will increase in the near future. These technologies will open up new avenues for precise genome editing that leaves no trace of transgenes in the genome-edited plants. Table 1.2 shows some recent instances of how different GenEd techniques were employed to create disease resistance in plants.

1.6 Conclusion and Future Prospects

Genome editing has been the most popular method for crop improvement and functional genomics in the modern era. CRISPR/Cas has been the most widely used gene-editing technique of the decade due to characteristics like simplicity, efficiency, integrity, and, in particular, multiplexing. It merely takes replacing the 20-nucleotide long guide sequence of sgRNA to target a new gene location. Between the 17th and 18th nucleotides of the target sequence (three nucleotides from the PAM), Cas9 effectively produces DSBs. Furthermore, providing a mix of sgRNAs substantially simplifies multiplexing. To combat unwanted off-target effects, careful use of nickase and alteration of the sgRNA structure result in more precise target identification. The application of CRISPR technology in many fields of biological inquiry, including biotechnology, genomics, transcriptomics, and proteomics, has been documented in a number of publications published in PubMed. With the use of genome engineering technologies such as CRISPR/Cas9, more S genes in many other plant species are necessary for the progressive development of disease resistance. To aid in the rapid advancement of this technology, government regulatory agencies are attempting to simplify many laws and regulations governing the production and use of transgenic crops, making them suitable for consumer use. The ZFN and TALEN systems need more time and effort. However, automated design of ZFN- and TALEN-expressing constructs is now possible, allowing for their accurate and economical commercial manufacturing. Furthermore, the fact that TALENs only produce breakage when the FokI domain dimerises, i.e., in pairs, boosts selectivity and minimises the possibility of off-target consequences. According to a recent study published in Nature Communications (Jain et al. 2021), TALEN is up to five times more effective than CRISPR-Cas9 in a highly compact form of DNA called heterochromatin. Cas9 is less effective in heterochromatin than TALEN, according to single-molecule imaging of genome-editing proteins, because Cas9 gets hampered by local searches on nonspecific sites in these areas. To sum up, each technology has advantages and disadvantages, and their use depends and varies on the specific design and necessity of the experiment, as well as which technologies are most suited for achieving the study aims. Above all, CRISPR technology has resulted in a variety of transgene-free modified crops, and CRISPR and the next new set of gene-editing tools will enable new ways to precise genome editing with no traces of transgenes left in genome-edited plants. Single molecule imaging of genome-editing proteins reveals that Cas9 is less efficient in heterochromatin than TALEN because Cas9 becomes encumbered by local searches on nonspecific sites in these regions. In conclusion, we would like to state that each technology has their pros and cons and their application depends and varies on particular design and need of experiment and also that are most appropriate to realise the research goals. Above all, CRISPR technology has rendered various modified crops which are transgene free, and so CRISPR and upcoming novel set of gene-editing tools will provide new approaches to achieve precise genome editing without any traces of transgenes remaining in genome-edited plants.

References

Abudayyeh OO, Gootenberg JS, Konermann S, Joung J, Slaymaker IM, Cox DB, Shmakov S, Makarova KS, Semenova E, Minakhin L, Severinov K, Regev A, Lander ES, Koonin EV, Zhang F (2016) C2c2 is a single-component programmable RNA-guided RNA-targeting CRISPR effector. Science (New York, N.Y.) 353(6299):aaf5573. https://doi.org/10.1126/science.aaf5573

Ainley WM, Sastry-Dent L, Welter ME, Murray MG, Zeitler B, Amora R, Corbin DR, Miles RR, Arnold NL, Strange TL, Simpson MA, Cao Z, Carroll C, Pawelczak KS, Blue R, West K, Rowland LM, Perkins D, Samuel P, Dewes CM et al (2013) Trait stacking via targeted genome editing. Plant Biotechnol J 11(9):1126–1134. https://doi.org/10.1111/pbi.12107

Ali Z, Abulfaraj A, Idris A et al (2015) CRISPR/Cas9-mediated viral interference in plants. Genome Biol 16:238. https://doi.org/10.1186/s13059-015-0799-6

Alphonse V, Antonysamy JM, Murugan K (2021) CRISPR and RNAi systems nano biotechnology approaches to plant breeding and protection nanobiotechnology of plant protection. Elsevier, pp 107–128

Aouida M, Eid A, Ali Z, Cradick TJ, Lee C, Deshmukh H, Atef A, Abusamra D, Gadhoum SZ, Merzaban JS et al (2015) Efficient fdCas9 synthetic endonuclease with improved specificity for precise genome engineering. PLoS One 10:e0133373. https://doi.org/10.1371/journal.pone.0133373

Arbuthnot P (2015) Chapter 3 - Engineering sequence-specific DNA binding proteins for antiviral gene editing. In: Gene therapy for viral infections. Academic Press, Amsterdam, pp 63–94. https://doi.org/10.1016/B978-0-12-410518-8.00003-X

Argast GM, Stephens KM et al (1998) I-PpoI and I-CreI homing site sequence degeneracy determined by random mutagenesis and sequential in vitro enrichment. J Mol Biol 280(3):345–353

Belfort M, Perlman PS (1995) Mechanisms of intron mobility. J Biol Chem 270(51):30237–30240. https://doi.org/10.1074/jbc.270.51.30237

Belfort M, Roberts RJ (1997) Homing endonucleases: keeping the house in order. Nucleic Acids Res 25(17):3379–3388

Blanvillain-Baufumé S, Reschke M, Solé M, Auguy F, Doucoure H, Szurek B, Meynard D, Portefaix M, Cunnac S, Guiderdoni E, Boch J, Koebnik R (2017) Targeted promoter editing for rice resistance to Xanthomonas oryzae pv. oryzae reveals differential activities for SWEET14-inducing TAL effectors. Plant Biotechnol J 15(3):306–317. https://doi.org/10.1111/pbi.12613

Boch J, Bonas U (2010) Xanthomonas AvrBs3 family-type III effectors: discovery and function. Annu Rev Phytopathol 48:419–436

Bortesi L, Fischer R (2015) The CRISPR/Cas9 system for plant genome editing and beyond. Biotechnol Adv 33(1):41–52

Briggs AW, Rios X, Chari R, Yang L, Zhang F, Mali P, Church GM (2012) Iterative capped assembly: rapid and scalable synthesis of repeat-module DNA such as TAL effectors from individual monomers. Nucleic Acids Res 40(15):e117. https://doi.org/10.1093/nar/gks624. Epub 2012 Jun 26. PMID: 22740649; PMCID: PMC3424587

Cermak T, Doyle EL, Christian M, Wang L, Zhang Y, Schmidt C, Baller JA, Somia NV, Bogdanove AJ, Voytas DF (2011) Efficient design and assembly of custom TALEN and other TAL effector-based constructs for DNA targeting. Nucleic Acids Res 39(12):e82. https://doi.org/10.1093/nar/gkr218

Chandrasegaran S, Carroll D (2016) Origins of programmable nucleases for genome engineering. J Mol Biol 428(5 Pt B):963–989. https://doi.org/10.1016/j.jmb.2015.10.014

Chandrasekaran J, Brumin M, Wolf D, Leibman D, Klap C, Pearlsman M, Sherman A, Arazi T, Gal-On A (2016) Development of broad virus resistance in non-transgenic cucumber using CRISPR/Cas9 technology. Mol Plant Pathol 17(7):1140–1153. https://doi.org/10.1111/mpp.12375

Chen W, Qian Y, Wu X, Sun Y, Wu X, Cheng X (2014) Inhibiting replication of begomoviruses using artificial zinc finger nucleases that target viral-conserved nucleotide motif. Virus Genes 48(3):494–501. https://doi.org/10.1007/s11262-014-1041-4

Chevalier BS, Stoddard BL (2001) Homing endonucleases: structural and functional insight into the catalysts of intron/intein mobility. Nucleic Acids Res 29(18):3757–3774. https://doi.org/10.1093/nar/29.18.3757

Christy B, Nathans D (1989) DNA binding site of the growth factorinducible protein Zif268". Proceedings of the National Academy of Sciences of the United States of America 86(22): 8737–8741. https://doi.org/10.1073/pnas.86.22.8737

Cho S, Kim S, Kim J et al (2013) Targeted genome engineering in human cells with the Cas9 RNA-guided endonuclease. Nat Biotechnol 31:230–232. https://doi.org/10.1038/nbt.2507

Choulika A, Perrin A, Dujon B, Nicolas JF (1995) Induction of homologous recombination in mammalian chromosomes by using the I-SceI system of Saccharomyces cerevisiae. Mol Cell Biol 15:1968–1973

Cristea S, Freyvert Y, Santiago Y, Holmes MC, Urnov FD, Gregory PD, Cost GJ (2013) In vivo cleavage of transgene donors promotes nuclease-mediated targeted integration. Biotechnol Bioeng 110(3):871–880. https://doi.org/10.1002/bit.24733

Curtin SJ, Zhang F, Sander JD, Haun WJ, Starker C, Baltes NJ, Reyon D, Dahlborg EJ, Goodwin MJ, Coffman AP, Dobbs D, Joung JK, Voytas DF, Stupar RM (2011) Targeted mutagenesis of duplicated genes in soybean with zinc-finger nucleases. Plant Physiol 156(2):466–473. https://doi.org/10.1104/pp.111.172981

de Pater S, Neuteboom LW, Pinas JE, Hooykaas PJ, van der Zaal BJ (2009) ZFN-induced mutagenesis and gene-targeting in Arabidopsis through agrobacterium-mediated floral dip transformation. Plant Biotechnol J 7(8):821–835. https://doi.org/10.1111/j.1467-7652.2009.00446.x

D'Halluin K, Vanderstraeten C, Stals E, Cornelissen M, Ruiter R (2008) Homologous recombination: a basis for targeted genome optimization in crop species such as maize. Plant Biotechnol J 6(1):93–102. https://doi.org/10.1111/j.1467-7652.2007.00305.x. Epub 2007 Nov 12

D'Halluin K, Vanderstraeten C, Van Hulle J, Rosolowska J, Van Den Brande I, Pennewaert A, D'Hont K, Bossut M, Jantz D, Ruiter R, Broadhvest J (2013) Targeted molecular trait stacking in cotton through targeted double-strand break induction. Plant Biotechnol J 11(8):933–941. https://doi.org/10.1111/pbi.12085

Ding Q, Regan SN, Xia Y, Oostrom LA, Cowan CA, Musunuru K (2013) Enhanced efficiency of human pluripotent stem cell genome editing through replacing TALENs with CRISPRs. Cell Stem Cell 12(4):393

Ding D, Chen K, Chen Y, Li H, Xie K (2018) Engineering introns to express RNA guides for Cas9- and Cpf1-mediated multiplex genome editing. Mol Plant 11:542–552. https://doi.org/10.1016/j.molp.2018.02.005

Doyle EL, Booher NJ, Standage DS, Voytas DF, Brendel VP, Vandyk JK, Bogdanove AJ (2012) TAL Effector-Nucleotide Targeter (TALE-NT) 2.0: tools for TAL effector design and target prediction. Nucleic Acids Res 40:W117–W122 [PMC free article] [PubMed] [Google Scholar]

Dujon B (1989) Group I introns as mobile genetic elements: facts and mechanistic speculations--a review. Gene 82(1):91–114. https://doi.org/10.1016/0378-1119(89)90034-6

Durai S, Mani M, Kandavelou K, Wu J, Porteus MH, Chandrasegaran S (2005) Zinc finger nucleases: custom-designed molecular scissors for genome engineering of plant and mammalian cells. Nucleic Acids Res 33(18):5978–5990. https://doi.org/10.1093/nar/gki912

Engler C, Kandzia R, Marillonnet S (2008) A one pot, one step, precision cloning method with high throughput capability. PLoS One 3(11):e3647. https://doi.org/10.1371/journal.pone.0003647

Feng Z, Zhang B, Ding W, Liu X, Yang D-L, Wei P, Cao F, Zhu S, Zhang F, Mao Y, Zhu J-K (2013) Efficient genome editing in plants using a CRISPR/Cas system. Cell Res 23(10):1229–1232. https://doi.org/10.1038/cr.2013.114

Feng Z, Mao Y, Xu N, Zhang B, Wei P, Yang D-L, et al. (2014) Multigeneration analysis reveals the inheritance, specificity, and patterns of CRISPR/Cas-induced gene modifications in Arabidopsis. Proc Natl Acad Sci USA 111:4632–4637

Fister AS, Landherr L, Maximova SN, Guiltinan MJ (2018) Transient expression of CRISPR/Cas9 machinery targeting TcNPR3 enhances defense response in *Theobroma cacao*. Front Plant Sci 9:268. https://www.frontiersin.org/article/10.3389/fpls.2018.00268

Fonfara I, Richter H, Bratovič M, Le Rhun A, Charpentier E (2016) The CRISPR-associated DNA-cleaving enzyme Cpf1 also processes precursor CRISPR RNA. Nature 532(7600):517–521. https://doi.org/10.1038/nature17945

Gaj T, Gersbach CA, Barbas CF 3rd (2013) ZFN, TALEN, and CRISPR/Cas-based methods for genome engineering. Trends Biotechnol 31(7):397–405. https://doi.org/10.1016/j.tibtech.2013.04.004

Gaj T, Sirk SJ, Shui SL, Liu J (2016) Genome-editing technologies: principles and applications. Cold Spring Harb Perspect Biol 8(12):a023754. https://doi.org/10.1101/cshperspect.a023754

Gao J, Wang G, Ma S, Xie X, Wu X, Zhang X, Wu Y, Zhao P, Xia Q (2015) CRISPR/Cas9-mediated targeted mutagenesis in Nicotiana tabacum. Plant Mol Biol 87(1–2):99–110. https://doi.org/10.1007/s11103-014-0263-0

Gomez MA, Lin ZD, Moll T, Chauhan RD, Hayden L, Renninger K, Beyene G, Taylor NJ, Carrington JC, Staskawicz BJ, Bart RS (2019) Simultaneous CRISPR/Cas9-mediated editing of cassava eIF4E isoforms nCBP-1 and nCBP-2 reduces cassava brown streak disease symptom severity and incidence. Plant Biotechnol J 17(2):421–434. https://doi.org/10.1111/pbi.12987

Guilinger JP, Thompson DB, Liu DR (2014) Fusion of catalytically inactive Cas9 to FokI nuclease improves the specificity of genome modification. Nat Biotechnol 32(6):577–582. https://doi.org/10.1038/nbt.2909

Han GZ (2018) Origin and evolution of the plant immune system. New Phytologist 222:70–83. https://doi.org/10.1111/nph.15596

Hogler and Timo (2012) Development of phytophthora resistant potato with increased yield. https://patents.google.com/patent/EP2535416A1/en

Holkers M, Maggio I, Liu J, Janssen JM, Miselli F, Mussolino C et al (2013) Differential integrity of TALE nuclease genes following adenoviral and lentiviral vector gene transfer into human cells. Nucleic Acids Res 41(5):e63–e63

Horvath P, Barrangou R (2010) CRISPR/Cas, the immune system of bacteria and archaea. Science (New York, N.Y.) 327(5962):167–170. https://doi.org/10.1126/science.1179555

Hsu PD, Scott DA, Weinstein JA, Ran FA, Konermann S, Agarwala V, Li Y, Fine EJ, Wu X, Shalem O, Cradick TJ, Marraffini LA, Bao G, Zhang F (2013) DNA targeting specificity of RNA-guided Cas9 nucleases. Nat Biotechnol 31(9):827–832. https://doi.org/10.1038/nbt.2647

Ishizaki T (2016) CRISPR/Cas9 in rice can induce new mutations in later generations, leading to chimerism and unpredicted segregation of the targeted mutation. Mol Breed 36(12):1–15

Iqbal Z, Iqbal MS, Ahmad A, Memon AG, Ansari MI (2020) New prospects on the horizon: Genome editing to engineer plants for desirable traits. Current Plant Biol 24:100171. https://doi.org/10.1016/j.cpb.2020.100171

Jain S, Shukla S, Yang C, Zhang M, Fatma Z, Lingamaneni M, Abesteh S, Lane ST, Xiong X, Wang Y, Schroeder CM, Selvin PR, Zhao H (2021) TALEN outperforms Cas9 in editing heterochromatin target sites. Nat Commun 12(1):606

Ji X, Zhang H, Zhang Y et al (2015) Establishing a CRISPR–Cas-like immune system conferring DNA virus resistance in plants. Nat Plants 1:15144. https://doi.org/10.1038/nplants.2015.144

Jia H, Orbovic V, Jones JB, Wang N (2016) Modification of the PthA4 effector binding elements in Type I CsLOB1 promoter using Cas9/sgRNA to produce transgenic Duncan grapefruit alleviating XccΔpthA4:dCsLOB1.3 infection. Plant Biotechnol J 14(5):1291–1301. https://doi.org/10.1111/pbi.12495

Jia H, Zhang Y, Orbović V, Xu J, White FF, Jones JB, Wang N (2017) Genome editing of the disease susceptibility gene CsLOB1 in citrus confers resistance to citrus canker. Plant Biotechnol J 15(7):817–823. https://doi.org/10.1111/pbi.12677

Jiang W, Bikard D, Cox D et al (2013) RNA-guided editing of bacterial genomes using CRISPR-Cas systems. Nat Biotechnol 31:233–239. https://doi.org/10.1038/nbt.2508

Jones JDG, Dangl JL (2006) The plant immune system. Nature 444(7117):323–329. https://doi.org/10.1038/nature05286

Joung JK, Sander JD (2013) TALENs: a widely applicable technology for targeted genome editing. Nat Rev Mol Cell Biol 14(1):49–55. https://doi.org/10.1038/nrm3486

Kamburova VS, Nikitina EV, Shermatov SE, Buriev ZT, Kumpatla SP, Emani C, Abdurakhmonov IY (2017) Genome editing in plants: an overview of tools and applications. Int J Agron 2017:7315351. https://doi.org/10.1155/2017/7315351

Khan Z, Khan SH, Mubarik MS, Ahmad A (2018) Targeted genome editing for cotton improvement, past, present and future trends in cotton breeding. IntechOpen. https://doi.org/10.5772/intechopen.73600

Kim HJ, Lee HJ, Kim H, Cho SW, Kim JS (2009) Targeted genome editing in human cells with zinc finger nucleases constructed via modular assembly. Genome Res 19(7):1279–1288. https://doi.org/10.1101/gr.089417.108. PMC 2704428

Kim YA, Moon H, Park CJ (2019) CRISPR/Cas9-targeted mutagenesis of Os8N3 in rice to confer resistance to *Xanthomonas oryzae* pv. *oryzae*. Rice (New York, N.Y.) 12(1):67. https://doi.org/10.1186/s12284-019-0325-7

Kis A, Hamar É, Tholt G, Bán R, Havelda Z (2019) Creating highly efficient resistance against wheat dwarf virus in barley by employing CRISPR/Cas9 system. Plant Biotechnol J 17(6):1004–1006. https://doi.org/10.1111/pbi.13077

Li T, Liu B, Spalding MH, Weeks DP, Yang B (2012) High-efficiency TALEN-based gene editing produces disease-resistant rice. Nat Biotechnol 30(5):390–392. https://doi.org/10.1038/nbt.2199

Li JF, Norville JE, Aach J, McCormack M, Zhang D, Bush J, Church GM, Sheen J (2013) Multiplex and homologous recombination-mediated genome editing in Arabidopsis and *Nicotiana benthamiana* using guide RNA and Cas9. Nat Biotechnol 31(8):688–691. https://doi.org/10.1038/nbt.2654

Li S, Shen L, Hu P, Liu Q, Zhu X, Qian Q, Wang K, Wang Y (2019) Developing disease-resistant thermosensitive male sterile rice by multiplex gene editing. J Integr Plant Biol 61(12):1201–1205. https://doi.org/10.1111/jipb.12774

Li H, Yang Y, Hong W, Huang M, Wu M, Zhao X (2020) Applications of genome editing technology in the targeted therapy of human diseases: mechanisms, advances and prospects. Signal Transduct Target Ther 5(1):1. https://doi.org/10.1038/s41392-019-0089-y

Liu J, Elmore JM, Coaker G (2009) Investigating the functions of the RIN4 protein complex during plant innate immune responses. Plant Signal Behav 4(12):1107–1110. https://doi.org/10.4161/psb.4.12.9944

Luo X, Li M, Su B (2016) Application of the genome editing tool CRISPR/Cas9 in non-human primates. Dong wu xue yan jiu 37(4):214–219. https://doi.org/10.13918/j.issn.2095-8137.2016.4.214

Maeder ML et al (2008) Rapid "open-source" engineering of customized zinc-finger nucleases for highly efficient gene modification. Mol Cell 31(2):294–301

Makarova KS, Wolf YI, Alkhnbashi OS, Costa F, Shah SA, Saunders SJ, Barrangou R, Brouns SJ, Charpentier E, Haft DH, Horvath P, Moineau S, Mojica FJ, Terns RM, Terns MP, White MF, Yakunin AF, Garrett RA, van der Oost J, Backofen R et al (2015) An updated evolutionary classification of CRISPR-Cas systems. Nat Rev Microbiol 13(11):722–736. https://doi.org/10.1038/nrmicro3569

Makino S, Sugio A, White F, Bogdanove AJ (2006) Inhibition of resistance gene-mediated defense in rice by *Xanthomonas oryzae* pv. *oryzicola*. Mol Plant-Microbe Interact 19(3):240–249. https://doi.org/10.1094/MPMI-19-0240

Malnoy M, Viola R, Jung M-H, Koo O-J, Kim S, Kim J-S, Velasco R, Nagamangala Kanchiswamy C (2016) DNA-free genetically edited grapevine and apple protoplast using CRISPR/Cas9 ribonucleoproteins. Front Plant Sci 7:1904. https://www.frontiersin.org/article/10.3389/fpls.2016.01904

Mao Y, Zhang H, Xu N, Zhang B, Gou F, Zhu JK (2013) Application of the CRISPR-Cas system for efficient genome engineering in plants. Mol Plant 6(6):2008–2011. https://doi.org/10.1093/mp/sst121

Maresca M, Lin VG, Guo N, Yang Y (2013) Obligate ligation-gated recombination (ObLiGaRe): custom-designed nuclease-mediated targeted integration through nonhomologous end joining. Genome Res 23(3):539–546. https://doi.org/10.1101/gr.145441.112

Martínez-Fortún J, Phillips DW, Jones HD (2017) Potential impact of genome editing in world agriculture. Emerg Top Life Sci 1(2):117–133. https://doi.org/10.1042/ETLS20170010

Maxmen A (2019) CRISPR might be the banana's only hope against a deadly fungus. Nature 574(7776):15. https://doi.org/10.1038/d41586-019-02770-7

Miao J, Guo D, Zhang J, Huang Q, Qin G, Zhang X, Wan J, Gu H, Qu L-J (2013) Targeted mutagenesis in rice using CRISPR-Cas system. Cell Res 23(10):1233–1236. https://doi.org/10.1038/cr.2013.123

Miller J, Tan S, Qiao G et al (2011) A TALE nuclease architecture for efficient genome editing. Nat Biotechnol 29:143–148. https://doi.org/10.1038/nbt.1755

Mitchell C, Arnold NL, Gopalan S, Meng X (2014) Precise genome modification in the Cantos C, Francisco P, Trijatmiko KR, Slamet-Loedin I, Chadha-Mohanty PK. Identification of "safe harbor" loci in indica rice genome by harnessing the property of zinc-finger nucleases to induce DNA damage and repair. Front Plant Sci 5:302

Mizuno S, Dinh TT, Kato K, Mizuno-Iijima S, Tanimoto Y, Daitoku Y, Hoshino Y, Ikawa M, Takahashi S, Sugiyama F, Yagami K (2014) Simple generation of albino C57BL/6J mice with G291T mutation in the tyrosinase gene by the CRISPR/Cas9 system. Mamm Genome 25(7–8):327–334. https://doi.org/10.1007/s00335-014-9524-0

Mohanraju P, Makarova KS, Zetsche B, Zhang F, Koonin EV, van der Oost J (2016) Diverse evolutionary roots and mechanistic variations of the CRISPR-Cas systems. Science (New York, N.Y.) 353(6299):aad5147. https://doi.org/10.1126/science.aad5147

Monaghan J, Zipfel C (2012) Plant pattern recognition receptor complexes at the plasma membrane. Curr Opin Plant Biol 15(4):349–357. https://doi.org/10.1016/j.pbi.2012.05.006

Naim F, Dugdale B, Kleidon J, Brinin A, Shand K, Waterhouse P, Dale J (2018) Gene editing the phytoene desaturase alleles of Cavendish banana using CRISPR/Cas9. Transgenic Res 27(5):451–460. https://doi.org/10.1007/s11248-018-0083-0

Nekrasov V, Staskawicz B, Weigel D, Jones JD, Kamoun S (2013) Targeted mutagenesis in the model plant Nicotiana benthamiana using Cas9 RNA-guided endonuclease. Nat Biotechnol, 31(8):691–693. https://doi.org/10.1038/nbt.2655

Nekrasov V, Wang CM, Win J et al (2017) Rapid generation of a transgene-free powdery mildew resistant tomato by genome deletion. Sci Rep 7:482

Oliva R, Ji C, Atienza-Grande G et al (2019) Broad-spectrum resistance to bacterial blight in rice using genome editing. Nat Biotechnol 37:1344–1350. https://doi.org/10.1038/s41587-019-0267-z

Oliver D, Yuan S, McSwiggin H, Yan W (2015) Pervasive genotypic mosaicism in founder mice derived from genome editing through pronuclear injection. PLoS One 10(6):e0129457. https://doi.org/10.1371/journal.pone.0129457

Ortigosa A, Gimenez-Ibanez S, Leonhardt N, Solano R (2019) Design of a bacterial speck resistant tomato by CRISPR/Cas9-mediated editing of SlJAZ2. Plant Biotechnol J 17:665–673. https://doi.org/10.1111/pbi.13006

Pan C, Ye L, Qin L et al (2016) CRISPR/Cas9-mediated efficient and heritable targeted mutagenesis in tomato plants in the first and later generations. Sci Rep 6:24765. https://doi.org/10.1038/srep24765

Paschon DE, Lussier S, Wangzor T et al (2019) Diversifying the structure of zinc finger nucleases for high-precision genome editing. Nat Commun 10:1133. https://doi.org/10.1038/s41467-019-08867-x

Peng A, Chen S, Lei T, Xu L, He Y, Wu L, Yao L, Zou X (2017) Engineering canker-resistant plants through CRISPR/Cas9-targeted editing of the susceptibility gene *CsLOB1* promoter in citrus. Plant Biotechnol J 15:1509–1519. https://doi.org/10.1111/pbi.12733

Peterson BA, Haak DC, Nishimura MT, Teixeira PJ, James SR, Dangl JL, Nimchuk ZL (2016) Genome-wide assessment of efficiency and specificity in CRISPR/Cas9 mediated multiple site targeting in arabidopsis. PLoS One 11(9):e0162169. https://doi.org/10.1371/journal.pone.0162169

Petolino JF (2015) Genome editing in plants via designed zinc finger nucleases. In Vitro Cell Dev Biol-Plant 51(1):1–8. https://doi.org/10.1007/s11627-015-9663-3

Prado JR, Segers G, Voelker T, Carson D, Dobert R, Phillips J et al (2014) Genetically engineered crops: from idea to product. Annu Rev Plant Biol 65:769–790. https://doi.org/10.1146/annurev-arplant-050213-04003

Puchta H (2005) The repair of double-strand breaks in plants: mechanisms and consequences for genome evolution. J Exp Bot 56:1–14. https://doi.org/10.1093/jxb/eri025

Puchta H, Fauser F (2013) Gene targeting in plants: 25 years later. Int J Dev Biol 57(6–8):629–637. https://doi.org/10.1387/ijdb.130194hp

Ramirez CL, Foley JE, Wright DA, Müller-Lerch F, Rahman SH, Cornu TI, Winfrey RJ, Sander JD, Fu F, Townsend JA, Cathomen T, Voytas DF, Joung JK (2008) Unexpected failure rates

for modular assembly of engineered zinc fingers. Nat Methods 5(5):374–375. https://doi.org/10.1038/nmeth0508-374
Ran Y, Liang Z, Gao C (2017) Current and future editing reagent delivery systems for plant genome editing. Sci China Life Sci 60(5):490–505. https://doi.org/10.1007/s11427-017-9022-1
Reyon D, Maeder ML, Khayter C, Tsai SQ, Foley JE, Sander JD, Joung JK (2013) Engineering customized TALE nucleases (TALENs) and TALE transcription factors by fast ligation-based automatable solid-phase high-throughput (FLASH) assembly. Curr Protoc Mol Biol, Chapter 12, Unit–12.16. https://doi.org/10.1002/0471142727.mb1216s103
Rodríguez-Leal D, Lemmon ZH, Man J, Bartlett ME, Lippman ZB (2017) Engineering quantitative trait variation for crop improvement by genome editing. Cell 171(2):470–480.e8
Roth N, Klimesch J, Dukowic-Schulze S, Pacher M, Mannuss A, Puchta H (2012) The requirement for recombination factors differs considerably between different pathways of homologous double-strand break repair in somatic plant cells. Plant J 72(5):781–790
Rouet P, Smih F, Jasin M (1994) Introduction of double-strand breaks into the genome of mouse cells by expression of a rare-cutting endonuclease. Mol Cell Biol 14:8096–8106
Sander JD, Dahlborg EJ, Goodwin MJ, Cade L, Zhang F, Cifuentes D, Curtin SJ, Blackburn JS, Thibodeau-Beganny S, Qi Y et al (2011) Selection-free zinc-finger-nuclease engineering by context-dependent assembly (CoDA). Nat Methods 8:67–69
Sandhya D, Jogam P, Allini VR et al (2020) The present and potential future methods for delivering CRISPR/Cas9 components in plants. J Genet Eng Biotechnol 18:25. https://doi.org/10.1186/s43141-020-00036-8
Santillán Martínez MI, Bracuto V, Koseoglou E et al (2020) CRISPR/Cas9-targeted mutagenesis of the tomato susceptibility gene *PMR4* for resistance against powdery mildew. BMC Plant Biol 20:284. https://doi.org/10.1186/s12870-020-02497-y
Sawyer E (2013) Editing genomes with the bacterial immune system. Scitable. Nature Publishing Group
Sera T (2005) Inhibition of virus DNA replication by artificial zinc finger proteins. J Virol 79(4):2614–2619. https://doi.org/10.1128/JVI.79.4.2614-2619.2005
Shukla VK, Doyon Y, Miller JC, DeKelver RC, Moehle EA, Worden SE, Mitchell JC, Arnold NL, Gopalan S, Meng X, Choi VM, Rock JM, Wu YY, Katibah GE, Zhifang G, McCaskill D, Simpson MA, Blakeslee B, Greenwalt SA, Butler HJ, Hinkley SJ, Zhang L, Rebar EJ, Gregory PD, Urnov FD (2009) Precise genome modification in the crop species Zea mays using zinc-finger nucleases. Nature 459(7245):437–441. Bibcode:2009Natur.459.437S. https://doi.org/10.1038/nature07992
Soyars CL, Peterson BA, Burr CA, Nimchuk ZL (2018) Cutting edge genetics: CRISPR/Cas9 editing of plant genomes. Plant Cell Physiol 59(8):1608–1620. https://doi.org/10.1093/pcp/pcy079
Stoddard BL (2011) Homing endonucleases: from microbial genetic invaders to reagents for targeted DNA modification. Structure (London, England: 1993) 19(1):7–15. https://doi.org/10.1016/j.str.2010.12.003
Stone CL, Frederick RD, Tooley PW, Luster DG, Campos B, Winegar RA et al (2018) Annotation and analysis of the mitochondrial genome of *Coniothyrium glycines*, causal agent of red leaf blotch of soybean, reveals an abundance of homing endonucleases. PLoS One 13(11):e0207062
Streubel J, Pesce C, Hutin M, Koebnik R, Boch J, Szurek B (2013) Five phylogenetically close rice SWEET genes confer TAL effector-mediated susceptibility to Xanthomonas oryzae pv. oryzae. New Phytol 200(3):808–819. https://doi.org/10.1111/nph.12411
Sun Y, Zhang X, Wu C, He Y, Ma Y, Hou H et al (2016) Engineering herbicide-resistant rice plants through CRISPR/Cas9-mediated homologous recombination of Acetolactate synthase. Mol Plant 9:628–631. https://doi.org/10.1016/j.molp.2016.01.001
Tang X, Zheng X, Qi Y, Zhang D, Cheng Y, Tang A, Zhang Y (2016) A single transcript CRISPR-Cas9 system for efficient genome editing in plants. Mol Plant 9(7):1088–1091

Tashkandi M, Ali Z, Aljedaani F, Shami A, Mahfouz MM (2018) Engineering resistance against tomato yellow leaf curl virus via the CRISPR/Cas9 system in tomato. Plant Signal Behav 13(10):e1525996. https://doi.org/10.1080/15592324.2018.1525996

Tegtmeier R, Pompili V, Singh J et al (2020) Candidate gene mapping identifies genomic variations in the fire blight susceptibility genes HIPM and DIPM across the Malus germplasm. Sci Rep 10:16317

Terns MP (2018) CRISPR-based technologies: impact of RNA-targeting systems. Mol Cell 72(3):404–412. https://doi.org/10.1016/j.molcel.2018.09.018

Thomazella PT, Brail Q, Dahlbeck D, Staskawicz B (2016) CRISPR-Cas9 mediated mutagenesis of a DMR6 ortholog in tomato confers broad-spectrum disease resistance. BioRxiv. https://doi.org/10.1101/064824

Townsend JA, Wright DA, Winfrey RJ, Fu F, Maeder ML, Joung JK, Voytas DF (2009) High-frequency modification of plant genes using engineered zinc-finger nucleases. Nature 459(7245):442–445. https://doi.org/10.1038/nature07845

Tripathi JN, Ntui VO, Ron M et al (2019) CRISPR/Cas9 editing of endogenous banana streak virus in the B genome of Musa spp. overcomes a major challenge in banana breeding. Commun Biol 2:46. https://doi.org/10.1038/s42003-019-0288-7

Upadhyay SK, Kumar J, Alok A, Tuli R (2013) RNA-guided genome editing for target gene mutations in wheat. G3 (Bethesda, Md.) 3(12):2233–2238. https://doi.org/10.1534/g3.113.008847

Vanyushin BF, Ashapkin VV (2011) DNA methylation in higher plants: past, present and future. Biochim Biophys Acta 1809(8):360–368. https://doi.org/10.1016/j.bbagrm.2011.04.006. Epub 2011 Apr 28. PMID: 21549230

Voytas DF, Gao C (2014) Precision genome engineering and agriculture: opportunities and regulatory challenges. PLoS Biol 12(6):e1001877. https://doi.org/10.1371/journal.pbio.1001877

Wada N, Ueta R, Osakabe Y et al (2020) Precision genome editing in plants: state-of-the-art in CRISPR/Cas9-based genome engineering. BMC Plant Biol 20:234. https://doi.org/10.1186/s12870-020-02385-5

Wang Y, Cheng X, Shan Q, Zhang Y, Liu J, Gao C, Qiu JL (2014) Simultaneous editing of three homoeoalleles in hexaploid bread wheat confers heritable resistance to powdery mildew. Nat Biotechnol 32(9):947–951. https://doi.org/10.1038/nbt.2969

Wang L, Chen L, Li R, Zhao R, Yang M, Sheng J, Shen L (2017) Reduced drought tolerance by CRISPR/Cas9-mediated SlMAPK3 mutagenesis in tomato plants. J Agric Food Chem 65(39):8674–8682. https://doi.org/10.1021/acs.jafc.7b02745

Wiedenheft B, Zhou K, Jinek M, Coyle SM, Ma W, Doudna JA (2009) Structural basis for DNase activity of a conserved protein implicated in CRISPR-mediated genome defense. Structure 17(6):904–912. https://doi.org/10.1016/j.str.2009.03.019. PMID: 19523907

Woo J, Kim J, Kwon S, et al. DNA-free genome editing in plants with preassembled CRISPR-Cas9 ribonucleoproteins. Nat Biotechnol 33:1162–1164 (2015). https://doi.org/10.1038/nbt.3389

Xie K, Yang Y (2013) RNA-guided genome editing in plants using a CRISPR–Cas system. Mol Plant 6(6):1975–1983. https://doi.org/10.1093/mp/sst119

Xie K, Minkenberg B, Yang Y (2015) Boosting CRISPR/Cas9 multiplex editing capability with the endogenous tRNA-processing system. Proc Natl Acad Sci 112(11):3570–3575. https://doi.org/10.1073/pnas.1420294112

Xu Z, Xu X, Gong Q, Li Z, Li Y, Wang S, Yang Y, Ma W, Liu L, Zhu B, Zou L, Chen G (2019) Engineering broad-spectrum bacterial blight resistance by simultaneously disrupting variable TALE-binding elements of multiple susceptibility genes in rice. Mol Plant 12(11):1434–1446. https://doi.org/10.1016/j.molp.2019.08.006

Yee J-K (2016) Off-target effects of engineered nucleases. FEBS J 283:3239–3248. https://doi.org/10.1111/febs.13760

Zaidi SSA, Vanderschuren H, Qaim M et al (2019) New plant breeding technologies for food security. Science 363:1390–1391

Zetsche B, Heidenreich M, Mohanraju P, Fedorova I, Kneppers J, DeGennaro EM, Winblad N, Choudhury SR, Abudayyeh OO, Gootenberg JS, Wu WY, Scott DA, Severinov K, van der Oost

J, Zhang F (2017) Multiplex gene editing by CRISPR-Cpf1 using a single crRNA array. Nat Biotechnol 35(1):31–34. https://doi.org/10.1038/nbt.3737

Zhang F, Maeder ML, Unger-Wallace E, Hoshaw JP, Reyon D, Christian M, Li X, Pierick CJ, Dobbs D, Peterson T, Joung JK, Voytas DF (2010) High frequency targeted mutagenesis in Arabidopsis thaliana using zinc finger nucleases. Proc Natl Acad Sci U S A 107(26):12028–12033. https://doi.org/10.1073/pnas.0914991107

Zhang H, Zhang J, Wei P et al (2014) The CRISPR/Cas9 system produces specific and homozygous targeted gene editing in rice in one generation. Plant Biotechnol J 12:797–807

Zhang Y, Bai Y, Wu G, Zou S, Chen Y, Gao C, Tang D (2017) Simultaneous modification of three homoeologs of TaEDR1 by genome editing enhances powdery mildew resistance in wheat. Plant J Cell Mol Biol 91(4):714–724. https://doi.org/10.1111/tpj.13599

Zhou Y, Zhu S, Cai C, Yuan P, Li C, Huang Y, Wei W (2014) High-throughput screening of a CRISPR/Cas9 library for functional genomics in human cells. Nature 509(7501):487–491. https://doi.org/10.1038/nature13166

Zhou J, Peng Z, Long J, Sosso D, Liu B, Eom JS, Huang S, Liu S, Vera Cruz C, Frommer WB, White FF, Yang B (2015) Gene targeting by the TAL effector PthXo2 reveals cryptic resistance gene for bacterial blight of rice. Plant J Cell Mol Biol 82(4):632–643. https://doi.org/10.1111/tpj.12838

Zhu H, Li C, Gao C (2020) Applications of CRISPR-Cas in agriculture and plant biotechnology. Nat Rev Mol Cell Biol 21(11):661–677. https://doi.org/10.1038/s41580-020-00288-9

Chapter 2
Synthetic Promoters in Regulating Disease Gene Expression

Badrinath Khadanga, Tsheten Sherpa, Jeky Chanwala and Nrisingha Dey

Abstract Precisely controlling the expression of a gene has a wide range of functions. These controls are primarily attributed to the gene's promoter region, where RNA polymerases bind and initiate the transcription. The promoter also consists of various *cis*-elements, which act as a regulatory region. Native/naturally occurring promoters are useful to some degree, but a synthetic promoter is preferred because of its smaller size, high transcriptional efficiency, and inducibility. To develop resistant plants against different phytopathogens, controlling the complex gene regulatory mechanism is essential. By reconstructing the *cis*-elements in the synthetic promoter, it is possible to express various pathogen resistance genes in the plant precisely. This expression of the resistance gene can drastically lower the pathogenesis of pests and therefore increase the survival of pest-infested plants. This chapter has discussed promoters and the development of pathogen-inducible synthetic promoters in depth. The presence of various *cis*-elements in the promoter seems to be the leading factor in controlling the regulation of the promoter. Therefore, by engineering these elements, different types of promoters such as inducible promoters, tissue-spatial specific promoters, and constitutive promoters can be developed and used to drive various pathogen resistance genes.

Keywords Synthetic promoter · *cis*-elements · phytopathogens · gene regulation

B. Khadanga · N. Dey (✉)
Division of Plant and Microbial Biotechnology, Institute of Life Sciences, Bhubaneswar, Odisha, India
e-mail: ndey@ils.res.in

T. Sherpa
Division of Plant and Microbial Biotechnology, Institute of Life Sciences, Bhubaneswar, Odisha, India

Regional Centre for Biotechnology, Faridabad, Haryana, India

J. Chanwala
Division of Plant and Microbial Biotechnology, Institute of Life Sciences, NALCO Square, Bhubaneswar, Odisha, India

© The Author(s), under exclusive license to Springer Nature Switzerland AG 2022

S. H. Wani et al. (eds.), *Transcription Factors for Biotic Stress Tolerance in Plants*, https://doi.org/10.1007/978-3-031-12990-2_2

2.1 Introduction

Globally, pathogen attacks has caused 15% losses in food production and has become a challenge in achieving food security. Other factors such as, climate change, food crisis, new pathogens, and weeds are also major threats to food crops. The extensive use of fertilizers, pesticides, and other chemicals leads to soil and water quality degradation that affects long-term agricultural production negatively. Additionally, the use of these chemicals also deteriorates the consumer's health. In this scenario, fulfilling the demand and improvement in food production has become a great challenge to meet the "no hunger" goal. Increasing population and decreasing landmass make this challenge tougher. Plant molecular farming may help in managing various environmental and biotic stress to improve agricultural production. Genetically modified stress-resistant plants should be developed to balance the quality and quantity of agricultural production. Employment of improved recombinant DNA technologies (rDT) in plant science has changed the background scenario of crop productivity in every aspects. The building of biological devices and different computational algorithms helps to boost plant molecular farming (Tito et al. 2018). Early prediction of molecular activities in a biological system is getting easier through bioinformatics-based software tools. The plant molecular farming emphasizes future goals to establish a sustainable agricultural production for upcoming generations. Expansion in rDT and synthetic biology multiplies the novel inventions for critical analysis and synthesis of new GMOs. The traditional breeding approach is a time-consuming and laborious process. To overcome traditional methods, understanding the depth of the plant genome at its molecular level is necessary. By understanding molecular arrangement, modern approaches can be implemented to increase the activities of native sequences.

Promoter is the fundamental element that initiates and regulates the transcription of a gene. A promoter contains different regulatory elements, which can enhance, or repress the transcription of a gene. Plant promoters are usually made up of different domains that are aligned in specific orientations and interact with various transcription factors. These domains are modular in nature and can be rearranged or interchanged to generate a novel gene-expressing module (Ranjan et al. 2011). These artificial promoters are called synthetic promoters. These synthetic promoters are usually more efficient than native promoters as these are designed in such a way that the spacing, copy number of *cis*-elements, and the interaction of transcription factors (TFs) with the promoter are optimum (Mehrotra et al. 2011). Additionally, synthetic promoters can be designed as an inducible, temporal-spatial, or constitutively expressing transcription module. These modules can have several implications: (1) during the gene-stacking approach, where multiple genes are expressed, different synthetic promoters can be used to avoid homology-dependent gene silencing1 (2) For metabolic engineering, different synthetic promoters with varying expressing capacity can be used to express a gene; (3) also, synthetic promoter use has a huge advantage in stress biology research as genes can be expressed using various stress-inducible promoters; and (4) molecular biology research requires expressing various genes in planta for studying its transcriptome, proteome, and metabolome profile. This expression of gene might require some extraordinary situations, such as inducible expression, temporal-spatial expression, or constitutive expression (Aysha et al. 2018). So for these instances, the use of synthetic promoter with defined expression is the key factor in driving the genes.

One of the essential objectives of plant biotechnology is to develop plants that are resistant to different stresses and nutritionally enriched. The non-transgenic-based approaches have been widely used in the past, such as classical breeding and marker-assisted breeding. These strategies have no doubt been a great success in elevating global crops yields, but complications such as (1) lack of R or S gene in the gene pool, (2) sexual incompatibility, (3) hybrid F1 sterility, (4) difficulty in segregating out undesirable traits, and (5) environmental pollution due to pesticides warrant us to search for other tactics (Bisht et al. 2019). Transgenic approaches for biotic stress management are an excellent alternative to the impending problems of agriculture. The success of Bt plants, viz. Bt maize, Bt brinjal, and Bt cotton, where the insecticidal gene from *Bacillus thuringienesis* is inserted into plants, is proof of the importance of transgenic plants in agronomy and pest management (Gassmann and Hutchison 2012). To express these various disease resistance genes, it was predicted that a strong constitutive promoter would suffice for the resistance against pathogens, but, disappointingly in many cases, the overexpression of resistance genes led to diminished growth and development (Rushton et al. 2002). It was then realized that the expression of the disease resistance gene would require a tightly controlled expression. Therefore, pathogen-inducible promoters would be of great advantage for expressing these genes. These pathogen-inducible promoters are expressed only in the presence of an inducer, hence controlling the amount of gene expression. Other types of promoters, such as spatial-temporal promoters expressed only during specific developmental stages and in particular tissue have also been valuable in expressing various genes. Recent advances in synthetic biology have led to the development of more tightly controlled and more specific promoters, enabling the expression of various types of disease resistance genes with minimal cost to plant fitness.

2.2 The Chemistry (Fundamental Structure) of Gene Regulation

Gene is responsible for every characteristics of living organism. The four nucleotide bases (A, T, G, and C) are arranged randomly and execute a specific character code. However, the expression of a gene is regulated by another stretch of DNA called promoter. A gene cassette has three parts: a promoter (regulating region), gene (coding region), and terminator (ending region). Promoter is a noncoding part of DNA, works as a regulatory component, and is present in the upstream of the gene. In the case of prokaryotic organisms, genes are mainly regulated by a single promoter (poly-cistronic), while in eukaryotic organisms, a single promoter is responsible for a expression of single gene (mono-cistronic). At the molecular level of the promoter, it is divided into three parts; mainly the core promoter domain containing TATAA box, the Initiator (Inr), Bre, DPE or DCE elements where various transcription factors binds and start transcription in the presence of RNA pol II. It is the minimum requirement for initiating transcription and is presents 20–40 bp upstream of transcription start site (TSS, +1). So it is also called a minimal promoter. The middle part of a promoter is called a proximal promoter, where various *cis-regulatory* elements (CREs) are present. CREs can be of various types such as silencer (decrease transcription rate), enhancer (increase transcription rate), or insulators. Under different environmental conditions, TFs bind to specific *cis-regulatory* element regions and regulate the gene function. The end portion of a promoter is called a distal promoter, which are present far from the core promoter region and it also can regulated the promoter activity. Hence, a promoter is a key element for gene regulation due to the presence of different regulatory modules in different regions. Based on gene expression, promoters can be categorized into constitutive, tissue-specific, and inducible (Dey et al. 2015). Reconstructing of a promoter can be done by manipulating endogenous/native promoter sequences as per desired gene expression. This newly synthesized promoter is called a synthetic promoter. As we know, CRE modules play an important role in gene expression in every aspect. The rearrangements of *cis*-modules can drive a transgene constitutively or in stress or in a specific tissue. *Cis–trans* element interaction and coordination between them inside the cell cause gene expression (Shrestha et al. 2018). The presence of various *cis*-motifs and their association with respective TFs can make a promoter activity more stronger and inducible. Most commonly found CREs are TATA-Box, CAAT-box, GC-motifs, WRKY, MYB, MYC, TGA, etc., which have a prominent role in gene regulations. From sequence analysis of naturally available biotic stress-inducible promoter, it is found that *cis-elements* like GCC, JERE, W1, W2-boxes, S, Gst1, and D are present in the upstream activating sequence (UAS) region. Understanding of these biotic stress-induced promoters can facilitate the development of novel stress-inducible promoters with strong gene driving capacity by tailoring native sequences. A synthetic promoter can also work as a bidirectional regulator with expression of two genes.

2.3 Construction of a Biotic Stress-Inducible Synthetic Promoter

Promoters are a prime need for a gene. Native promoters tend to drive genes at a normal rate due to positive and negative regulators, and the length of promoter is also long. Different studies have shown increased driving efficiency in modified promoters (negative regulators removed) than compared to native full-length promoters. Mainly plant infecting pararetroviral promoters are used to generate synthetic promoters due to them having similar *cis*-modulatory elements resembling plant promoters and strong expression. Such synthetically developed promoter fragments have the capability for over-expression of a gene. The enhanced CaMV35S synthetic promoter is widely used for driving transgenes in different plant systems. A list of developed novel plant synthetic promoters and their expression types is given in Table 2.1.

Construction of synthetic promoters using molecular techniques like domain hybridization, DNA shuffling, and CRISPR-mediated gene editing enhance the promoter's specificity and strength. It has been seen that increasing the *cis*-elements' copy number inside a promoter makes the promoter more effective for driving of the

Table 2.1 Various biotic-stress inducible synthetic promoters expressed in plants

Promoter name	Source	Inducer	Species tested	References
4X CCTC	CaMV35S minimal promoter with Potato Pi transporter 3 (StPT3) promoter regions	Fungus inducible	Potato and lotus	Lota et al. (2013)
4X SARE; 4X ERE; 4X PR1; 4X JAR	CaMV35S minimal promoter with hormone/pathogen-inducible elements	Bacterial; hormonal inducible	Tobacco; *Arabidopsis*	Liu et al. (2013)
4X GCC	CaMV35S minimal promoter with *Arabidopsis* PDF1.2 promoter	Jasmonic acid	*Arabidopsis*	Van der Does et al. (2013)
2 X W2/2 X S/2 X D, 4 X W2/4 X S	CaMV35S minimal promoter with *cis*-elements containing W1, W2, GCC, JERE, S, Gst1, and D	Pathogen and wound inducible	*Arabidopsis*	Rushton et al. (2002)
CMPG1	CaMV35S minimal promoter with a dimerized form of the F element of CMPG1 promoter	Elicitor (Pep25) responsiveness (bacterial derived)	*Arabidopsis*, *Petroselinum crispum*	Heise et al. (2002)
(SP), SP-EE, SP-FF, and SP-FFEE	CaMV35S minimal promoter with E17 and F *cis*-acting pathogen-inducible region	Fungal elicitors, salicylic and jasmonic acid inducible	*Brassica napus*	Shokouhifar et al. (2011)
EFCFS-HS-1, EFCFS-HS-2, EFCFS-HS-3	Figwort mosaic virus CP and UAS region with additional Dof-1 (AAAG) *cis*-elements	Salicylic and jasmonic acid inducible	*Nicotiana tabacum*	Ranjan and Dey (2012)
SP-DDEE	Parsley D, E17 elements + minimal promoter	Chitin and fungal elicitor inducible	*Brassica napus*	Moradyar et al. (2016)

gene. Domain hybridization is another method to develop a synthetic promoter. It is a molecular shuffling between the enhancer region (UAS) and the core promoter region (CP). The DNA shuffling is another method for synthetic promoter construction, where desired promoter fragments are mixed, cut down by *DNAase*I, and allow them to be annealed by PCR reaction. A library of synthetic fragments will be made and the desired promoter is screened. An advanced CRISPR-Cas9 genome editing tool is a boon for molecular biology. By using this approach, directly targeting CREs through sg-RNA/dCas9 for regulating the activity of promoter is an attractive technique.

Normally constitutive promoter for disease gene regulation is not usually required as the constitutive expression of a defense gene may cause a negative impact on other cells and increase cytotoxicity. So, making a biotic stress-inducible synthetic promoter for disease gene regulation is a favorable option. For this the *cis*-architecture of the synthetic promoter should contain specific CREs like W-box [(T)TGAC(C/T)], D-box (GGAACC), GCC-box (AGCCGCC), JERE (AGACCGCC), DRE (TACCGACAT), and S-box (AGCCACC) which are biotic stress inducible CREs. Under pathogen attack or wound in plants or any other biotic stress, this promoter will induce overexpressing the pathogen-resistant gene and protect from pathogen attacks (Muthusamy et al. 2017).

2.4 Role of Synthetic Promoter and TFs for Gene Regulation

As discussed, the promoter is the key element necessary for the control and drive of a gene. The *cis*-acting elements present on the promoter define the promoter type. The activity of the promoter depends upon *cis–trans* interaction. *Trans*-acting elements are known as TFs, which are nuclear proteins that binds to specific *cis*-motifs. It is a molecular interaction of these DNA–protein that regulates and initiates the promoter. Under different stresses, stimuli/signals are formed which initiates the expression of specific TF's. These TFs in turn interact with *cis*-elements present on the promoter, and the *cis–trans* interaction is jointly established for a successful expression of the gene (Shrestha et al. 2018).

In case of synthetic promoter, the arrangement of different *cis*-elements can led to a differential expression of the gene depending on different environmental conditions or in which cells it has to be expressed. The fate of gene expression can be optimized with the synthesis of novel promoters with assembling of various CREs. The distance between the CREs, core promoter, and other essential segments like enhancers, silencers, and associated TFs also plays vital role for construction of synthetic promoters. In vivo transgene expression at the region of interest (ROI) requires a well-arranged *cis*-modules in the promoter. CREs present in the synthetic promoter are activated after strong attachment with specific TFs. These TFs not only boost the transcription rate but also control other regulatory elements associated with expressed genes. The transcription frequency is regulated by the *cis–trans* interaction in response to environmental stimuli. Shrestha et al. 2018 proposed that the perfect combination of *cis–trans* engineering can enhance the environmental-specific plant gene regulation. Computational biology software, binding-site estimation suite of tools (BEST), would be a platform for the prediction and evaluation of novel plant CREs under biotic and abiotic stresses. Not only synthetic promoters but also synthetic TFs can control gene expression. Considering *cis*-sequences present in the promoter region, synthetic TFs can be designed for binding with DNA binding domains (DBDs) and regulating transgene expression in plants.

A new approach of synthetic TFs for activation of native gene and transgene in plant system also emphasizes the transcription level. Synthetic TFs, e.g., zinc finger transcription factors (ZN-TFs) and transcription activator-like effector transcription factors (TALE-TFs), are implemented to improve the transcription rate in genetically modified plants. Synthetic ZN-TFs domain was used to bind with respective gene promoters for activation of genes, β-ketoacyl-acyl-carrier-protein synthase II (*KASII*) in *Brassica napus*, and *APETALA1* in *Arabidopsis thaliana*. Positively strong gene expression was found in respective transgenic plants. The study proves that the synthetic zinc finger protein (from ZN-TFs) successfully binds to the promoter site of the gene and regulates its expression. Such a type of molecular engineering facilitates plant synthetic biology. Synthetic TALE-TF domain can easily be edited for a specific DNA binding domain, and it has been used as a transcriptional repressor in *A. thaliana* (Mahfouz et al. 2012). Therefore, such types of molecular tools have a vital role in plant genetic engineering.

2.5 The Presence of CREs (*Cis*-Regulatory Elements) Inducing Strength and Effectiveness

CREs or *cis*-regulatory elements are the noncoding segment of DNA that plays a central role in gene transcription regulation. It is the building block of the promoter and comprises multiple binding sites for transcription factors. These TFs then in turn function as activators, repressors, and insulators of transcription and can also directly associate with general transcription factors (GTFs) and other chromatin architectural proteins (Noonan and McCallion 2010). The modular nature of the *cis*-elements is one of the main reasons for the varied expression pattern of different promoters, and therefore identifying and understanding CREs and their interaction with different regulatory proteins (TFs) is of utmost importance in unraveling the complex gene regulatory mechanisms.

Table 2.2 shows different types of pathogen-related CREs that have been identified as W-Box, GCC-like elements, etc., which associate with various families of TFs such as WRKYs, bZIPs, bHLH, Dofs, ERFs, and Mybs and have been found to play major roles in the plant's defense-related signaling (Chen et al. 2002). By combining various CREs described above, many synthetic promoters were developed, which were induced by different plant defense-associated signaling molecules such as salicylic acid (SA), ethylene (ET), and jasmonic acid (JA) (Mazarei et al. 2008).

Table 2.2 Pathogen-inducible CREs and their respective transcription factors

CRE Motif name	CRE sequence	Transcription factor	Inducer	References
W box	(C/T)TGAC(T/C)	WRKY	Fungus, bacteria, oomycetes, salicylic acid	Eulgem et al. (1999), Kumar et al. (2012), Li et al. (2004)
JERE (jasmonic acid responsive element)	AGACCACC	AP2/ERF	Jasmonic acid, fungus, bacteria, yeast elicitors	Gurr and Rushton (2005), Rushton et al. (2002)
SARE	TTCGACCTCC	Unknown	Salicylic acid	Shah and Klessig (1996)
S box	AGCCACC	AP2/ERF	Oomycete, fungal, and bacterial elicitors	Kirsch et al. (2000), Rushton et al. (2002)
GT1 element	G(T/A)AA(T/A)	GT1	Various pathogens; salicylic acid inducible	Park et al. (2004), Zhou (1999)
GCC box	AGCCGCC	AP2/ERF	Jasmonic acid inducible	Van der Does et al. (2013)
as-1	TGACG	bZIP (TGA/OBF)	Salicylic acid inducible	Garretón et al. (2002), Sarkar et al. (2018)
Dof-binding site	AAAG	Dof	Salicylic and jasmonic acid	Ranjan and Dey (2012), Yanagisawa (2004)
MRE	A(A/C)C(A/T)A(A/C)C	Myb	Fungal elicitor	Gurr and Rushton (2005), Rushton and Somssich (1998)

In the same study, PR1 and NPR1 *cis*-elements containing promoters were tested in transgenic tobacco, and they were found to be induced by *alfalfa mosaic virus* (AMV) infection. So using a combination of specific CRE in a promoter, temporal and spatial features of gene expression can be controlled.

The presence of defined *cis*-elements in a synthetic promoter can be very useful in studying the precise function of that element. The study by Rushton et al. (2002) shed some important light on the copy number of various pathogen-inducible CREs and its implications. The copy number of pathogen-related CREs such as S box, D box, and W box from various promoters of parsley (*Petroselinum crispum*) was increased and was found to have a significant difference in the gene expression rate. Increasing the number of CREs from one to eight was found to significantly increase the transcription rate, which is probably caused by the increase in the binding site of TFs (Fig. 2.1). These findings were tested in vivo, and the results were consistent with the transient assay data. Although the best copy number for use in synthetic promoters was found to be two, as it showed the highest signal: noise ratio.

Another important factor that determines the strength and specificity of the promoter is the spacing between CREs and the minimal promoter. The study by

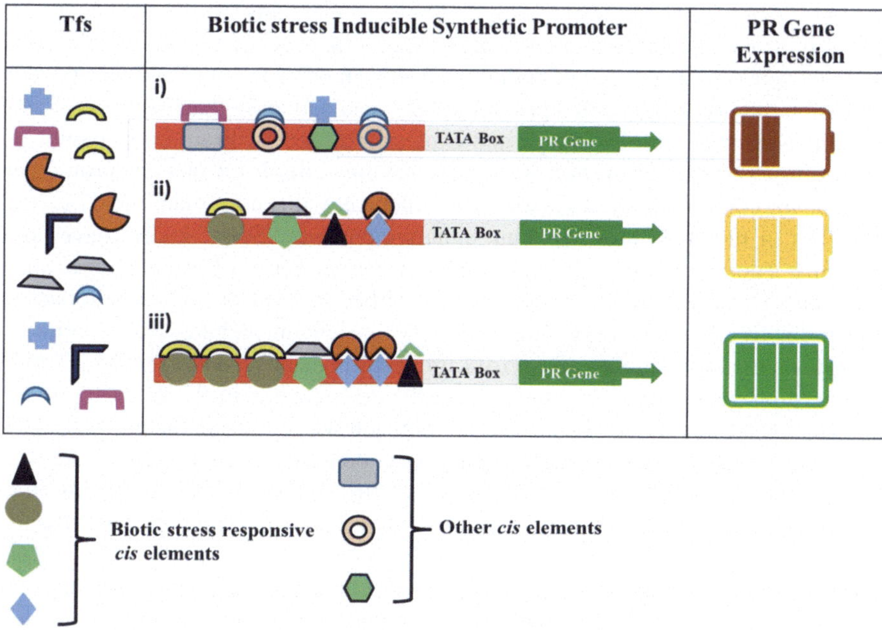

Fig. 2.1 Schematic representation of synthetic promoter inducibility in the presence of biotic stress-responsive *cis*-elements. (i) Synthetic promoter is not showing any significant PR gene expression due to the absence of biotic stress-responsive *cis*-elements. (ii) Synthetic promoter with moderate PR gene expression in the presence of single copy number biotic stress-responsive *cis*-elements. (iii) Increasing copy number of biotic stress-responsive *cis*-elements; synthetic promoter activity induced significantly with PR gene expression

Krawczyk et al. (2002) found that the as-1 element was most effective in binding to TGA factors (as-1 binding protein) when the spaces between two TGACG motifs were 10–12 bp. It appears that if the CREs are spaced too close to one another, they lose their ability to bind to TFs. This is probably due to the fact that a better spaced CRE will have a favorable situation for the binding of protein factors. The importance of spacing between different CREs and the minimal promoter was also analyzed using pathogen-induced CRE (ACGT). The presence of ACGT significantly enhanced the expression of promoter twofold to threefold when placed 100 bp upstream of the TATA box, whereas two ACGT separated by five nucleotides enhanced expression by sixfold when placed 50 bp upstream of the TATA box (Mehrotra et al. 2005). When the CRE is placed too close to the minimal promoter, the activity of the promoters drastically reduces. This may be because, as the CRE is close to the minimal promoter, the TFs binding to it have to compete with the general transcription factors that are involved in the formation of the pre-initiation complex. Therefore, it is best to have CRE located a minimum of 50 bp upstream of the minimal promoter region.

Another criterion that has been observed while trying to make good synthetic promoters is that those promoters that have a diverse *cis*-elements are more effective than those that have multiple copies of the same *cis*-elements (Rushton et al. 2002). In a biotic stress-inducible promoter, high expression is needed in an infection site and low expression is needed in noninfected sites in order to avoid any nonspecific expression. It was observed that promoter that contained more than one type of *cis*-elements was better able to induce and express transgene with a lower amount of nonspecific expression. This may be because having multiple *cis*-element promoters that contain combinations of *cis*-elements interacts with more than one TFs, and these TFs in turn lead to a different signaling pathway. Some studies have also found that a combination of different CREs in specific series operates as one functional unit. This property of a promoter was studied in PMT promoters in tobacco. PMT promoters contain a GAG fragment in its *cis*-element, and they can be induced by jasmonate and wounding. The GAG fragment consists of a G-box, an AT-rich spacer region, and a GCC-like box. When the *cis*-elements in the GAG fragment were studied separately or when even one fragment was removed, the promoter lost its activity. Only when the promoter contained G box-AT rich region-GCC-like box in a specific series did the promoter show strong expression, hence proving that these three segments act as one functional unit to induce the gene expression (Sears et al. 2014).

Understanding CREs and their signaling mechanism is not only important in designing "designer promoter," but it is also important in understanding the regulatory network of various genes. With better knowledge and innovations, more robust and more effective ways can be developed for better management of plant diseases.

2.6 Plant Defense Mechanism Under Biotic Stress

The defense mechanism of the plant is inherited. Generally, the plant comes under two types of stresses, i.e., biotic and abiotic stresses. Plants defend these stresses through the impletion of different proteins, enzymes, secondary metabolites, and morphological and structural barriers. The stress developed in the plant due to infection caused by biotic agents like microorganisms (virus, bacteria, and fungi), insects, weeds, pests, nematodes, and intraspecific competition is called biotic stress. These biotic factors severely affect and damage the plant, so they are called phytopathogens. During biotic stress, the plant adapts some physiological mechanisms to protect itself. Like vertebrates, the plant has different mechanisms to fight various pathogens. It defends the biotic stress through a defense system. At the preliminary stage of pathogen attack, the plant releases reactive oxygen species (ROS) and destroys the pathogens that are overspread. On the later stage, the plant protects itself by introducing proteins, enzymes, chemical compounds, different structural barriers, etc. These provide resistance and strength to plants to overcome biotic stress. Some hormonal signaling molecules like abscisic acid (ABA), salicylic acid (SA), jasmonic acid (JA), MeJA, and ethylene have a crucial role in response to biotic stress. Under biotic stress, pathogenesis-related (PR) proteins are released to increase plant resistance against pathogens (Madani et al. 2019).

Plant defense mechanisms range from anatomy, physiology, biochemistry, genetics, development, and evolution according to their molecular dynamics. Genetic regulation is very crucial to protect plants from biotic stress. A number of PR genes are expressed through different signaling pathways to resist biotic stress. Mainly PR gene-expressed proteins have leucine-rich repeats (LRR). Therefore, it is a signature for knowing pathogenesis-related proteins. The plant has its own mechanism to resist different pathogens by employing physiological, chemical, molecular, and genetical processes. In the plant innate immune system, a conserved recognition molecular structure known as plant pathogen recognition receptors (PRRs) detects the type of biotic agents that infect the plant through pathogen-associated molecular patterns (PAMP) or microbes-associated molecular patterns (MAMP). After plant–pathogen interaction, the PAMP-triggered immunity (PTI) system activates to fight against it. The PTI system includes MAP kinase cascade to activate the transcriptional regulators associated with PTI-related gene expression. On the other side, infecting microbes produce virulence chemicals to inhibit the action of PTI and spread disease inside the host cell. The toxic chemicals from pathogens, also known as effector, are recognized by the plant innate immune system through intracellular immune receptors. These intracellular immune receptors, called resistant proteins, identify the type of effector. Due to this effector, plant induces an immunity, called effector-triggered immunity (ETI). The induction of ETI is superficial against the pathogens and expresses the resistant gene (*R* gene). These pathogenesis-related (PR) proteins are one of the key compounds of plant innate immunity protected against infected cells and inhibit the spreading of infection. These endogenous PR proteins have a hallmark of LRR. Systemic acquired resistance (SAR) is one type of plant immunity system that develops under primary infection by pathogens. Such type of immunity shows strong resistance against a number of consequent pathogenic infections for a longer period. Molecular signals (signals from infected host cell/salicylic acid) control the activity of SAR (Han and Jung 2013).

2.7 Plant Defense Genes and Their Functions

The plant defense system is regulated by two types of genes, namely, *R* gene (resistant gene) and *S* gene (susceptible gene). Mainly the *R* gene is involved in pathogen resistance and defending against it. Effector triggered immunity (ETI) employs the expression of the *R* gene and production of a specific NB-LRR (nucleotide-binding leucine-rich repeats) class of protein, which can recognize the respective pathogen effectors through a gene-for-gene system. Proteins from *R* genes have been divided into five classes considering their structure and functions:

1. Serine–threonine protein kinase R protein (e.g., *pto*), which acts in signal transduction.
2. Extracellular transmembrane R protein (e.g., *Xa21* of rice), which is involved in receiving and transmitting of kinase-like protein signals.

3. Cytoplasmic R protein-1 (e.g., *N1* gene of tobacco, *L6* gene of flax, *RPP5* of *Arabidopsis*), which acts as a receptor and plays an important role in transfer of TFs into the nucleus. It contains a NB-LRR conserved domain and TIR (Toll like receptor) that helps in TF translocation from the cytoplasm to the nucleus and activates the hypersensitive response (HR)-related genes.
4. Cytoplasmic R protein-2 (e.g., *RPS2*, *RPM1*) also contains NB-LRR domain, but instead of TIR, it has coiled leucine zipper domain.
5. Extracellular cell membrane R protein (e.g., *cf2-cf9* genes of tomatoes) having a common LRR, present outside of the cell membrane but connected to the membrane by a transmembrane anchor. The extracellular receiving of effector molecules such as R proteins plays key role and activates the corresponding gene expression (Edition).

The constant arms race between the pathogen and host has led to the evolution of new strategies by pathogens to increase compatibility with host and to evade or suppress the host's immune response. In contrast to the *R* gene, there is another type of gene called the susceptibility gene or *S* gene. These plant gene promotes or facilitates the infection of the pathogens. Various *S* genes have been identified and categorized mainly into three types: (a) genes that allow compatibility with pathogen and promote host–pathogen recognition, (b) genes that inhibit immune response of host, and (c) genes that sustain the required nutritional or structural needs (van Schie and Takken 2014). *S* gene-mediated resistance has been generated for various cultivars, by silencing or interfering these genes. In a recent study, promoter engineering using the CRISPR/Cas9 of *S* gene present in rice called SWEET gene was done (Fig. 2.2). The gene encodes a sucrose transporter gene, and many species of the *Xanthomonas* were found to upregulate this gene to increase the pathogen's nutritional requirements by transporting sugars to the infected sites (Chen et al. 2010). A species of *Xanthomonas* called *Xanthomonas oryzae* pv. *oryzae* (*Xoo*) causes rice bacterial blight disease, which is a very devastating disease in rice fields. Xoo after infection has been found to release effector molecules called TALE (transcriptional activator like effector), which goes and binds to the region on the promoter called the effector binding region (EBE) on different SWEET family of genes. This binding in turn upregulates the gene. Using CRISPR/Cas9, the EBE region of the promoter of three SWEET genes, *OsSWEET11*, *OsSWEET13*, and *OsSWEET14*, was mutated, and these mutants were found to be resistant to a wide variety of *Xoo* strains collected from various regions of China and Philippians (Xu et al. 2019). Another similar study was done in citrus plants where the EBE was mutated in the promoter region of CsLOB1 and the *S* gene promoted the growth of pathogens and formation of erumpent pastule. These mutants were found to be resistant to *Xanthomonas citri subso. Citri* (*Xcc*) causes serious disease called citrus canker (Jia et al. 2017; Peng et al. 2017). Study on these types of relationship between pathogen and host can most certainly pave the way to a wide variety of pathogen-resistant plants.

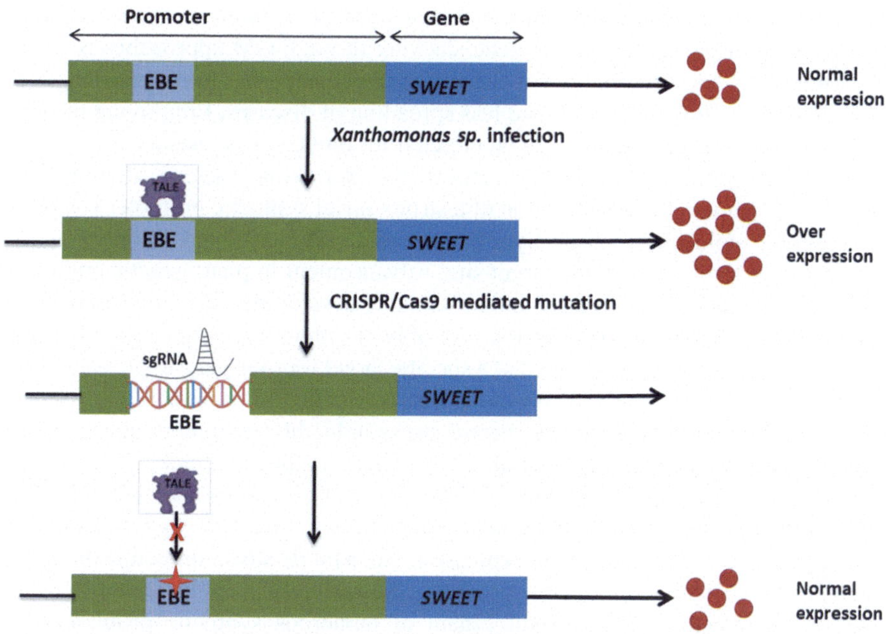

Fig. 2.2 Gene editing through CRISPR-Cas9 approaches. Mutation in the EBE (effector binding element) of ***SWEET*** gene promoter using CRISPR-Cas9 leads to pathogen-released elicitor (TALE) not recognizing the EBE region, and this leads to disease resistance

2.8 Synthetic Promoter as a Powerful Tool for Plants

Environmental and climatic changes are increasing day by day and imposing an effect on other living organisms. Due to these changes, many pathogenic organisms became more powerful in overcoming PR genes, and there may be a high infection rate. In a study, it is found that plants infected with the pathogen *Phytophthora* are resistant to different PR genes. Such type changes are a threat to future days (Jacobsen et al. 2009). Considering the present challenges, some efforts must have taken to overcome the future scenarios like (a) resistance capacity increase, (b) increased resilience of PR genes, and (c) enlarging the resistant variety. Using traditional breeding approaches for selected PR genes, developing a stress-resistant candidate is time consuming. While transferring PR genes through the crossing process, it is difficult to express the trait in that variety due to the low recombination of undesired traits linked to PR genes.

With the advancement of plant science, traditional approaches can be replaced by genetically modified (GM) techniques. Interestingly, selection of desired PR gene candidates from various plants and isolating clones particularly inside a suitable gene cassette (T-DNA) for plant expression are performed. After that, by employing *Agrobacterium-mediated* transformation, the desired gene is translocated into the plant cell. In this way, more than one PR gene can be transferred into the plant genome. This polyculture can make the plant more resistant against a broad spectrum of pathogens. PR gene stacking through GM approaches is a reliable strategy for developing new resistant variety considering future challenges to overcome the pathogenic attack and less spreading of diseases. Employing the GM approach, valuable crop plants can be targeted for biotic stress resistance.

Alongside, promoters are the key factor for the expression of a gene. Manipulating the native promoter, the construction of a strong novel synthetic promoter is a powerful tool for molecular biology. Synthetic promoters are designed according to the expression of transgenes in the target site. Advancement in plant genetic engineering greatly impacts food security and sustainable development for crop production to meet future challenges. Mainly, the assembly of different *cis*-regulatory elements (CREs) in a promoter can be aimed at a specific target for transgene expression. Due to such positivity in synthetic promoters, stress-inducible promoters are used to control different genes in plants. Biotic stress-inducible promoters are activated through pathogen attacks (Bisht et al. 2019). Under pathogen attack, PR genes are expressed to defend against pathogens. The pathogens release some chemicals like elicitors or some hormones and secondary metabolites, which are signals for induction of promoters to regulate gene expression. So, a biotic stress-inducible promoter can be called a chemical inducible promoter. Accordingly, plant biotechnology has achieved a milestone in the development of numerous synthetic promoters for pathogen resistance.

A pathogen-inducible synthetic promoter named Hv-Ger4c strongly driving transgene *Ta-Lr34res* (encodes for ATP-binding cassette transporter protein in wheat) in transgenic barley showed strong resistance against powdery mildew and leaf rust. Several chemically inducible synthetic promoters were reported for gene expression or suppression through the growth of a plant. The activity of another inducible synthetic promoter known as probenazole inducible promoter was observed in the *Arabidopsis* plant. The probenazole (PBZ) inducible promoter synthesized by tailoring of different CREs has the potential to drive a gene under multiple signaling pathways like salicylic acid, MAPK, ethylene, jasmonic acid, and calcium. Some CREs are triggered after a pathogen attack. So, such pathogen-inducible CREs can be added upstream of the synthetic promoter to make a resilient synthetic promoter against pathogen infection. E17 and F, two CREs, were added upstream of the minimal CaMV35S promoter for the development of a pathogen-inducible synthetic promoter and its efficacy was checked. The resultant promoters pGEE, pGFF, and pGFFEE showed effective expression of *the GUS* gene (a reporter gene) under hormonal and fungal treatments in *Brassica napus*. Pathogen-inducible synthetic promoter SP-DDEE (containing D, E17 *cis-elements*) strongly inhibits the growth of the phytopathogenic fungus *Sclerotinia scletrotiorum* (Ali and Kim 2019). The genetically engineered promoter becomes an effective candidate for guiding PR genes and improving the plant immune system against broad-spectrum pathogens without affecting other ecological substances. Both synthetic promoter and synthetic TFs introduced into the plant genome will help control gene expression at the molecular level. In transgenic plants containing pathogen-inducible promoters activated against the pathogen attack or in the presence of certain chemical substances used for plant protection, the plants are made resistant against a wide range of phytopathogens. Stable transgenic plants can meet future challenges and provide sustainable development in agriculture.

2.9 Conclusion

Upcoming climate change and the evolution of new pathogens are a threat to the agriculture field. New molecular technologies in plant genetic engineering are suitable tools for combating these threats. Advancements on developing novel synthetic promoters and TFs for various stress resistance may increase food production and security. Synthetic promoters can be a suitable promising candidate for translational research in plant synthetic biology. By assembling different CREs in a single promoter, it can be developed into an inducible promoter expressed under various stresses through *cis–trans* engineering approaches. Our goal in the agriculture field is to make a single plant with multiple transgene expression systems by using various synthetic promoters. Specific synthetic promoters can be made by reconstructing, re-arranging, manipulating, or using different molecular techniques to generate a robust, specific, and effective synthetic promoter for plant genetic engineering.

The use of new chemicals and pesticides into agricultural lands has led to increased toxicity in the environment. It decreases the soil quality and also affects the soil microbiota. Also due to the extensive use of these chemicals on crops for protection from phytopathogens, the pathogens has started to become resistant to these chemicals. However, crop diseases is increasing day by day with the generation of broad spectrum resistance of pathogens. Biotic stress-resistant crops for sustainable agricultural development can be developed through molecular farming. The successful implication of synthetically developed molecular tools for the protection of crop plants will improve the production rate of the crops. The toxicity dangers related with plant molecular farming is less and therefore the value of production, and food security will be more. Thus plant synthetic biology can have an effective impact on eliminating broad-spectrum resistance pathogens and promoting food security.

A perfect strategy for designing and constructing synthetic promoters and TFs could be implemented for pathogen-resistant transgene regulation, synthesis of recombinant proteins, and other metabolic activities in planta. Development and accurate engagement of such advanced synthetic molecular devices for sustainable agriculture would be a reward for global food crises.

References

Ali S, Kim W-C (2019) A fruitful decade using synthetic promoters in the improvement of transgenic plants. Front Plant Sci 10:1433

Aysha J, Noman M, Wang F, Liu W, Zhou Y, Li H, Li X (2018) Synthetic promoters: designing the cis regulatory modules for controlled gene expression. Mol Biotechnol 60(8):608–620

Bisht DS, Bhatia V, Bhattacharya R (2019) Improving plant-resistance to insect-pests and pathogens: the new opportunities through targeted genome editing. In: Seminars in cell & developmental biology. Elsevier, pp 65–76

Chen W, Provart NJ, Glazebrook J, Katagiri F, Chang HS, Eulgem T, Mauch F, Luan S, Zou G, Whitham SA, Budworth PR, Tao Y, Xie Z, Chen X, Lam S, Kreps JA, Harper JF, Si-Ammour A, Mauch-Mani B, Heinlein M, Kobayashi K, Hohn T, Dangl JL, Wang X, Zhu T (2002) Expression profile matrix of Arabidopsis transcription factor genes suggests their putative functions in response to environmental stresses. Plant Cell 14(3):559–574. https://doi.org/10.1105/tpc.010410

Chen L-Q, Hou B-H, Lalonde S, Takanaga H, Hartung ML, Qu X-Q, Guo W-J, Kim J-G, Underwood W, Chaudhuri B, Chermak D, Antony G, White FF, Somerville SC, Mudgett MB, Frommer WB (2010) Sugar transporters for intercellular exchange and nutrition of pathogens. Nature 468(7323):527–532. https://doi.org/10.1038/nature09606

Dey N, Sarkar S, Acharya S, Maiti IB (2015) Synthetic promoters in planta. Planta 242(5):1077–1094

Edition. F GEORGE N. AGRIOS

Eulgem T, Rushton PJ, Schmelzer E, Hahlbrock K, Somssich IE (1999) Early nuclear events in plant defence signalling: rapid gene activation by WRKY transcription factors. EMBO J 18:4689–4699. https://doi.org/10.1093/emboj/18.17.4689

Garretón V, Carpinelli J, Jordana X, Holuigue L (2002) The as-1 promoter element is an oxidative stress-responsive element and salicylic acid activates it via oxidative species. Plant Physiol 130:1516–1526. https://doi.org/10.1104/pp.009886

Gassmann AJ, Hutchison WD (2012) Bt crops and insect pests: past successes, future challenges and opportunities. Taylor & Francis. GM Crops Food 2012;3(3):139. https://doi.org/10.4161/gmcr.21778

Gurr SJ, Rushton PJ (2005) Engineering plants with increased disease resistance: what are we going to express? Trends Biotechnol 23:275–282. https://doi.org/10.1016/j.tibtech.2005.04.007

Han S-W, Jung HW (2013) Molecular sensors for plant immunity; pattern recognition receptors and race-specific resistance proteins. J Plant Biol 56(6):357–366

Heise A, Lippok B, Kirsch C, Hahlbrock K (2002) Two immediate-early pathogen-responsive members of the *AtCMPG* gene family in *Arabidopsis thaliana* and the W-box-containing elicitor-response element of *AtCMPG1*. Proc Natl Acad Sci 99(13):9049–9054. https://doi.org/10.1073/pnas.132277699

Jacobsen E, der Vossen V, In E (2009) Plant disease resistance: breeding and transgenic approaches. In: Schaechter M (ed) Encyclopedia of microbiology. Elsevier, Oxford

Jia H, Zhang Y, Orbović V, Xu J, White FF, Jones JB, Wang N (2017) Genome editing of the disease susceptibility gene CsLOB1 in citrus confers resistance to citrus canker. Plant Biotechnol J 15(7):817–823. https://doi.org/10.1111/pbi.12677

Kirsch C, Takamiya-Wik M, Schmelzer E, Hahlbrock K, Somssich IE (2000) A novel regulatory element involved in rapid activation of parsley ELI7 gene family members by fungal elicitor or pathogen infection. Mol Plant Pathol 1:243–251. https://doi.org/10.1046/j.1364-3703.2000.00029.x

Krawczyk S, Thurow C, Niggeweg R, Gatz C (2002) Analysis of the spacing between the two palindromes of activation sequence-1 with respect to binding to different TGA factors and transcriptional activation potential. Nucleic Acids Res 30(3):775–781

Kumar D, Patro S, Ghosh J, Das A, Maiti IB, Dey N (2012) Development of a salicylic acid inducible minimal sub-genomic transcript promoter from Figwort mosaic virus with enhanced root- and leaf-activity using TGACG motif rearrangement. Gene 503:36–47. https://doi.org/10.1016/j.gene.2012.04.053

Li J, Brader G, Palva ET (2004) The WRKY70 transcription factor: a node of convergence for jasmonate-mediated and salicylate-mediated signals in plant defense. Plant Cell 16:319–331. https://doi.org/10.1105/tpc.016980

Liu W, Mazarei M, Rudis MR, Fethe MH, Peng Y, Millwood RJ, Schoene G, Burris JN, Stewart CN Jr (2013) Bacterial pathogen phytosensing in transgenic tobacco and Arabidopsis plants. Plant Biotechnol J 11(1):43–52. https://doi.org/10.1111/pbi.12005

Lota F, Wegmüller S, Buer B, Sato S, Bräutigam A, Hanf B, Bucher M (2013) The cis-acting CTTC–P1BS module is indicative for gene function of LjVTI12, a Qb-SNARE protein gene that is required for arbuscule formation in Lotus japonicus. Plant J 74(2):280–293. https://doi.org/10.1111/tpj.12120

Madani B, Mirshekari A, Imahori Y (2019) Physiological responses to stress. In: Postharvest physiology and biochemistry of fruits and vegetables. Elsevier, pp 405–423

Mahfouz MM, Li L, Piatek M, Fang X, Mansour H, Bangarusamy DK, Zhu J-K (2012) Targeted transcriptional repression using a chimeric TALE-SRDX repressor protein. Plant Mol Biol 78(3):311–321

Mazarei M, Teplova I, Hajimorad MR, Stewart CN (2008) Pathogen phytosensing: plants to report plant pathogens. Sensors 8(4). https://doi.org/10.3390/s8042628

Mehrotra R, Kiran K, Chaturvedi CP, Ansari SA, Lodhi N, Sawant S, Tuli R (2005) Effect of copy number and spacing of the ACGT and GT cis elements on transient expression of minimal promoter in plants. J Genet 84(2):183–187. https://doi.org/10.1007/bf02715844

Mehrotra R, Gupta G, Sethi R, Bhalothia P, Kumar N, Mehrotra S (2011) Designer promoter: an artwork of cis engineering. Plant Mol Biol 75(6):527–536

Moradyar M, Motallebi M, Zamani MR, Aghazadeh R (2016) Pathogen-induced expression of chimeric chitinase gene containing synthetic promoter confers antifungal resistance in transgenic canola. In Vitro Cell Dev Biol Plant 52(2):119–129

Muthusamy S, Sivalingam P, Sridhar J, Singh D, Haldhar S, Kaushal P (2017) Biotic stress inducible promoters in crop plants-a review. J Agric Ecol 4:14–24

Noonan JP, McCallion AS (2010) Genomics of long-range regulatory elements. Annu Rev Genomics Hum Genet 11(1):1–23. https://doi.org/10.1146/annurev-genom-082509-141651

Park HC, Kim ML, Kang YH, Jeon JM, Yoo JH, Kim MC, Park CY, Jeong JC, Moon BC, Lee JH, Yoon HW, Lee SH, Chung WS, Lim CO, Lee SY, Hong JC, Cho MJ (2004) Pathogen- and NaCl-induced expression of the SCaM-4 promoter is mediated in part by a GT-1 box that interacts with a GT-1-like transcription factor. Plant Physiol 135:2150–2161. https://doi.org/10.1104/pp.104.041442

Peng A, Chen S, Lei T, Xu L, He Y, Wu L, Yao L, Zou X (2017) Engineering canker-resistant plants through CRISPR/Cas9-targeted editing of the susceptibility gene CsLOB1 promoter in citrus. Plant Biotechnol J 15(12):1509–1519. https://doi.org/10.1111/pbi.12733

Ranjan R, Dey N (2012) Development of vascular tissue and stress inducible hybrid–synthetic promoters through Dof-1 motifs rearrangement. Cell Biochem Biophys 63:235–245. https://doi.org/10.1007/s12013-012-9359-9

Ranjan R, Patro S, Kumari S, Kumar D, Dey N, Maiti IB (2011) Efficient chimeric promoters derived from full-length and sub-genomic transcript promoters of Figwort mosaic virus (FMV). J Biotechnol 152(1–2):58–62

Rushton PJ, Somssich IE (1998) Transcriptional control of plant genes responsive to pathogens. Curr Opin Plant Biol 1:311–315. https://doi.org/10.1016/1369-5266(88)80052-9

Rushton PJ, Reinstädler A, Lipka V, Lippok B, Somssich IE (2002) Synthetic plant promoters containing defined regulatory elements provide novel insights into pathogen- and wound-induced signaling. Plant Cell 14:749–762. https://doi.org/10.1105/tpc.010412

Sarkar S, Das A, Khandagale P, Maiti IB, Chattopadhyay S, Dey N (2018) Interaction of Arabidopsis TGA3 and WRKY53 transcription factors on Cestrum yellow leaf curling virus (CmYLCV) promoter mediates salicylic acid-dependent gene expression in planta. Planta 247:181–199. https://doi.org/10.1007/s00425-017-2769-6

Sears MT, Zhang H, Rushton PJ, Wu M, Han S, Spano AJ, Timko MP (2014) NtERF32: a non-NIC2 locus AP2/ERF transcription factor required in jasmonate-inducible nicotine biosynthesis in tobacco. Plant Mol Biol 84(1):49–66. https://doi.org/10.1007/s11103-013-0116-2

Shah J, Klessig DF (1996) Identification of a salicylic acid-responsive element in the promoter of the tobacco pathogenesis-related β-1,3-glucanase gene, PR-2d. Plant J 10:1089–1101. https://doi.org/10.1046/j.1365-313X.1996.10061089.x

Shokouhifar F, Zamani MR, Motallebi M, Mousavi A, Malboobi MA (2011) Construction and functional analysis of pathogen-inducible synthetic promoters in Brassica napus. Biol Plant 55(4):689. https://doi.org/10.1007/s10535-011-0169-5

Shrestha A, Khan A, Dey N (2018) Cis–trans engineering: advances and perspectives on customized transcriptional regulation in plants. Mol Plant 11(7):886–898

Tito R, Vasconcelos HL, Feeley KJ (2018) Global climate change increases risk of crop yield losses and food insecurity in the tropical Andes. Glob Chang Biol 24(2):e592–e602

Van der Does D, Leon-Reyes A, Koornneef A, Van Verk MC, Rodenburg N, Pauwels L, Goossens A, Körbes AP, Memelink J, Ritsema T, Van Wees SCM, Pieterse CMJ (2013) Salicylic acid suppresses jasmonic acid signaling downstream of SCFCOI1-JAZ by targeting GCC promoter motifs via transcription factor ORA59. Plant Cell 25:744–761. https://doi.org/10.1105/tpc.112.108548

van Schie CCN, Takken FLW (2014) Susceptibility genes 101: how to be a good host. Annu Rev Phytopathol 52(1):551–581. https://doi.org/10.1146/annurev-phyto-102313-045854

Xu Z, Xu X, Gong Q, Li Z, Li Y, Wang S, Yang Y, Ma W, Liu L, Zhu B, Zou L, Chen G (2019) Engineering broad-spectrum bacterial blight resistance by simultaneously disrupting variable TALE-binding elements of multiple susceptibility genes in rice. Mol Plant 12(11):1434–1446. https://doi.org/10.1016/j.molp.2019.08.006

Yanagisawa S (2004) Dof domain proteins: plant-specific transcription factors associated with diverse phenomena unique to plants. Plant Cell Physiol 45:386–391. https://doi.org/10.1093/pcp/pch055

Zhou DX (1999) Regulatory mechanism of plant gene transcription by GT-elements and GT-factors. Trends Plant Sci 4:210–214. https://doi.org/10.1016/s1360-1385(99)01418-1

Chapter 3
Transcription Factors Associated with Defense Response Against Fungal Biotrophs

Thirunarayanan Permal, Tulasi Korra, Sushree Suparna Manopatra, K. Manikandan, and Jyostrani Prathan

Abstract Transcription factors act as transcriptional regulators found exclusive to plants playing a key role in signal transduction pathways. They controlled a wide range of biological functions from receptors to signaling networks of various genes, and also help in crosstalk between stress responses. This chapter's focus is on the important unregulates and downregulates the defense response to smut, rust, and powdery mildew. This knowledge provides insights to improve resistance of the invading pathogens.

Keywords Transcription factors · Fungal biotrophs · Rust · Powdery mildew · Smut

3.1 Introduction

Transcription factors play a major role in the gene expression corresponding to both internal and external signals and associate with biological processes such as transcription, post- transcription, translation, and post -translation (Zhang et al. 2020).

T. Permal · K. Manikandan
Department of Mycology and Plant Pathology, Institute of Agricultural Sciences, Banaras Hindu University, Varanasi, Uttar Pradesh, India

T. Korra (✉)
Department of Plant Pathology, Odisha University of Agriculture and Technology, Bhubaneswar, India

S. S. Manopatra
Department of Plant Pathology, IARI, New Delhi, India

J. Prathan
Department of Plant physiology, RPCAU, Samstipur, India

© The Author(s), under exclusive license to Springer Nature Switzerland AG 2022
S. H. Wani et al. (eds.), *Transcription Factors for Biotic Stress Tolerance in Plants*, https://doi.org/10.1007/978-3-031-12990-2_3

The transcription factors family is in fungal biotrophs (e.g., WRKY, NAC, ZN finger, bZIP, C_2H_2, GATA, VELVET, and HOX).

On attack and invasion, pathogens lead to ignition of various signaling genes and enzymes that pave the way for the host to combat the attack and further destroy the host cells. Among these immunity-inducing pathways—Pathogen Associated Molecular Pattern (PAMP), Damage Associated Molecular Pattern (DAMP), Microbe Associated Molecular Pattern (MAMP), Effector Triggered Immunity (ETI), and Pathogen Triggered Immunity (PTI)—are the most vital ones (Spoel and Dong 2012). The host plant receives the signal of pathogen invasion, and it activates a network of defense signaling pathways, which in turn consider certain significant transcriptional reprograming that regulate the large series of defense genes (Li et al. 2016a, b). This significant transcriptional reprograming requires a functioning modulated system of the various transcription factors.

A transcription factor (TF) is the entity that has a significant role to play in defense-regulation pathways in response to various stress conditions; they do this by acting as "master switches" of a series of genes initiated by unique regulatory signaling networks (Liu et al. 2014). Defense responses involve cross-talking of several signaling pathways in reaction to biotic stresses that take several different genes into consideration. The genes governing transcriptional programing are of utmost importance.

The TFs involved in defense regulation are categorized into six major families, as follows: MYB (Myeloblastosis Related Proteins), bHLH (Basic Helix-Loop-Helix), AP2 (Apetala2)/ERF (Ethylene Response Factor), WRKY, NAC (i.e., CUC, NAM, and ATAF) and (bZIP (Basic Leucine Zipper Domain) (Zanetti et al. 2017). Many genetic studies so far have characterized the role of these families of TFs in the immune responses against pathogens that induce the activation of defense enzymes and the production of anti-microbial or anti-fungal secondary metabolites. Each TF family has a specific binding domain (i.e., helix-turn-helix, zinc finger, and bZIP). Previously mentioned domains bound to DNA *cis*-elements were found to respond to environmental stress.

Out of the TFs families, NAC proteins include a greater number of TFs, which is unusual to plants, mainly rice and Arabidopsis. This group of proteins are comprised of two parts—namely, terminal-N and C- terminal region. Puranik et al. (2012) estimated that the terminal-N, also known as the NAC domain, is highly conserved. This takes place in DNA-binding and dimer oligomerization, homodimerization, or heterodimerization with other NAC proteins. Jensen et al. (2010) explained that the C-terminal is less conserved and more diverse, acting as the transcription regulatory domain. NAC transcriptional factors act as master regulators of secondary cell-wall synthes that is mediated through a NAC–MYB signaling cascade (Huang et al. 2015). Voitsik et al. (2013) explained that two NAC transcription factors (i.e., ZmNAC41 and ZmNAC100) are transcriptional-induced hemibiotrophic and later ensure necrotroph colonization of corn leaves by *Colletotrichum graminicola*. ZmNAC41 mutants do not regulate the post-infection; ZmNAC100 induction would not occur in such interactions (Wang et al. 2009).

The bHLH transcription factors are associated with synthesis of other secondary metabolites (e.g., flavonoids) in response to stress and fungal interaction with the host (Nemesio-Gorriz et al. 2017). According to Zander et al. (2010)), class II TGA transcription factors play a vital role in introducing PDF1.2 transcription post-infection with *Pseudomonas syringae* or *Botrytis cinerea,* or in the Jasmonic acid (JA)/Ethylene (ET) induction plant, establishing a molecular link that connects the salicylic acid (SA) to the signaling networks. The expression of localized PR proteins (e.g., PR3, PDF 1, and PR4) regulates the ethylene and JA-dependent pathway.

Transcription factors are bound to the GCC box individually; the Ethylene-Responsive Element Binding Proteins (EREBPs) were first isolated from tobacco and ethylene-induced and were required in the expression of the GCC box containing PR genes (Suzuki et al. 1998). Gu et al. (2002) estimated that the resistance against bacterial speck disease in the tomato is governed by the *Pto*-resistant gene, which encodes a Ser/Thr protein kinase (Martin et al. 1994). Similarly, Zhou et al. (1997) found the three types of *ERF* transcription factors—that is, *Pti5, Pti4*, and *Pti6*—from tomato crops by virtue of their specific combined interaction with *Pto* kinase in the yeast-2 hybrid assay method.

bZIP proteins act as a master regulator of the SA-mediated signaling pathway, the TGA proteins mediated modulation; Group D consists of a class of *Arabidopsis* bZIP proteins with responses to biotic stress. Furthermore, they controlled the subcellular localization and functioning of genes during defense responses against biotrophic pathogens (Jakoby et al. 2002). The AP_2 domian specific to binding to DNA acts synergistically with other factors and the gene expression on the plant-defense response (Büttner and Singh 1997).

The WRKY proteins play a key role in various molecular events (e.g., senescence, seed development, seed germination, or seed dormancy and stresses) in plants. The WRKY family plays a role in pathogen infection. WRKY factors and their interactions can play an important role in signaling, chromatin remodelization, transcription, and cellular events. WRKY transcription factors are nodes for interplay between SA, ethylene signaling, JA pathways, and they are involved in plant defense by these signaling pathways.

Cross-talk between the previously mentioned signaling pathways appear as an vital regulator in plant disease resistance. Yet, SA signals are more prone to resistance responses against biotrophic pathogens. Exogenous application of synthetic chemicals, environmental conditions, and biotic pathogens trigger the expression of WRKY genes against multiple pathogens (Rushton et al. 2010).

Resistant genes correspond to temperature-reliant resistance. In the wheat plant temperature plays a major role against the biotrophic fungal pathogen stripe rust caused by *Puccinia striformis.* High temperature stimuli-induced expressions are relevant WRKY transcription factors in wheat plant seedlings resistant to *Pst.* When the wheat plant seedlings are exposed to high temperature treatment, *TaWRKY49*-silenced and non-silenced *TaWRKY62* leaves after *Pst* post-inoculation, O_2^-, H_2O_2 are high at 24 h. The accumulated level is high in HT-treated non-silenced and attend to the excess by way of the anti-oxidant enzyme gene (*TaPOD).* It enhanced the hypersensitive response (HR) cell death that triggers microbial pathogens (Wang et al. 2017). Exogenous application of SA induces the expression of PtrWRKY73 in *Populus trichocarpa* and comes under WRKY factor group-1. It is like tobacco AtWRKY33 and NtWRKY1. PtrWRKY73 is accumulated in sprouts, roots, leaves, and stems. The transcription activatation of the genes is related to Camalexin and Indole-glucosinolate. Biosynthesis is activated by the double mutant *atwrky18* and *atwrky40* (Duan et al. 2015).

In tea plant the disease-resistant gene, *CsWRKY14*, showed resistance against *Exobasidium vexans* (blister blight) fungal biotroph. The resistant cultivar exhibited higher activities of anti-oxidant enzymes and less H_2O_2 accumulation (Liu et al. 2021).

Some of the WRKY transcription factors' expression is negatively regulating the resistance mechanism. The phytohormones (i.e., SA, JA, and ET) have several ways on WRKY of TFs. For example, the WRKY-70 transcriptional factor where JA-responsive act as repressor genes and SA-induced ones are activated genes. These phytohormones use mutually antagonistic pathways. When there is an expression of WRKY transcription factors by a JA-dependent pathway, it suppresses the resistance response against biotrophic pathogens. PR-1 expression and disease resistance is reduced by overexpression of WRKY38 or WRKY62. The WRKY38 and WRKY62 are negative regulators (Jiang et al. 2016).

3.2 Role of TF Defense Response in Powdery Mildew and Rust

Transcriptional factors contribute the main role in plant defense recognized by stress signals and control downstream defense gene expression (Jan et al. 2021). Zhang et al. (2021) noted that wheat line TcLr14b showed the presence of *TaNAC*35 gene expression during virulent pathogen, *Puccinia triticina*, resulting in reduced haustorial mother-cell formation and also mycelia growth. Zhang et al. (2018) studied the wheat stripe rust pathogen, *Ta*NAC2, transcription factor associated with early suppression of the biotroph pathogen mycelia growth by induced H_2O_2 production, which enhanced defense response in wheat plant (Zhang et al. 2021).

The NAC transcription factor *Ta* NAC069 displayed resistance against the leaf rust pathogen in wheat by actuating the pathogenesis-related genes or by reducing the ROS-related genes (Wang et al. 2018). *Ta* NAC30 transcription factor expression of obligate pathogen wheat stripe rust shows the increased defense responses by enhancing the H_2O_2 in the wheat plant and localizing the nucleus and participate role in the transcription activator (Xia et al. 2010). The TaNAC4 transcription factor associated with the stripe rust of wheat pathogen by activating the transcription activity in the initiation of a defense-signaling pathway against the biotroph pathogen infection in wheat.

The *TaWRKY45* gene expressed the multiple disease resistance in wheat powdery mildew and leaf rust (Bahrini et al. 2011). *Ta*NAC6s transcription factor in the powdery mildew pathogen on wheat exhibited transcriptional activation activity via localized sub-cellular in the nucleus involved in basal broad-spectrum resistance and reduced the Haustorium index of the pathogen. The *WRKY3* transcription factor repressor defends against barley powdery mildew (Han et al. 2020; Zhou et al. 2018). Transcription factor *Hv*NAC6 expressed a strong basal defenses against the penetration of virulent powdery mildew fungus into barley (Jensen et al. 2007).

Transcription factors are proteins involved in activation or repression of the target genes' transcription process by binding with specific sequences. Besides the resistant gene (R gene), TFs (e.g., MYB, NAC, and WRKY) may also help plants to overcome pathogenic and non-pathogenic attack. Among all eukaryotic organisms, MYB transcription factors are most common and this protein family is large in higher plants. Based on adjacent repeats of the MYB domain classified four sub-families—(1) R_1MYB, (2) R_2R_3-MYB, (3) 3R-MYB, and (4) 4R-MYB. Most of the plant MYB transcription factors belong to the R_2R_3 type, and these are associated with the process of growth regulation.

The MYB proteins are substantially related to different kinds of functions regulating secondary metabolism (e.g., morphogenesis, epidermal wax, and the biotic and abiotic stress responses). Overexpression of MYB proteins increase the structural defense responses, such as wax layer formation, and promote the hypersensitive response related to plant disease resistance. TaLHY is a disease resistance-related MYB transcription factor that has been discovered in wheat plant that fights against the stripe rust pathogen. The loss of function of TaLHY in wheat plants reduces its immune capability against stripe rust pathogen. *TaMYB4,* a member of R_2R_3-MYB family of gene, is localized in the nucleus of wheat plants. During incompatible interaction of wheat and *Pst, TaMYB4* significantly up-regulated and *TaMYB4* was promoted by the hormones. Silencing of *TaMYB4* expression reduced the immune capability of wheat varieties and incompatible *Pst* race.

NAC transcription factors are most widely present in the plant kingdom, which contains an *mi*R164 complementary site, and acts as a role transcriptional regulator of plant growth and stress. For stripe rust pathogen, the TF factor *TaNAC21/22* acts as a negative regulator. Reduction of NAC transcription factor TaNAC21/22 shows increased stripe rust resistance. The NAM domain is localized in the nucleus. When the host plant is infected with virulent race *Pst* rust fungus, the expression level is high in $TaNAC_{30}$. In compatible wheat-*Pst* interaction, $TaNAC_{30}$ negatively regulates plant resistance. H_2O_2 production significantly increases in the $TaNAC_{30}$-silenced plants and enhances the resistance mechanism against *Pst.* During incompatible interaction of wheat and *Pst,* there is an overexpression of $TaNAC_4$ and $TaNAC_8$. The *Lr 19* rust-resistant gene was identified from *Agropryon elongatum.* In the wheat crop, *TcLr19* is a transcription factor induced by *Pt* (*Puccunia triticina*). During wheat-*Pst* compatible interaction of $TaNAC_2$ is involved in resistance reaction by generating H_2O_2.

The WRKY transcription factors play an extremely important role in plant defense regulation. It is involved in several physiochemical processes; in this way, it controls the expression of several genes under biotic- and abiotic-stress conditions. Various WRKY genes are involved in different kinds of defense mechanisms, either by up-regulating or down-regulating the target gene expression. In plant defense systems, some of the WRKY transcription factors act as negative regulators, whereas others act as positive regulators by being associated with specific regulatory pathways. The wheat cultivar contains the resistance gene, Lr28, infected with a virulent race of leaf rust pathogen, resulting in 146-fold increase of TFs. TaWRKY1B is comparable to a susceptible cultivar.

3.3 Role of TF Defense Response in Smut

Smut fungi include these varities: rice false smut, sugarcane whip smut, sorghum smut, maize smut, ragi smut, cumbu smut, and smut of barley (Wani et al. 2021). According to plant databases, there are TFs in rice (2389), Saccharum spp. (672), and sugarcane (39), WRKY (44), and NAC have been identified. Rice false smut, for undergoing mechanisms of resistance, were performed on RNA-*Seq*, and investigated transcriptional modulation in resistance (IR28) is 1405 and susceptible (HXZ) varieties is 1066 differentially expressed genes (DEGs). The pattern of expressed genes indicates changes in the pathogenic behavior of the pathogen (Yang et al. 2021). Though, the home-box TF gene (UVHOX2) regulated the chamldyospore production by inserted mutant B-766.

For further confirmation of regulatory function, CRISPR/Cas9 and agrobacterium-mediated transformation was used as solid evidence to stoppage of formation of clamydospore and decreased virulence (Xu et al. 2021). Similarly, the SUN-family protein, UvSUN1, reduces the development and virulence of smut fungi and production of toxic compounds and TF, UvMsn2, enhances the development and virulence in rice smut fungi, where Msn2 is involved in the stress response, mitochondrial morphology, and conidiogenesis (Gasch et al. 2000). Futhermore, they were characterized as a zinc finger TF in *Ustilaginoidea virens* (*UvMsn*$_2$)—a homologous of *MoMsn*$_2$ from the paddy blast. *UvMsn2* mutant also regulates stress response and virulence in smut fungus.

The genome broadly identified the transcriptional family in smut fungi—that is, basic leucine zipper (bZIP) and the GATA-binding TF family (Yu et al. 2019). bZIP has a role in plant growth and development of biotic- and abiotic-stress. The total of 28 bZIP, 17 of them were up-regulated during the initial infection period, and 11 were examined under H_20_2 stress. Transcription factors (e.g., Uvprol, UvCom1, and UvHox2) play a major role in conidial formation responses to *U.virens*-plant infection (Chen et al. 2020). The rice TF OsMYB55 induces the elevated temperature for heat-stress response, and the same results were reported for the Arabidopsis (AtMYB68) gene by Li et al. (2016a). For cold treatment, OsMYB4 TF acts as an ABA-independent cold response and accumulates in proline, which imparts resistance in several plants and similar results have been reported in barley for seed germination and Arabidopsis, either freezing and cold tolerance (Soltész et al. 2013). Transgenic rice overexpressing TF (*OsMYB*$_3$*R*-$_2$) enhances the cold tolerance and effectively improves under the stress condition.

On the other hand, high expression of TF *OsNAC045* up-regulated two stress responsive genes (i.e., LEA3 and PM1) and increased salt along with drought tolerance; this is also true with the *OsNAC2* gene being cold tolerant (Vannini et al. 2006). Another TF family is the NAC factor, consisting of more than 26 NAC genes that rise under biotic stress in rice. The WRKY factor was also demonstrated under biotic stress in rice; OsWRKY genes control plants under stress at 420 C and cold at 40 C (Qu et al. 2013). Furthermore, the other TFs (e.g., OsWRKY76, OsWRKY13, OsWRKY45–2, OsWRKY13, and OsWRKY42) positively regulated resistance to pathogens and can associate with MAP kinases; thus, they represent key components of plant defense signaling. This transcriptional network may help to respond quickly and efficiently to deter plant pathogens (Pandey and Somssich 2009). Rice smut fungus, Zn2Cys6 (TF family), and the orthologues Pro1 have been reported for regulatory function in sexual development, sporulation, and hyphal growth.

Another disease, sugarcane smut, is caused by *Sporisorium scitamineum*. The WRKY transcription factor family is important to regulate the phytopathogens. ScWRKY was a suppressed smut-susceptible variety at early stages (0–72 h) after inoculation. Transcriptional factor SsPrf1 via the two signaling pathways—Histidine kinase SlnI and cAMP/PKA—regulated sugarcane smut virulence and mating of the smut fungi (Cai et al. 2014). The corn plant conserved transcription factor (ZTF1) was involved in vegetative growth of the fungi and virulence in smut fungus of maize. The maize zinc finger transcriptional factor, Mzr 1 Cys_2His_2-type, acts as a transcriptional activator that regulates the fungal gene expression of smut disease infection. Transcriptional factor *Prf*1 through the two signaling pathways—MAPK and PKA—regulated pathogencity. The overexpression of maize homolougous protein, ZmDREB2A, induced stress responsive genes, whereas ectopic expression on *zea mays* transcription factors, ZmMYB30, promoted salt stress (Chen et al. 2017); it also confers drought-resistance genes in Arabidopsis transgenic plants.

The maize smut fungi, bZIP factor, regulates CL synthesis (Teichmann et al. 2010). Corn smut, APSES transcriptional factor StuAp, Efg1p, Phd1p, Sok2p, Asm1p, characterized under phylum basidiomycotina and regulated the conidial morphology and virulence. The StuA-homolog, Ust1, plays a role in conserved and distinct functions during cell differentiation in corn smut fungus, *Ustilago maydis*. Identification of a new transcriptional factor, *Fox 1*, expressed during development of the pathogen and deletion of *Fox 1* impaired tumor development and reduced virulence (Zahiri et al. 2010). The smut fungi of maize, the zinc finger transcriptional factor Rbf1, triggers to control the pathogencity. This Rbf1 is a regulator for sexual development.

On the other hand, Ros 1 under the WOPR family and mutant ros 1 induced infection in later stages (Heimel et al. 2010). *zfp 1*, a putative gene modulates the pathogenic development at various stages, but the deletion of *zfp 1* stops the pathogenic growth and reduces infection efficiency. The maize smut fungus, C_2H_2, with the orthologue PacC is a regulatory function (i.e., pH response regulator, SM biosynthesis, and carbohydrate metabolism). Velvet Ve A and Vel C act in multiple regulatory functions (i.e., sproluation, sexual development, abiotic stress tolerance, cell wall integrity). Pac 2, with the orthologue Worl, regulates hyphal growth development and effectors regulation (Cheung et al. 2021).

3.4 Conclusion

This chapter provides a global overview of the role of the transcriptional factors (TFs) for mounting defense responses against smut, rust, and powdery mildew in various crops. TFs were identified with several families with transcriptional networks and regulated infection periods for fungus. The overexpression of TFs unregulated or down-regulated by pathogen attack and validated through transcriptional analysis were studied and applied transgenic crops were reviewed to improve resistance against plant pathogens. Interestingly, the study applied genetic engineering of crops against plant pathogens, and function analysis done to determine modes of action and target sites.

References

Bahrini I, Ogawa T, Kobayashi F, Kawahigashi H, Handa H (2011) Overexpression of the pathogen-inducible wheat TaWRKY45 gene confers disease resistance to multiple fungi in transgenic wheat plants. *Breed Sci 61*(4):319–326

Büttner M, Singh KB (1997) Arabidopsis thaliana ethylene-responsive element binding protein (AtEBP), an ethylene-inducible, GCC box DNA-binding protein interacts with an ocs element binding protein. *Proc Natl Acad Sci 94*(11):5961–5966

Cai R, Zhao Y, Wang Y, Lin Y, Peng X, Li Q et al (2014) Overexpression of a maize WRKY58 gene enhances drought and salt tolerance in transgenic rice. Plant Cell, Tissue and Organ Culture (PCTOC) *119*(3):565–577

Chen X, Lu X, Shu N, Wang D, Wang S, Wang J, Ye W (2017) GhSOS1, a plasma membrane Na+/H+ antiporter gene from upland cotton, enhances salt tolerance in transgenic Arabidopsis thaliana. *PLoS One 12*(7):e0181450

Chen X, Hai D, Tang J, Liu H, Huang J, Luo C et al (2020) UvCom1 is an important regulator required for development and infection in the rice false smut fungus Ustilaginoidea virens. *Phytopathology 110*(2):483–493

Cheung HK, Donaldson ME, Storfie ER, Spence KL, Fetsch JL, Harrison MC, Saville BJ (2021) Zfp1, a putative Zn (II) 2Cys6 transcription factor, influences Ustilago maydis pathogenesis at multiple stages. *Plant Pathol 70*(7):1626–1639

Duan Y, Jiang Y, Ye S, Karim A, Ling Z, He Y, Luo K (2015) PtrWRKY73, a salicylic acid-inducible poplar WRKY transcription factor, is involved in disease resistance in *Arabidopsis thaliana*. *Plant Cell Rep 34*(5):831–841

Gasch AP, Spellman PT, Kao CM, Carmel-Harel O, Eisen MB, Storz G, Botstein D, Brown PO (2000) Genomic expression programs in the response of yeast cells to environmental changes. *Mol Biol Cell 11*((12)):4241–4257

Gu YQ, Wildermuth MC, Chakravarthy S, Loh YT, Yang C, He X, Martin GB (2002) Tomato transcription factors Pti4, Pti5, and Pti6 activate defense responses when expressed in Arabidopsis. *Plant Cell 14*(4):817–831

Han X, Zhang L, Zhao L, Xue P, Qi T, Zhang C, Shen QH (2020) SnRK1 phosphorylates and destabilizes WRKY3 to enhance barley immunity to powdery mildew. *Plant Communications 1*(4):100083

Heimel K, Scherer M, Vranes M, Wahl R, Pothiratana C, Schuler D, Kämper J (2010) The transcription factor Rbf1 is the master regulator for b-mating type controlled pathogenic development in Ustilago maydis. *PLoS Pathog 6*(8):e1001035

Huang D, Wang S, Zhang B, Shang-Guan K, Shi Y, Zhang D, Zhou Y (2015) A gibberellin-mediated DELLA-NAC signaling cascade regulates cellulose synthesis in rice. *Plant Cell 27*(6):1681–1696

Jakoby M, Weisshaar B, Dröge-Laser W, Vicente-Carbajosa J, Tiedemann J, Kroj T, Parcy F (2002) *bZIP* transcription factors in Arabidopsis. *Trends Plant Sci 7*(3):106–111

Jan R, Asaf S, Numan M, Kim KM (2021) Plant secondary metabolite biosynthesis and transcriptional regulation in response to biotic and abiotic stress conditions. *Agronomy 11*(*5*):968

Jensen MK, Rung JH, Gregersen PL, Gjetting T, Fuglsang AT, Hansen M, Collinge DB (2007) The Hv NAC6 transcription factor: a positive regulator of penetration resistance in barley and Arabidopsis. *Plant Mol Biol 65*(1):137–150

Jensen MK, Kjaersgaard T, Nielsen MM, Galberg P, Petersen K, O'shea C, Skriver K (2010) The Arabidopsis thaliana NAC transcription factor family: structure–function relationships and determinants of ANAC019 stress signalling. *Biochem J 426*(2):183–196

Jiang M, Liu QE, Liu ZN, Li JZ, He CM (2016) Over-expression of a WRKY transcription factor gene BoWRKY6 enhances resistance to downy mildew in transgenic broccoli plants. *Australas Plant Pathol 45*(3):327–334

Li B, Meng X, Shan L, He P (2016a) Transcriptional regulation of pattern-triggered immunity in plants. *Cell Host Microbe 19*(5):641–650

Li C, Tang Z, Wei J, Qu H, Xie Y, Xu G (2016b) The OsAMT1. 1 gene functions in ammonium uptake and ammonium–potassium homeostasis over low and high ammonium concentration ranges. *J Genet Genomics 43*(11):639–649

Liu B, Ouyang Z, Zhang Y, Li X, Hong Y, Huang L, Song F (2014) Tomato NAC transcription factor SlSRN1 positively regulates defense response against biotic stress but negatively regulates abiotic stress response. *PLoS One 9*(7):e102067

Liu S, Zhang Q, Guan C, Wu D, Zhou T, Yu Y (2021) Transcription factor WRKY14 mediates resistance of tea plants (*Camellia sinensis* (L.) O. Kuntze) to blister blight. *Physiol Mol Plant Pathol 115*:101667

Martin GB, Frary A, Wu T, Brommonschenkel S, Chunwongse J, Earle ED, Tanksley SD (1994) A member of the tomato Pto gene family confers sensitivity to fenthion resulting in rapid cell death. *Plant Cell 6*(11):1543–1552

Nemesio-Gorriz M, Blair PB, Dalman K, Hammerbacher A, Arnerup J, Stenlid J, Elfstrand M (2017) Identification of Norway spruce MYB-bHLH-WDR transcription factor complex members linked to regulation of the flavonoid pathway. *Front Plant Sci 8*:305

Pandey SP, Somssich IE (2009) The role of WRKY transcription factors in plant immunity. *Plant Physiol 150*(4):1648–1655

Puranik S, Sahu PP, Srivastava PS, Prasad M (2012) NAC proteins: regulation and role in stress tolerance. *Trends Plant Sci 17*(6):369–381

Qu AL, Ding YF, Jiang Q, Zhu C (2013) Molecular mechanisms of the plant heat stress response. Biochem Biophys Res Commun 432(2):203–207

Rushton PJ, Somssich IE, Ringler P, Shen QJ (2010) WRKY transcription factors. *Trends Plant Sci 15*(5):247–258

Soltész A, Smedley M, Vashegyi I, Galiba G, Harwood W, Vágújfalvi A (2013) Transgenic barley lines prove the involvement of TaCBF14 and TaCBF15 in the cold acclimation process and in frost tolerance. *J Exp Bot 64*(7):1849–1862

Spoel SH, Dong X (2012) How do plants achieve immunity? Defence without specialized immune cells. *Nat Rev Immunol 12*(2):89–100

Suzuki K, Suzuki N, Ohme-Takagi M, Shinshi H (1998) Immediate early induction of mRNAs for ethylene-responsive transcription factors in tobacco leaf strips after cutting. *Plant J 15*(5):657–665

Teichmann B, Liu L, Schink KO, Bölker M (2010) Activation of the ustilagic acid biosynthesis gene cluster in *Ustilago maydis* by the C2H2 zinc finger transcription factor Rua1. *Appl Environ Microbiol 76*(8):2633–2640

Vannini C, Iriti M, Bracale M, Locatelli F, Faoro F, Croce P, Genga A (2006) The ectopic expression of the rice Osmyb4 gene in Arabidopsis increases tolerance to abiotic, environmental and biotic stresses. *Physiol Mol Plant Pathol 69*(1–3):26–42

Voitsik AM, Muench S, Deising HB, Voll LM (2013) Two recently duplicated maize NAC transcription factor paralogs are induced in response to *Colletotrichum graminicola* infection. *BMC Plant Biol 13*(1):1–16

Wang XE, Basnayake BVS, Zhang H, Li G, Li W, Virk N, Song F (2009) The Arabidopsis ATAF1, a NAC transcription factor, is a negative regulator of defense responses against necrotrophic fungal and bacterial pathogens. *Mol Plant-Microbe Interact 22*(10):1227–1238

Wang J, Tao F, Tian W, Guo Z, Chen X, Xu X, Hu X (2017) The wheat WRKY transcription factors TaWRKY49 and TaWRKY62 confer differential high-temperature seedling-plant resistance to *Puccinia striiformis* f. sp. *tritici*. *PLoS One 12*(7):e0181963

Wang B, Wei J, Song N, Wang N, Zhao J, Kang Z (2018) A novel wheat NAC transcription factor, TaNAC30, negatively regulates resistance of wheat to stripe rust. *J Integr Plant Biol 60*(5):432–443

Wani SH, Anand S, Singh B, Bohra A, Joshi R (2021) WRKY transcription factors and plant defense responses: latest discoveries and future prospects. *Plant Cell Rep 7*:1–15

Xia N, Zhang G, Liu XY, Deng L, Cai GL, Zhang Y, Kang ZS (2010) Characterization of a novel wheat NAC transcription factor gene involved in defense response against stripe rust pathogen infection and abiotic stresses. *Mol Biol Rep 37*(8):3703–3712

Xu Y, Wu S, Yu Z, Moeketsi EK, Yang Z, Zhang Z, Zhang H (2021) Transcription factor UvMsn2 is important for vegetative growth, conidiogenesis, stress response, mitochondrial morphology and pathogenicity in the rice false smut fungus Ustilaginoidea virens. Phytopathology Research 3(1):1–10

Yang YN, Kim Y, Kim H, Kim SJ, Cho KM, Kim Y, Park OK (2021) The transcription factor ORA59 exhibits dual DNA binding specificity that differentially regulates ethylene-and jasmonic acid-induced genes in plant immunity. *Plant Physiol 187*(4):2763–2784

Yu J, Yu M, Song T, Cao H, Pan X, Yong M, Liu Y (2019) A homeobox transcription factor UvHOX2 regulates chlamydospore formation, conidiogenesis, and pathogenicity in Ustilaginoidea virens. *Front Microbiol 10*:1071

Zahiri A, Heimel K, Wahl R, Rath M, Kämper J (2010) The *Ustilago maydis* forkhead transcription factor Fox1 is involved in the regulation of genes required for the attenuation of plant defenses during pathogenic development. *Mol Plant-Microbe Interact 23*(9):1118–1129

Zander M, La Camera S, Lamotte O, Métraux JP, Gatz C (2010) *Arabidopsis thaliana* class-II TGA transcription factors are essential activators of jasmonic acid/ethylene-induced defense responses. *Plant J 61*(2):200–210

Zanetti ME, Rípodas C, Niebel A (2017) Plant NF-Y transcription factors: key players in plant-microbe interactions, root development and adaptation to stress. Biochim Biophys Acta (BBA)-Gene Regulatory Mechanisms 1860(5):645–654

Zhang XM, Zhang Q, Pei CL, Li X, Huang XL, Chang CY, Kang ZS (2018) TaNAC2 is a negative regulator in the wheat-stripe rust fungus interaction at the early stage. *Physiol Mol Plant Pathol 102*:144–153

Zhang N, Yuan S, Zhao C, Park RF, Wen X, Yang W, Liu D (2020) TaNAC35 acts as a negative regulator for leaf rust resistance in a compatible interaction between common wheat and Puccinia triticina. *Mol Gen Genomics 296*(2):279–287

Zhang Y, Geng H, Cui Z, Wang H, Liu D (2021) Functional analysis of wheat NAC transcription factor, TaNAC069, in regulating resistance of wheat to leaf rust fungus. *Front Plant Sci 12*:604797. https://doi.org/10.3389/fpls.2021.604797

Zhou J, Tang X, Martin GB (1997) The Pto kinase conferring resistance to tomato bacterial speck disease interacts with proteins that bind a cis-element of pathogenesis-related genes. *EMBO J 16*(11):3207–3218

Zhou W, Qian C, Li R, Zhou S, Zhang R, Xiao J, Cao A (2018) TaNAC6s are involved in the basal and broad-spectrum resistance to powdery mildew in wheat. *Plant Sci 277*:218–228

Chapter 4
Transcription Factors Associated with Defense Response Against Fungal Necrotrophs

L. Mathan, Namo Dubey, Swati Verma, and Kunal Singh

Abstract Among phytopathogenic fungal and oomycetes, the necrotrophic mode of nutrition uptake (including hemibiotrophs that switch to necrotrophic mode later in their life cycle) is the most widespread method of deriving nutrition. These pathogens kill the plant tissue at the site of infection. Plants employ various methods of defense against them, including the utilization of a number of transcription factors that regulate various hormonal pathways and help in necessary downstream cross talk. These transcription factors may act as positive as well as negative regulators under stress to modulate plant immunity. Cataloging, summarizing, and extensive analysis of these transcription factors and regulators may increase our understanding of their role against harmful fungal necrotrophs.

Keywords Plant defense · Fungal necrotrophs · Phytopathogenic fungi · Transcription factors

---

L. Mathan and Namo Dubey are joint first authors and contributed equally.

L. Mathan
CSIR-Institute of Himalayan Bioresource Technology, Palampur, India

N. Dubey · K. Singh (✉)
CSIR-Institute of Himalayan Bioresource Technology, Palampur, India

Academy of Scientific and Innovative Research (AcSIR), Ghaziabad, India
e-mail: kunal@ihbt.res.in

S. Verma
CSIR-Institute of Himalayan Bioresource Technology, Palampur, India

College of Horticulture and Forestry Thunag, Dr. Yashwant Singh Parmar University of Horticulture and Forestry, Solan, HP, India

© The Author(s), under exclusive license to Springer Nature Switzerland AG 2022

S. H. Wani et al. (eds.), *Transcription Factors for Biotic Stress Tolerance in Plants*, https://doi.org/10.1007/978-3-031-12990-2_4

4.1 Introduction

Stresses caused by various environmental factors, including biotic and abiotic agents, affect plant health and their reproductive potential. If such plants are widely cultivated as crops for human consumption or benefit, then decreases in harvest and yield lead to economic losses, a decrease in industrial output in the case of cash crop, hunger, and sometimes even famine (Fry and Goodwin 1997; Goss et al. 2014). Many microbes and related organisms act as biotic factors to cause stress by either colonizing the plants as pathogens or deriving nutrition directly like aphids and insects. Among these biotic factors, phytopathogenic fungi and filamentous oomycetes are one of the most prominent and adversely affecting microbes. A few of them choose a parasitic pathogenic lifestyle where they do not kill the plant, yet keep deriving nutrition from living tissue and are called biotrophs. Most of the other oomycetes and fungal pathogens derive nutrition by ultimately causing necrosis and cell death at the site of infection and colonization and are known as necrotrophs (Roth et al. 2021). These necrotrophic fungal pathogens are highly devastating to cultivated plants and agriculture.

In the last few decades, many researchers have tried to understand the mechanism of plant–phytopathogenic fungus interaction at molecular and cellular levels to use the genetic knowledge obtained in breeding programs. These efforts have identified a few major classes of genes and proteins involved in such biotic stress, including nucleotide-binding site–leucine rich repeat (NBS-LRR) proteins (Dubey and Singh 2018), pathogenesis-related (PR) proteins (Ali et al. 2018), and many transcription factors (Burke et al. 2020). Yet, collective and summarized literature on the role of transcription factors in fungal biotic stress is almost negligible. Furthermore, literature on transcription factors' role in plant defense and interaction against necrotrophic phytopathogenic fungi is completely lacking. The present effort has tried to fill this gap.

4.2 Transcription Factors and Their Role in Fungal Stress Signaling

Transcription factors (TFs) are proteins that contain an activation domain (AD) and one or more DNA-binding domains (DBDs). The DBD attaches to sequence-specific cis-acting regions in target gene promoters. The DBD is generally well conserved and is used to categorize TFs into classes and super-classes (Stegmaier et al. 2004), but the activation domain is significantly less well conserved. Based on the differences in activation domain and binding domain amino acid composition and structure, more than 60 transcription factor families have been classified to date in plants (Hong 2016). They are involved in the regulation of various stresses, including host–pathogen interaction. TFs cause significant gene expression reprogramming in both the host and the pathogen and connect various defense pathways through cross talk into a complex regulatory network. Though they have been found in high numbers in each plant, members of a few major families of proteins have been associated more prominently in plant–fungus interaction (Alves et al. 2014), including families of MYB (myeloblastosis-related proteins), NAC (NAM, ATAF, and CUC), WRKY (WRKYGQK repeat amino acid containing proteins), and bZIP (basic leucine zipper domain). All these families of transcription factors are discussed in detail with a focus on their involvement in plant–fungal–necrotroph interaction.

4.2.1 Role of WRKY Transcription Factor

Many homologs and orthologs of WRKY gene family members have been implicated in fungal biotic stress responses (Alves et al. 2014). In *Arabidopsis thaliana*, the WRKY protein family has been explored for their role in plant–fungal–necrotroph interaction along with other defense pathways like AtWRKY75, AtWRKY33, and AtWRKY40. AtWRKY75 has been linked to a variety of stress-related biological processes, including oxalic acid-mediated stress tolerance, defensive responses, and the unfolded protein response. AtWRKY75 regulates leaf senescence by increasing SA production and inhibiting H_2O_2 sequestration. It has been studied that WRKY75-SA-ROS loop acts as a age-related moiety in ET-induced leaf senescence (Guo et al. 2017). Other WRKY transcription factors, including AtWRKY10, AtWRKY33, and AtWRKY40, employ a defense response signaling cascade driven by jasmonic acid (JA) and ethylene (ET) to protect the plant from necrotrophic fungus (Zheng et al. 2006). AtWRKY33 overexpression inhibits the SA-mediated signaling pathway, making it vulnerable to *P. syringae* (Zheng et al. 2006). In contrast, BnWRKY33 participates in *Brassica napus* defense against fungal necrotroph *Sclerotinia sclerotiorum* by activating the salicylic acid (SA)- and jasmonic acid-mediated immunity by decreasing H_2O_2 accumulation (Wang et al. 2014). Transgenic *Arabidopsis* plants overexpressing the PtWRKY40 from *Populus trichocarpa* showed a greater level of transcript accumulation of jasmonic acid signaling pathway genes including *VSP2*, *PDF1.2*, and *PR3*, as well as enhanced resistance to the necrotrophic fungus *Botrytis cinerea* (Karim et al. 2015). Two members of the WRKY protein family, AtWRKY3 and AtWRKY4, have also shown to behave differently for separate pathogens. Single and double mutants of two proteins viz. atwrky3, atwrky4, and atwrky3::atwrky4 become more susceptible to *B. cinerea* while unaffected for *Pseudomonas syringae* infection. When the two genes overexpressed under constitutive promoter, then though AtWRKY3 had no effects on pathogen survival, the AtWRKY4 overexpressed lines showed susceptible behavior under *P. syringae* infection without any differential behavior against *B. cinerea* (Lai et al. 2008). In *Vitis vinifera*, *VvWRKY2* got induced in leaves by damage and *Plasmopara viticola* infection. When *VvWRKY2* was expressed under constitutive promoter in tobacco, the transgenic lines were less vulnerable to three fungal diseases that infected different parts of the plant: *Botrytis cinerea* (leaves), *Alternaria tenuis* (seeds), and *Pythium* spp. (roots) (Mzid et al. 2007). The example of AtWRKY4 shows that members of the WRKY transcription factor family act as negative regulators in plant immunity.

4.2.2 Role of AP2/ERF Transcription Factor Family

Ethylene-responsive element-binding proteins/ethylene-responsive factors (ERFs) are highly conserved proteins having an AP2 (Apetala-2) domain. These proteins bind to the GCC box and DRE (dehydration-responsive element) at the target promoter site. Many members of the protein family are reported for their role in plant stress, including pathogen resistance and osmotic stress like NtOPBP1, NtTsi1, NtTSRF1, SlTERF1, and CaERFLP1/PF1. For example, *Arabidopsis* resistance against *Botrytis cinerea* was improved in transgenic lines employing overexpressed *AcERF2* from *Atriplex canescens* (Sun et al. 2018). The transgenic lines provided immunity against pathogens by upregulating multiple pathogenesis-related (PR) genes involved in plant defense like PR1, PR2, and PR5. In *Arabidopsis*, via expression analysis methodology, AtERF4 has been shown as a negative regulator to *F. oxysporum* interaction and JA-responsive pathway genes while AtERF2 was found to positively modulate the defense pathways under *F. oxysporum* attack (McGrath et al. 2005). Other ERF transcription factors, including AtERF5 and AtERF6, play antagonistic roles in their expression, rendering them resistant to fungal necrotrophs via the JA/ET pathway but susceptible to the bacterial pathogen *Pseudomonas* via the SA signaling pathway. In other plants, TaERF3 protects *Triticum aestivum* against *Blumeria graminis* in the early stages of pathogen infection via SA signaling. *TaERF3* expression also provides defense to wheat plants against *Fusarium gaminearum* and *Rhizoctonia cerealis* in the later stages of fungal infection via ET/JA pathways (Zhang et al. 2007). TaERF3 also plays a major role in providing plant immunity to *T. aestivum* plants against multiple pathogens employing different hormone signaling.

4.2.3 Role of NAC Transcription Factors

NAC TFs carry the NAC domain (NAM, no apical meristem; ATAF, *Arabidopsis* transcription activation factor; CUC, cup-shaped cotyledon) in their N-terminal region that functions as a DNA binding domain. Many NAC proteins are implicated in *Arabidopsis* in the control of plant defense against necrotrophic fungal pathogens such as *Botrytis cinerea*, *Fusarium oxysporum*, and *Alternaria brassicicola*. Such resistance involved but not limited to include members of NAC transcription factor, AtATAF1, AtANAC055, AtANAC019, AtATAF2, AtANAC072, AtNTL9/CBNAC (calmodulin-binding NAC protein), and AtANAC042/JUB1 (Tsuda and Somssich 2015). Work on *AtATAF1* revealed its role as a negative regulator of plant defense against necrotrophic pathogens. When *AtATAF1* overexpressing transgenic lines were generated and tested against fungal necrotrophs like *B. cinerea* and *A. brassicicola*; their immunity was compromised (Wu et al. 2009; Wang et al. 2009). Similarly, in Cotton plants, resistance to *Verticilium dahliae* and *B. cinerea* was reduced by GhATAF1 overexpression, a homolog of ATAF1, which was linked with suppression of JA-mediated signaling and promotion of SA-mediated signaling (He et al. 2016). Similar to *AtATAF1*, the AtATAF2 overexpression enhanced sensitivity to *Fusarium oxysporum* and decreased defense gene expression in transgenic *Arabidopsis* plants. Double mutants of the transcription factors anac019 and anac055 have also been demonstrated to act as negative regulators and mutants being more resistant to *B. cinerea* in *Arabidopsis* (Zheng et al. 2012). In contrast, ANAC042/JUB1 mutations resulted in increased sensitivity to *A. brassicicola* and decreased camalexin accumulation, showing that ANAC042/JUB1 regulates camalexin synthesis and therefore impacts immunity against *A. brassicicola* in *Arabidopsis* plants as a positive regulator (Saga et al. 2012).

Rice resistance to the necrotrophic fungus *Magnaporthe oryzeae* is promoted by the transcription factors OsNAC6, OsNAC111, OsNAC58, OsNAC066, OsNAC60, OsNAC122, and OsNAC131 (Park et al. 2017). In wheat, a NAC-like transcription factor—TaNACL-D1—positively regulates plant immunity against *F. graminearum* attack. The TaNACL-D1 protein interacts with a unique orphan protein—TaFROG (Fusarium Resistance Orphan Gene)—and forms a complex to provide immunity against Fusarium Head Blight (FHB). *TaNACL-D1* overexpression in wheat cultivar-fielder plants made them more resistant to FHB than wild-type plants. In tomato plants, inhibiting *SlSRN1* enhanced sensitivity to *B. cinerea* (Liu et al. 2014). Silencing of two membrane-localized NAC TFs from the NTL (NAC with a transmembrane (TM) motif1-like) subfamily of potato plants, *StNTP1* and *StNTP2*, increased *Nicotiana benthamiana* sensitivity to *Phytophthora infestans* infection (McLellan et al. 2013). Cotton plants that were silenced for *GbNAC1* had reduced *Verticillium dahliae* resistance, whereas *Arabidopsis* plants that were overexpressed for GbNAC1 had better *V. dahliae* resistance (Wang et al. 2016). *Artemisia annua AaNAC1* or grapevine *VvNAC1* overexpression in tobacco or *Arabidopsis* plants increased resistance to *B. cinerea* (Le Hénanff et al. 2013).

4.2.4 Role of bZIP Transcription Factor

The bZIP family of transcription factors plays a prominent role in plant defense, as discussed by Amorim et al. (2017). As the name suggests, bZIP constitutes of leucine zipper as DNA-binding domain (DBD) with ACGT as a core element at the binding site, and based on flanking sequences, it has been categorized as C-box (GACGTC), G-box (CACGTG), and A-box (TACGTA). All three group bZIP TFs play distinct but important roles in pathogen defense.

Gene expression analysis revealed the putative positive role of *ZmbZIP21* and *ZmbZIP65* in *Zea mays* against *Colletotrichum graminicola* and *TcRT42C09* and

TcRT57A09 in *Theobroma cacao* against *Moniliophthora perniciosa* as their expression was found significantly upregulated under pathogen attack (Alves et al. 2013). Similarly, RNA-Seq-based gene expression analysis revealed many members of bZIP TFs getting differentially expressed in *Gossypium barbadense* cv. 7124 under *V. dahliae* attack. Gene characterization work on bZIP member *GmbZIP15* in soybean showed the involvement of the gene in providing resistance against *Phytophthora sojae* and *Sclerotinia sclerotiorum.* The transgenic soybean plants overexpressing GmbZIP15 modulate defense response against fungal pathogens by regulating antioxidant enzymes and phytohormone signaling (Zhang et al. 2021). The RcTGA positively controls rose (*Rosa chinensis*) resistance to *Botrytis cinerea* (Gao et al. 2021).

4.2.5 Role of MYB Transcription Factor

Myeloblastosis-related domain-containing proteins, also known as MYB transcription factors, play a prominent role in plant defense and immunity. In plants, two major subfamilies belong to R2R3 and R1R2R3 types (Hong 2016). Many R2R3-type MYBs are plant specific and involved in response to abiotic and biotic stresses (Zhou et al. 2009). Certain R2R3-type MYBs are critical components of the plant pathosystem's defense against fungal infections. For example, *AtMYB30* from *A. thaliana* is considered a master regulator in plant defense against multiple pathogens, including fungal necrotrophs (Raffaele and Rivas 2013). When *AtMYB30* was overexpressed, transgenic lines of *Arabidopsis* and tobacco became more resistant to hemibiotrophic–necrotrophic fungal pathogen *Cercospora nicotianae.* These overexpressing transgenic lines were found to generate intense hypersensitive responses against pathogens (Vailleau et al. 2002). AtMYB44 transcription factors act as antagonists that regulate SA-mediated response by interacting with AtWRKY70 and promote resistance against biotrophic pathogens, which in turn downregulate the JA-mediated regulation, ultimately leading to increased susceptibility toward necrotrophic infection (Shim et al. 2013). AtMYB108/BOS1 (Botrytis-Susceptible1) shows the resistance to *B. cinerea* and *A. brassicicola* in the *Arabidopsis* plant (Mengiste et al. 2003). In another study, the expression of *AtMYB46* induces necrotrophic infection in the *Arabidopsis* plant (Ramírez et al. 2011). In *Hevea brasiliensis*, overexpression of *HbMYB1* exhibits disease resistance against *B. cinerea,* which results from the suppression of hypersensitive response (Peng et al. 2011).

In grapes, VvMYB44 acts as a positive regulator of *B. cinerea* resistance in an NPR1-dependent way. These findings imply that 2R-type VvMYB44 may have a beneficial function in BABA (β-aminobutyric acid)-induced priming defense in grape berries, potentially reducing soluble sugar intake during postharvest storage. Antisense ASR1 fruits were also shown to be more sensitive to *Botrytis cinerea* (Guadalupe Dominguez et al. 2021).

4.2.6 *Role of bHLH (Basic Helix–Loop–Helix) Transcription Factors*

The bHLH transcription factor family comprised the N-terminal basic helix–loop–helix domain, as the name suggests. In *Arabidopsis*, AtFAMA, a bHLH transcription factor that interacts with MED8, provides plant immunity to *B. cinerea* (Li et al. 2018). In mutant lines of fama-1 and fama-2, the susceptibility toward *B. cinerea* increased. This was accompanied by downregulation of defense-related genes in the mutant. When the transgenic lines were generated using *AtFAMA* under constitutive promoter, the resistance to *B. cinerea* increased. AtFAMA act as a positive regulator of plant defense in *A. thaliana.* In contrast, four bHLH members viz. AtbHLH3, AtbHLH13, AtbHLH14, and AtbHLH17 were revealed to be acting as transcriptional repressors in response to phytopathogens including *B. cinerea* (Song et al. 2013). Loss-of-function mutants of these genes displayed reduced JA-mediated defense response.

4.2.7 *Role of Other Transcription Factors*

Lateral organ boundaries (LOB) domain-containing transcription factors are responsible for *Fusarium* disease development because it acts as a susceptible factor. In previous studies, knockout of LOB domain-containing protein20 (LBD20) were found to be resistant to *F. oxysporum*. The lbd20 mutant also showed increased expression of the JA-regulated defense genes Vegetative storage protein 2 (VSP2) and Thionin2.1 (THI2.1) (Thatcher et al. 2012). A total of 43 LBD genes in *Arabidopsis* have been identified and classified into two categories. The LBD proteins, including LBD20, contain LOB domains that are most similar to the defining LOB member, having a cysteine-repeat motif, a conserved glycine residue, and a leucine zipper sequence (Majer and Hochholdinger 2011).

4.3 Role of Transcription Factors in the Regulation of Plant Hormone Signaling in Response to Fungal Necrotrophs

It is recognized that the interplay of different signaling pathways, such as those mediated by the hormones jasmonic acid (JA), salicylic acid (SA), and ethylene (ET), have a role in plant resistance to necrotrophs. The following section summarizes the role of transcription factors in several hormonal signaling pathways.

4.3.1 Transcription Factors Involved in JA Signaling

Jasmonates (JA) and ethylene (ET) are defensive signaling chemicals that control immunity against necrotrophs and herbivorous predators. Knockout of the Jasmonic acid gene in *Arabidopsis* shows the susceptibility to necrotrophic fungal disease. For instance, redox-responsive TF 1 (RRTF1) overexpression enhanced resistance to the fungus *Botrytis cinerea*, whereas rrtf1 mutant plants were more vulnerable (Li et al. 2021a, b). Transcription factors such as WRKY, bHLH, and TGA are responsible for the positive and negative regulation of the JA signaling pathway (Fig. 4.1).

The COI1 (Coronatine insensitive 1) gene, according to the findings, is responsible for resistance to necrotrophic bacteria such as *A. brassicicola* as well as necrotrophic fungi such as *B. cinerea*. COI1, a SCFCOI1 component, recognizes JA signals and binds to JAZs, which are then ubiquitinated and eliminated via the 26S proteasome pathway. According to Chen et al. (2021), JAZ-regulated AtWRKY75 interacts with the JA-sensitive gene ORA59 and activates the JA signaling pathway. AtWRKY57 acts as a negative feedback loop to control protective reaction. It has the potential to increase the expression of JAZ1 and JAZ5. AtWRKY33 inhibits JAZ1, and JAZ5 functions as a repressor, controlling the protective response. NINJA (Novel interactor of JAZ), one of the JAZ interactors, recruits TPL (TOPLESS) and TPL-related protein that blocks the MYC and bHLH transcription factors and turns off the JA-responsive gene.

JAZ-targeted transcription activators and repressors (AtbHLH3, AtbHLH13, AtbHLH14, and AtbHLH17) are released to control their target genes such as TAT1 (tyrosine and tryptophan amino acid transporter) and DFR (dihydroflavonol reductase) antagonistically and cooperatively, potentially affecting expression of JA-sensitive genes critical for different JA responses (Song et al. 2013). JAZ repressors interact with AtMYC2, AtMYC3, and AtMYC4, as well as AtbHLH003, AtbHLH013, and AtbHLH017 in the absence of JA-Ile. JAZs are a repressive complex that also comprises NINJA and TOPLESS. The COI1/JAZ co-receptor recognizes hormone in response to a stimulus, and the proteasome degrades JAZ proteins. TFs such as AtMYC2, AtMYC3, and AtMYC4 increase transcription after being released from JAZ, whereas AtbHLH003, AtbHLH013, and AtbHLH017 inhibit it. Because both sets of proteins (MYCs and bHLHs) compete for the G-box, the activity balance between these two sets of TFs will govern the output response. AtMYC2 also increases AtbHLH017 expression, which leads to an increase in the effective repressor and a reduction in transcriptional activation over time (Table 4.1).

4.3.2 The Ethylene Cascade

Plants produce ethylene to defend themselves from necrotrophic diseases like *B. cinerea*. In the previous studies, the pepr1/pepr2 double mutant was found to be less sensitive to ET and more susceptible to *B. cinerea*, indicating that it participates in ET-mediated defenses against fungal necrotrophs. Similarly, the *Arabidopsis* mutants viz. ein 2 (ethylene-insensitive) and constitutive triple response (ctr1) are vulnerable to necrotrophic fungi such as *Fusarium* and *Pythium*. Signaling cascades such as MAPK signaling pathways are activated in response to necrotroph infection, and they control ACS (acetate scavenging acetyl-CoA synthetase) transcriptionally and post-translationally. AtMPK3/AtMPK6 promotes ET synthesis post-translationally by phosphorylating rate-limiting ACS isoforms (ACS2 and ACS6) to prevent ubiquitin-mediated degradation (Meng et al. 2013). The AtWRKY33 transcription factor controls the expression of the ACS2 and ACS6 genes via the AtMPK3/AtMPK6 cascade. MAPK signaling pathways that positively regulate ACS are activated in plants to increase ET synthesis. As a result, the detection of a pathogen or related PAMPs initiates a complicated series of signaling cascades that end in the production of ethylene and jasmonic acid.

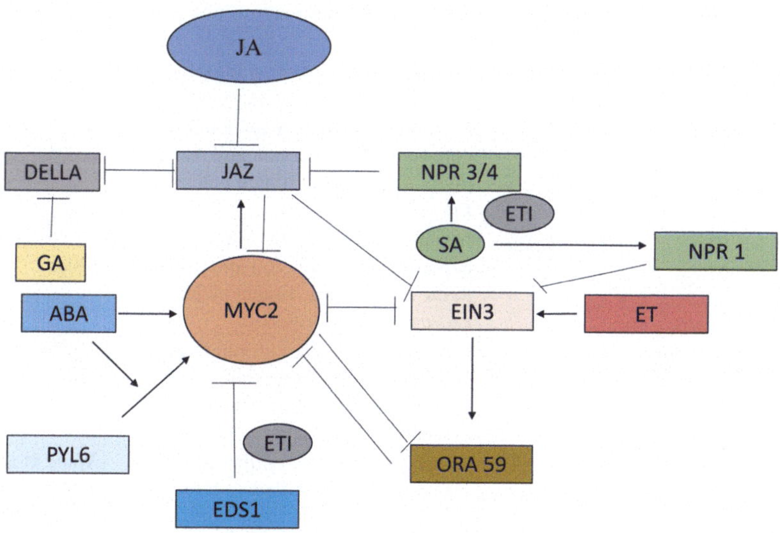

Fig. 4.1 The Jasmonate Zim Domain (JAZ) Protein regulates cross talks across jasmonic acid (JA) hormone signaling pathways in plant growth and stress responses. DELLAs interact with JAZ repressors, allowing MYC2 to be released from JAZ repression and promoting defensive responses via MYC2 activation. MYC2 is also the most critical element in JA-gibberellin (GA) cross talk. In plant resistance, JAZ inhibition of EIN promotes ET and JA signaling synergy, whereas MYC2 inhibition of EIN facilitates ethylene (ET) and JA signaling antagonism. EIN promotes the ethylene response factor ORA59 (Octadecanoid-responsive *Arabidopsis* 59). In SA signaling, NPR1 paralogues NPR3 and NPR4 function as SA receptors, regulating hormone signaling-mediated defense mechanisms. Other TFs such as PYL-6 (Pyrabactin resistance1-Like proteins) stimulate MYC2 and hence initiate the JA signaling cascade. It is worth noting that the lines with bars and arrows signify negative and positive regulatory actions, respectively

Table 4.1 A list of transcription factors that are involved against fungal necrotrophs

S.No.	Transcription factor	Name of the plant	Function	References
1.	AtWRKY33	*Arabidopsis*	Induce JA/ET pathway against *B. cineria* and *A. brassicicola*	Zheng et al. (2006)
2.	BnWRKY15	*Brassica napus*	Prevent WRKY33 transcription activation and increase disease susceptibility	Liu et al. (2018)
3.	AtWRKY75	*Arabidopsis*	Activator of JA-mediated pathway against necrotrophic pathogen	Chen et al. (2021)
4.	PtWRKY40	*Populus trichocarpa*	Overexpression increasing JA pathway and resistance against *B. cineria*	Karim et al. (2015)
5.	AtWRKY3	*Arabidopsis*	Positive role in JA/ET-regulated resistance to necrotrophic pathogens	Lai et al. (2008)
6.	AtWRKY4	*Arabidopsis*	Induce JA/ET-regulated resistance to necrotrophs and overexpression promoting biotrophic bacterial infection	Lai et al. (2008)
7.	VvWRKY2	*Vitis vinifera*	Resistance against necrotrophs	Mzid et al. (2007)
8.	AcERF2	*Atriplex canescens*	Enhanced resistance to pathogen	Sun et al. (2018)
9.	AtERF2	*Arabidopsis*	Positively regulate JA-responsive defense gene expression and resistance to the necrotrophic fungal pathogen *Fusarium oxysporum*	McGrath et al. (2005)
10.	AtERF4	*Arabidopsis*	Negatively regulate JA-responsive defense gene expression and resistance to the necrotrophic fungal pathogen *Fusarium oxysporum*	McGrath et al. (2005)
11.	AtERF5 and AtERF6	*Arabidopsis*	Defense against necrotrophic fungus and constitutive expression suppressing SA pathway and susceptible to *P. syringae* infection	Moffat et al. (2012)
12.	AtATAF1	*Arabidopsis*	Overexpression susceptible to *B. cineria* and *A. brassicicola* infection	Wang et al. (2009), Wu et al. (2009)
13.	AtATAF1	*Cotton*	Overexpression leading to repression of JA responsive pathway and negative regulator in resistance to *B. cineria*	He et al. (2016)
14.	AtATAF2	*Arabidopsis*	Overexpression increasing the *Fusarium oxysporum* infection	Delessert et al. (2005)
15.	TaNACL-D1	*Wheat*	Increased resistance against *Fusarium* head blight	Perochon et al. (2019)
16.	SlSRN1	*Tomato*	Silencing enhancing resistance to *B. cineria*	Liu et al. (2014)
17.	GmbZIP15	*Soybean*	Overexpression enhancing resistance against *S. sclerotiorum*	Zhang et al. (2021)

(continued)

Table 4.1 (continued)

S.No.	Transcription factor	Name of the plant	Function	References
18.	RcTGA	*Rose*	Resistance against *B. cineria*	Gao et al. (2021)
19.	AtMYB108	*Arabidopsis*	Resistance against *B. cineria* and *A. brassicicola*	Mengiste et al. (2003)
20.	VvMYB44	*Grape*	Positive regulator of *B. cineria*	Li et al. (2021a, b)
21.	AtbHLH13	*Arabidopsis*	Act as a transcriptional repressor and modulate the expression of JA-responsive gene	Song et al. (2013)
22.	TaERF3	Wheat	Against *B. graminis* in the early stages via SA signaling and *F. gaminearum* and *R. cerealis* in the later stages via ET/JA pathways	Zhang et al. (2007)

4.4 The JA and ET Pathways

JA does not function alone; it is part of a complicated signaling network that includes other plant hormone signaling pathways. Plant stress response is controlled in an antagonistic or cooperative manner by JA and ET. Ethylene insensitive 3 (EIN3) is a homolog of EIN2. EIN3-like 1 (EIL1) and JAZs-MYC2 in the ethylene signaling pathway and JAZs-MYC2 in the jasmonic acid signaling pathway enable cross talk between the ET and JA signaling pathways (Zhang et al. 2014). To resist herbivorous insects, exogenous JA causes JAZ breakdown, and the release of MYC2 controls the ORA59/ERF1 expression and the wound-responsive VSP2 gene expression (Verhage et al. 2011).

JAZ suppresses EIL2/transcriptional EIN3 activity in the ET signaling pathway and activates downstream ORA59/ERF1, which targets and enhances the synthesis of Plant defensin 1.2 (PDF1.2), allowing plants to withstand necrotrophic and hemibiotrophic pathogen infection (Zhu et al. 2011). MeJA promoted the activity of 1-aminocyclopropane-1-carboxylic acid oxidase (ACO), leading to higher ET synthesis and ET augmented the induction of allene oxidase synthase (AOS), which facilitated the first step in the biosynthesis of JA (Hudgins and Franceshi 2004). This indicates that JA and ET signaling pathways are interconnected, and both are triggered by their productions.

4.5 Transcription Factors of Fungal Necrotrophs and Their Role in Virulence and Pathogenicity

As transcription factors are involved in plant defense against various pathogens, including phytopathogenic fungal necrotrophs, they are also reported to have a role in providing crucial assistance to pathogenicity behavior of fungi and regulation of release of various toxins, effectors, and virulence factors. To have a counterview of the role of TFs from fungi, they are also discussed here in brief.

The reports on the TF repertoire of phytopathogenic fungus are limited. The majority of the research utilized *Saccharomyces cerevisiae* and *Schizosaccharomyces pombe* (Spitz and Furlong 2012). Because of the improved accessibility and availability of whole genome and transcriptome data, genome-wide TF identification has shown to be extremely useful in gaining insights into an organism's TF repertoire. *Tuber melanosporum*, a mycorrhizal fungus, was used in a genome-wide search and functional identification of TFs. Researchers used a functional analysis combined with the bioinformatics method in yeast and transcriptome profiling to identify TFs in *T. melanosporum. T. melanosporum* has 57 potential TF homologs, 102 previously known TF homologs, and 42 possible tuber-specific TFs (Montanini et al. 2011). A yeast screen confirmed approximately one-fifth of the *T. melanosporum* TFs predicted in silico. Furthermore, 29 transcription factors were shown to be increased in ectomycorrhizae or fruiting bodies. Other databases and studies on genome-wide TF analysis are available for mice (Zheng et al. 2008), humans (Fulton et al. 2009), plants (Mochida et al. 2010), *Drosophila melanogaster*, and rats (Adryan and Teichmann 2006). The Animal Transcription Database, on the other hand, is a comprehensive TF database for 50 different animal species spanning from *Caenorhabditis elegans* to humans (Zhang et al. 2012). TRANSFAC and its module TRANSCompel can be used to study eukaryotic transcriptional gene regulation (Matys et al. 2006). The DBD database contains 930 completely sequenced prokaryotes, eukaryotes, and eukaryotes prokaryotes genomes with projected TF repertoires (Wilson et al. 2008). In the case of phytopathogenic fungi, there are presently insufficient resources to investigate transcription factors.

4.6 Postulated Transcription Factors in Necrotroph

Pyrenophora tritici-repentis and *Cochliobolushetero strophus*, two closely related necrotrophic fungi, contain 366 and 362 predicted TFs, respectively, which are similar to *A. rabiei*'s 381 putative TFs. A total of 389 putative TFs in *Botrytis cinerea* were predicted. In *Sclerotinia sclerotiorum* and *Parastagonospora nodorum*, 431 and 435 putative TFs, respectively, were predicted. Surprisingly, the number of projected putative TFs in biotrophic fungi such as *Puccinia graminis* f. sp. *tritici* (287), *Blumeria graminis* f. sp. *hordei* (224), and *Ustilago maydis* (272) was lower than predicted in necrotrophic fungi. Furthermore, the hemibiotrophic fungi *Fusarium oxysporum* f. sp. *lycopersici* (566), *Mycosphaerella graminicola* (623), and symbiont *Laccaria bicolor* (613) exhibited higher number of TFs. *Neurospora crassa* (403) and *Magnaporthe oryzae* (378), both saprotrophs, displayed putative TFs similar to those predicted for *A. rabiei*.

4.7 Gene Expression Reprogramming Facilitated by TFs

When a pathogen infects a host plant, both the virus and the host undergo significant gene expression reprogramming, which is aided by transcription factors (TFs) (Verma et al. 2013). TFs appear to have a key role in filamentous fungal proliferation, growth, and virulence, according to various studies. In filamentous fungus and yeast, bZIP (the basic leucine zipper) TF Activating Protein 1 (AP-1) serves as a transcriptional activator in response to oxidative stress (Reverberi et al. 2008) and has numerous functions as a redox regulator (Karin et al. 1997). Pap1 in *S. pombe*, Yap1 in *S. cerevisiae*, Kap1 in *Kluyveromyces lactis*, and Cap1 in *Candida albicans* are all members of the yeast AP-1 transcription factor family (Toone et al. 2001). Yap1 orthologs were later discovered in *Aspergillus fumigatus*, *Aspergillus nidulans*, and *Aspergillus parasiticus*, and they have been demonstrated to play an important role in cellular defense against oxidative stress (Reverberi et al. 2008). Catalase B expression in the necrotrophic phytopathogenic fungus *B. cinerea* was regulated by the bZIP TF (BcAtf1), but it did not contribute to osmotic or oxidative stress tolerance (Temme et al. 2012). MST12 (a homolog of the yeast TF Ste12) is required for *M. oryzae* host penetration and colonization (Park et al. 2002). A few additional specialized TFs, such as *Cochliobolus carbonum* ccSNF1, control the expression of cell wall-degrading enzymes in plants (Tonukari et al. 2002). Other specific transcription factors, such as *Cochliobolus carbonum* ccSNF1, regulate the development of cell wall-destroying enzymes in plants (Tonukari et al. 2002). MYT3, a Myb-like transcription factor, affects pathogenicity and sexual development in *Fusarium graminearum*. Similarly, the *Alternaria brassicicola* AbPf2 TF plays an important role in pathogenesis while having little influence on other cellular processes. Because TFs play such a crucial role in the life cycle of fungi, their presence or absence may give opportunities or impose constraints on the natural environment of fungal species. PnPf2, a Zn2Cys6 transcription factor in *Parastagonospora nodorum*, regulates NE gene expression and is required for wheat pathogenicity. As a result, TFs are critical for the survival and completion of life.

4.8 Conclusion

Since transcription factors play an essential role in plant life, their absence or presence may provide a starting point for understanding how plants respond to pathogens. The current study lays the groundwork for future research into the role of transcription factors in the battle against fungal necrotrophs. This will help researchers understand how regulatory mechanisms developed in fungal necrotrophs. Using the knowledge already available, the role of TFs in conferring resistance or vulnerability to illness caused by fungal necrotrophs will be decoded. The considerable knowledge gained will assist in the development of strategies and techniques for controlling destructive diseases.

Acknowledgments The authors are thankful to the Director of CSIR-IHBT for providing a working facility. ND is grateful to the University Grant Commission of India for financial support in the form of a fellowship. This manuscript's IHBT communication number is 10092.

References

Adryan B, Teichmann SA (2006) FlyTF: a systematic review of site-specific transcription factors in the fruit fly *Drosophila melanogaster*. Bioinformatics 22(12):1532–1533

Amorim LL, da Fonseca Dos Santos R, Neto JPB, Guida-Santos M, Crovella S, Benko-Iseppon AM (2017) Transcription factors involved in plant resistance to pathogens. Curr Protein Pept Sci 18(4): 335–351

Ali S, Ganai BA, Kamili AN, Bhat AA, Mir ZA, Bhat JA, Tyagi A, Islam ST, Mushtaq M, Yadav P, Rawat S (2018) Pathogenesis-related proteins and peptides as promising tools for engineering plants with multiple stress tolerance. Microbiol Res 212:29–37

Alves MS, Dadalto SP, Gonçalves AB, De Souza GB, Barros VA, Fietto LG (2013) Plant bZIP transcription factors responsive to pathogens: a review. Int J Mol Sci 14(4):7815–7828

Alves MS, Dadalto SP, Gonçalves AB, De Souza GB, Barros VA, Fietto LG (2014) Transcription factor functional protein-protein interactions in plant defense responses. Proteomes 2(1):85–106

Burke R, Schwarze J, Sherwood OL, Jnaid Y, McCabe PF, Kacprzyk J (2020) Stressed to death: the role of transcription factors in plant programmed cell death induced by abiotic and biotic stimuli. Front Plant Sci 11:1235

Chen L, Zhang L, Xiang S, Chen Y, Zhang H, Yu D (2021) The transcription factor WRKY75 positively regulates jasmonate-mediated plant defense to necrotrophic fungal pathogens. J Exp Bot 72(4):1473–1489

Delessert C, Kazan K, Wilson IW, Straeten DV, Manners J, Dennis ES, Dolferus R (2005) The transcription factor ATAF2 represses the expression of pathogenesis-related genes in Arabidopsis. Plant J 43(5):745–757

Dubey N, Singh K (2018) Role of NBS-LRR proteins in plant defense. In: Molecular aspects of plant-pathogen interaction. Springer, Singapore, pp 115–138

Fry WE, Goodwin SB (1997) Resurgence of the Irish potato famine fungus. Bioscience 47(6):363–371

Fulton DL, Sundararajan S, Badis G, Hughes TR, Wasserman WW, Roach JC, Sladek R (2009) TFCat: the curated catalog of mouse and human transcription factors. Genome Biol 10(3):1–4

Gao P, Zhang H, Yan H, Wang Q, Yan B, Jian H, Tang K, Qiu X (2021) RcTGA1 and glucosinolate biosynthesis pathway involvement in the defence of rose against the necrotrophic fungus *Botrytis cinerea*. BMC Plant Biol 21(1):1–7

Goss EM, Tabima JF, Cooke DE, Restrepo S, Fry WE, Forbes GA, Fieland VJ, Cardenas M, Grünwald NJ (2014) The Irish potato famine pathogen *Phytophthora infestans* originated in central Mexico rather than the Andes. Proc Natl Acad Sci 111(24):8791–8796

Guadalupe Dominguez P, Conti G, Duffy T, Insani M, Alseekh S, Asurmendi S, Fernie AR, Carrari F (2021) Multi-omics analyses reveal the roles of the ASR1 transcription factor in tomato fruits. J Exp Bot 72(18):6490–6509

Guo P, Li Z, Huang P, Li B, Fang S, Chu J, Guo H (2017) A tripartite amplification loop involving the transcription factor WRKY75, salicylic acid, and reactive oxygen species accelerates leaf senescence. Plant Cell 29(11):2854–2870

He X, Zhu L, Xu L, Guo W, Zhang X (2016) GhATAF1, a NAC transcription factor, confers abiotic and biotic stress responses by regulating phytohormonal signaling networks. Plant Cell Rep 35(10):2167–2179

Hong JC (2016) General aspects of plant transcription factor families. InPlant transcription factors. Academic Press, pp 35–56

Hudgins JW, Franceschi VR (2004) Methyl jasmonate-induced ethylene production is responsible for conifer phloem defense responses and reprogramming of stem cambial zone for traumatic resin duct formation. Plant Physiol 135(4):2134–2149

Karim A, Jiang Y, Guo L, Ling Z, Ye S, Duan Y, Li C, Luo K (2015) Isolation and characterization of a subgroup IIa WRKY transcription factor PtrWRKY40 from *Populus trichocarpa*. Tree Physiol 35(10):1129–1139

Karin M, Liu ZG, Zandi E (1997) AP-1 function and regulation. Curr Opin Cell Biol 9(2):240–246

Lai Z, Vinod KM, Zheng Z, Fan B, Chen Z (2008) Roles of Arabidopsis WRKY3 and WRKY4 transcription factors in plant responses to pathogens. BMC Plant Biol 8(1):1–3

Le Hénanff G, Profizi C, Courteaux B, Rabenoelina F, Gérard C, Clément C, Baillieul F, Cordelier S, Dhondt-Cordelier S (2013) Grapevine NAC1 transcription factor as a convergent node in developmental processes, abiotic stresses, and necrotrophic/biotrophic pathogen tolerance. J Exp Bot 64(16):4877–4893

Li X, Yang R, Chen H (2018) The Arabidopsis thaliana Mediator subunit MED8 regulates plant immunity to *Botrytis Cinerea* through interacting with the basic helix-loop-helix (bHLH) transcription factor FAMA. PLoS One 13(3):e0193458

Li C, Cao S, Wang K, Lei C, Ji N, Xu F, Jiang Y, Qiu L, Zheng Y (2021a) Heat shock protein HSP24 is involved in the BABA-induced resistance to fungal pathogen in postharvest grapes underlying an NPR1-dependent manner. Front Plant Sci 12:292

Li J, Meng Y, Zhang K, Li Q, Li S, Xu B, Georgiev MI, Zhou M (2021b) Jasmonic acid-responsive RRTF1 transcription factor controls DTX18 gene expression in hydroxycinnamic acid amide secretion. Plant Physiol 185(2):369–384

Liu B, Ouyang Z, Zhang Y, Li X, Hong Y, Huang L, Liu S, Zhang H, Li D, Song F (2014) Tomato NAC transcription factor SlSRN1 positively regulates defense response against biotic stress but negatively regulates abiotic stress response. PLoS One 9(7):e102067

Liu F, Li X, Wang M, Wen J, Yi B, Shen J, Ma C, Fu T, Tu J (2018) Interactions of WRKY 15 and WRKY 33 transcription factors and their roles in the resistance of oilseed rape to *Sclerotinia* infection. Plant Biotechnol J 16(4):911–925

Majer C, Hochholdinger F (2011) Defining the boundaries: structure and function of LOB domain proteins. Trends Plant Sci 16(1):47–52

Matys V, Kel-Margoulis OV, Fricke E, Liebich I, Land S, Barre-Dirrie A, Reuter I, Chekmenev D, Krull M, Hornischer K, Voss N (2006) TRANSFAC® and its module TRANSCompel®: transcriptional gene regulation in eukaryotes. Nucleic Acids Res 34(suppl_1):D108–D110

McLellan H, Boevink PC, Armstrong MR, Pritchard L, Gomez S, Morales J, Whisson SC, Beynon JL, Birch PR (2013) An RxLR effector from *Phytophthora infestans* prevents re-localisation of two plant NAC transcription factors from the endoplasmic reticulum to the nucleus. PLoS Pathog 9(10):e1003670

McGrath KC, Dombrecht B, Manners JM, Schenk PM, Edgar CI, Maclean DJ, Scheible WR, Udvardi MK, Kazan K (2005) Repressor-and activator-type ethylene response factors functioning in jasmonate signaling and disease resistance identified via a genome-wide screen of Arabidopsis transcription factor gene expression. Plant Physiol 139(2):949–959

Meng X, Xu J, He Y, Yang KY, Mordorski B, Liu Y, Zhang S (2013) Phosphorylation of an ERF transcription factor by Arabidopsis MPK3/MPK6 regulates plant defense gene induction and fungal resistance. Plant Cell 25(3):1126–1142

Mengiste T, Chen X, Salmeron J, Dietrich R (2003) The BOTRYTIS SUSCEPTIBLE1 gene encodes an R2R3MYB transcription factor protein that is required for biotic and abiotic stress responses in Arabidopsis. Plant Cell 15(11):2551–2565

Mochida K, Yoshida T, Sakurai T, Yamaguchi-Shinozaki K, Shinozaki K, Tran LS (2010) Legume TFDB: an integrative database of *Glycine max*, *Lotus japonicus* and *Medicago truncatula* transcription factors. Bioinformatics 26(2):290–291

Moffat CS, Ingle RA, Wathugala DL, Saunders NJ, Knight H, Knight MR (2012) ERF5 and ERF6 play redundant roles as positive regulators of JA/Et-mediated defense against *Botrytis cinerea* in Arabidopsis. PloS Oone 7(4):e35995

Montanini B, Levati E, Bolchi A, Kohler A, Morin E, Tisserant E, Martin F, Ottonello S (2011) Genome-wide search and functional identification of transcription factors in the mycorrhizal fungus *Tuber melanosporum*. New Phytol 189(3):736–750

Mzid R, Marchive C, Blancard D, Deluc L, Barrieu F, Corio-Costet MF, Drira N, Hamdi S, Lauvergeat V (2007) Overexpression of VvWRKY2 in tobacco enhances broad resistance to necrotrophic fungal pathogens. Physiol Plant 131(3):434–447

Park G, Xue C, Zheng L, Lam S, Xu JR (2002) MST12 regulates infectious growth but not appressorium formation in the rice blast fungus *Magnaporthe grisea*. Mol Plant Microbe Interact 15(3):183–192

Park SR, Kim HS, Lee KS, Hwang DJ, Bae SC, Ahn IP, Lee SH, Kim ST (2017) Over-expression of rice NAC transcription factor OsNAC58 on increased resistance to bacterial leaf blight. J Plant Biotechnol 44(2):149–155

Peng SQ, Wu KX, Huang GX, Chen SC (2011) HbMyb1, a Myb transcription factor from *Hevea brasiliensis*, suppresses stress induced cell death in transgenic tobacco. Plant Physiol Biochem 49(12):1429–1435

Perochon A, Váry Z, Malla KB, Halford NG, Paul MJ, Doohan FM (2019) The wheat SnRK1α family and its contribution to *Fusarium* toxin tolerance. Plant Sci 288:110217

Raffaele S, Rivas S (2013) Regulate and be regulated: integration of defense and other signals by the AtMYB30 transcription factor. Front Plant Sci 4:98

Ramírez V, Agorio A, Coego A, García-Andrade J, Hernández MJ, Balaguer B, Ouwerkerk PB, Zarra I, Vera P (2011) MYB46 modulates disease susceptibility to *Botrytis cinerea* in Arabidopsis. Plant Physiol 155(4):1920–1935

Reverberi M, Zjalic S, Ricelli A, Punelli F, Camera E, Fabbri C, Picardo M, Fanelli C, Fabbri AA (2008) Modulation of antioxidant defense in *Aspergillus parasiticus* is involved in aflatoxin biosynthesis: a role for the Ap yapA gene. Eukaryot Cell 7(6):988–1000

Roth MG, Shao D, Smith DL, Kabbage M (2021) Effectors of plant necrotrophic fungi. Front Plant Sci 12:995

Saga H, Ogawa T, Kai K, Suzuki H, Ogata Y, Sakurai N, Shibata D, Ohta D (2012) Identification and characterization of ANAC042, a transcription factor family gene involved in the regulation of camalexin biosynthesis in Arabidopsis. Mol Plant Microbe Interact 25(5):684–696

Shim JS, Jung C, Lee S, Min K, Lee YW, Choi Y, Lee JS, Song JT, Kim JK, Choi YD (2013) AtMYB44 regulates WRKY 70 expression and modulates antagonistic interaction between salicylic acid and jasmonic acid signaling. Plant J 73(3):483–495

Song S, Qi T, Fan M, Zhang X, Gao H, Huang H, Wu D, Guo H, Xie D (2013) The bHLH subgroup IIId factors negatively regulate jasmonate-mediated plant defense and development. PLoS Genet 9(7):e1003653

Spitz F, Furlong EE (2012) Transcription factors: from enhancer binding to developmental control. Nat Rev Genet 13(9):613–626

Stegmaier P, Kel AE, Wingender E (2004) Systematic DNA-binding domain classification of transcription factors. Genome Inform 15(2):276–286

Sun X, Yu G, Li J, Liu J, Wang X, Zhu G, Zhang X, Pan H (2018) AcERF2, an ethylene-responsive factor of *Atriplex canescens*, positively modulates osmotic and disease resistance in *Arabidopsis thaliana*. Plant Sci 274:32–43

Temme N, Oeser B, Massaroli M, Heller J, Simon A, González Collado I, Viaud M, Tudzynski P (2012) BcAtf1, a global regulator, controls various differentiation processes and phytotoxin production in *Botrytis cinerea*. Mol Plant Pathol 13(7):704–718

Thatcher LF, Powell JJ, Aitken EA, Kazan K, Manners JM (2012) The lateral organ boundaries domain transcription factor LBD20 functions in *Fusarium* wilt susceptibility and jasmonate signaling in Arabidopsis. Plant Physiol 160(1):407–418

Tonukari NJ, Scott-Craig JS, Waltonb JD (2002) The *Cochliobolus carbonum* SNF1 gene is required for cell wall–degrading enzyme expression and virulence on maize. Plant Cell 12(2):237–247

Toone WM, Morgan BA, Jones N (2001) Redox control of AP-1-like factors in yeast and beyond. Oncogene 20(19):2336–2346

Tsuda K, Somssich IE (2015) Transcriptional networks in plant immunity. New Phytol 206(3):932–947

Vailleau F, Daniel X, Tronchet M, Montillet JL, Triantaphylides C, Roby D (2002) A R2R3-MYB gene, AtMYB30, acts as a positive regulator of the hypersensitive cell death program in plants in response to pathogen attack. Proc Natl Acad Sci 99(15):10179–10184

Verhage A, Vlaardingerbroek I, Raaijmakers C, Van Dam N, Dicke M, Van Wees S, Pieterse CM (2011) Rewiring of the jasmonate signaling pathway in Arabidopsis during insect herbivory. Front Plant Sci 2:47

Verma S, Nizam S, Verma PK (2013) Biotic and abiotic stress signaling in plants. In: Stress signaling in plants: genomics and proteomics perspective, vol 1. Springer, New York, pp 25–49

Wang XE, Basnayake BV, Zhang H, Li G, Li W, Virk N, Mengiste T, Song F (2009) The Arabidopsis ATAF1, a NAC transcription factor, is a negative regulator of defense responses against necrotrophic fungal and bacterial pathogens. Mol Plant-Microbe Interact 22(10):1227–1238

Wang Z, Fang H, Chen Y, Chen K, Li G, Gu S, Tan X (2014) Overexpression of BnWRKY33 in oilseed rape enhances resistance to *Sclerotinia sclerotiorum*. Mol Plant Pathol 15(7):677–689

Wang W, Yuan Y, Yang C, Geng S, Sun Q, Long L, Cai C, Chu Z, Liu X, Wang G, Du X (2016) Characterization, expression, and functional analysis of a novel NAC gene associated with resistance to verticillium wilt and abiotic stress in cotton. G3: Genes, Genomes, Genetics 6(12):3951–3961

Wilson D, Charoensawan V, Kummerfeld SK, Teichmann SA (2008) DBD—taxonomically broad transcription factor predictions: new content and functionality. Nucleic Acids Res 36(suppl_1):D88–D92

Wu Y, Deng Z, Lai J, Zhang Y, Yang C, Yin B, Zhao Q, Zhang L, Li Y, Yang C, Xie Q (2009) Dual function of Arabidopsis ATAF1 in abiotic and biotic stress responses. Cell Res 19(11):1279–1290

Zhang Z, Yao W, Dong N, Liang H, Liu H, Huang R (2007) A novel ERF transcription activator in wheat and its induction kinetics after pathogen and hormone treatments. J Exp Bot 58(11):2993–3003

Zhang HM, Chen H, Liu W, Liu H, Gong J, Wang H, Guo AY (2012) AnimalTFDB: a comprehensive animal transcription factor database. Nucleic Acids Res 40(D1):D144–D149

Zhang X, Zhu Z, An F, Hao D, Li P, Song J, Yi C, Guo H (2014) Jasmonate-activated MYC2 represses ethylene insensitive3 activity to antagonize ethylene-promoted apical hook formation in Arabidopsis. Plant Cell 26(3):1105–1117

Zhang M, Liu Y, Li Z, She Z, Chai M, Aslam M, He Q, Huang Y, Chen F, Chen H, Song S (2021) The bZIP transcription factor GmbZIP15 facilitates resistance against *Sclerotinia sclerotiorum* and *Phytophthora sojae* infection in soybean. iScience 24(6):102642

Zheng Z, Qamar SA, Chen Z, Mengiste T (2006) Arabidopsis WRKY33 transcription factor is required for resistance to necrotrophic fungal pathogens. Plant J 48(4):592–605

Zheng G, Tu K, Yang Q, Xiong Y, Wei C, Xie L, Zhu Y, Li Y (2008) ITFP: an integrated platform of mammalian transcription factors. Bioinformatics 24(20):2416–2417

Zheng XY, Spivey NW, Zeng W, Liu PP, Fu ZQ, Klessig DF, He SY, Dong X (2012) Coronatine promotes *Pseudomonas syringae* virulence in plants by activating a signaling cascade that inhibits salicylic acid accumulation. Cell Host Microbe 11(6):587–596

Zhou J, Lee C, Zhong R, Ye ZH (2009) MYB58 and MYB63 are transcriptional activators of the lignin biosynthetic pathway during secondary cell wall formation in Arabidopsis. Plant Cell 21(1):248–266

Zhu Z, An F, Feng Y, Li P, Xue L, Mu A, Jiang Z, Kim JM, To TK, Li W, Zhang X (2011) Derepression of ethylene-stabilized transcription factors (EIN3/EIL1) mediates jasmonate and ethylene signaling synergy in Arabidopsis. Proc Natl Acad Sci 108(30):12539–12544

Chapter 5
Role of Plant Transcription Factors in Virus Stress

Bipasha Bhattacharjee and Vipin Hallan

Abstract Plants get exposed to various types of microorganisms in real time, which positively or negatively regulates their growth and development. On contact or sensing foreign biotic agents, intricate molecular and physiological changes are triggered within the plant system, which leads to the activation of a host of pathways relating to defense and morphogenesis. These mechanisms involve complex roles of Transcription Factors (TFs) and their interactome between co-factors and *cis*-regulatory genomic elements. Mostly involving multi-domain families, transcription factors prompt multifaceted responses during pathogen stress leading to activation of diverse pathways progressing to the production of several metabolites, defense proteins, plant hormones, and transcriptional and posttranscriptional gene modification (TGM/PTGM). By activating or repressing downstream signaling pathways, transcription factor family members interact directly or indirectly to affect the defense response. As a result, understanding how plant viruses and TFs interact and decoding changes in the defense pathway are required before crops may be engineered to withstand biotic stressors. A few families of TFs like bZIP, ERF/AP2, MYB, NAC, and WRKY have comprehensively been studied for biotic stress mitigation response in general and plant virus stress in particular, and through this chapter, we shall be elucidating the roles of such plant transcription factors in gene expression regulation during plant–virus attack and overall plant stress response.

Keywords Plant virus · Transcription Factor · Biotic Stress · Defense

B. Bhattacharjee · V. Hallan (✉)
Academy of Scientific and Innovative Research (AcSIR), Ghaziabad, India

Plant Virology Laboratory, Division of Biotechnology, CSIR-Institute of Himalayan Bioresource Technology, Palampur, India
e-mail: hallan@ihbt.res.in

© The Author(s), under exclusive license to Springer Nature Switzerland AG 2022
S. H. Wani et al. (eds.), *Transcription Factors for Biotic Stress Tolerance in Plants*, https://doi.org/10.1007/978-3-031-12990-2_5

5.1 Introduction

Plant stressors, including environmental (abiotic) and living (biotic) stresses, comprise mainly pressures that affect all types of plants globally, leading to significant crop losses and an eventual loss of livelihood. Abiotic stresses comprise salt, drought, heat, and cold stress, whereas biotic stresses involve microorganisms like bacteria, protozoa, fungi, and viruses and others such as insects and nematodes (Van Verk et al. 2009). Such stresses have been a part of a continual evolution and therefore have been a complex part of plant–microbe interaction ecosystem for eons. Plants modulate cellular, biochemical, molecular, and physiological feedback in a comeback to various biotic and abiotic stresses. These modulation types are affected by transcription activators and repressors encoding genes, which are very common in affecting downstream stress-responsive genes that regulate various metabolic and defense-related processes within the plant system (Tolosa and Zhang 2020). A hypersensitive response (HR) is mounted after transcriptional activation/repression that leads to the limitation of pathogen invasion and multiplication, leading to subsequent programmed cell death (PCD) (Coll et al. 2011). In this type of response, the molecular changes in plants involve protein phosphorylation/dephosphorylation through calcium-dependent or mitogen-activated kinases (CDPK/MAPK), reactive oxygen species (ROS) production, plasma membrane depolarization and channel activity modulation, modification of host transcriptional regulation, and plant cell wall changes, among others (Meng and Zhang 2013). Therefore, plant transcription reprogramming is essential for defense response regulation (Fan et al. 2014) and has been reported to participate in an adjustment between development and immunity (Lozano-Durán et al. 2013) to ensure proper cellular resource allotment for the survival of the host plant (Malinovsky et al. 2014). Hormonal signaling and transcriptional regulation of gene expression are therefore very common methods of host defense mounting of expression response under any form of stress.

Plant transcriptional regulation is very well established in many plant species. The defense transcription factors consist of two types of mode of action, i.e., DNA-binding TFs of the families of Apetela2/ethylene-responsive element (AP2/ERF), basic leucine zipper (bZIP), myeloblastosis-related protein (MYB), WRKY family of TFs, and NAC TFs comprising no apical meristem (NAM), *Arabidopsis* transcription activator factor (ATAF1/2) and cup-shaped cotyledon (CUC), and corresponding proteins that interact and regulate these TFs through a multitude of molecular interactions (Eulgem 2005; Erpen et al. 2018). A DNA-binding domain (DBD) of transcription factors accurately detects the target DNA sequence, forming a transcriptional complex, and so modulating gene expression (Ikeda and Ohme-Takagi 2009).

Plant viruses are agronomically important as they cause the destruction of crops and are difficult to control. These are obligate parasites that only get activated when within the host tissue. Chlorosis, necrosis, vein clearing, and wilting are all symptoms of viral infections in agricultural plants, and they influence the physiology and morphology of the plants (Maisonneuve et al. 2018; Lei et al. 2017). A deeper understanding of the defense and counter-defense mechanisms utilized by plants and pathogenic viruses is necessary in such situations because both viruses and host species have developed specific methods to support their survival and expansion. The role of TFs is widely elucidated in viral stress response. Therefore, in the following sections, we aim to curate and analyze the roles of different plant transcription factors and their regulation in the case of defense response against plant virus invasion and infection in plants.

5.2 Plant Virus and Virus-Like Pathogens

Single-stranded (ss) DNA viruses, double-stranded (ds) DNA viruses, ss positive sense RNA viruses, double-stranded RNA (dsRNA) viruses, and ss negative sense RNA viruses are among the 16 families and three orders of plant viruses that produce viral stress (Lefkowitz et al. 2018). Viroids have free RNA molecules and lack coat protein, are alike viruses, and replicate within the host (Marwal and Gaur 2020). These pathogens proliferate within the cells using plasmodesmata and trigger varying degrees of resistance mechanisms within the plant system, thereby modulating the genetic makeup and molecular dynamics of the plant–pathogen cosplay (Boualem et al. 2016). To halt viral replication, plants use responses to hormones, gene silencing, metabolite level management, protein degradation via ubiquitin proteasome pathway (UPS), immunological receptor signaling, and PAMP-triggered immunity. Plant hormones, such as abscisic acid, auxin, brassinosteroids, cytokinin, ethylene, gibberellin, jasmonic acid, salicylic acid, and ROS, have a role in virus defense (Calil and Fontes 2017). Diverse plant transcription factors show major roles in damage tolerance against virus attacks. We categorize the various TF factor families and genes associated, which have been reported to be regulated under different kinds of virus stress in this chapter.

5.3 AP2/ERF Transcription Factor Family

5.3.1 Classification and Constitution AP2/ERF Family

This family of TFs is among the most widespread in plants and has diverse roles in growth modulation, plant development, hormone responses, and environmental and microorganism stress responses. The APETALA2 (AP2)/Ethylene Responsive Element Binding Factor (EREB) domain typically has about 70 amino acids participating in the interaction with the DNA elements responsible for stress regulation (Feng et al. 2005; Sakuma et al. 2002; Nakano et al. 2006). APETALA2 (AP2), Related to Abscisic Acid Insensitive 3/Viviparous 1 (RAV), Dehydration-Responsive Element Binding proteins (DREBs) (subgroups A1 to A6), and Ethylene-Responsive Factors (ERFs) are four key sub-families of AP2/ERFs belonging to the subgroup five to ten (Sakuma et al. 2002; Nakano et al. 2006). Two amino acids, namely, aspartate (position 14) and alanine (position 9), take part in the *cis*-element binding (Sakuma et al. 2002).

Tightly controlled gene regulation is essential for stress tolerance mechanisms (Feng et al. 2005), and it has been shown that AP2/ERF family constitutively express in low levels, although expression can be enhanced or repressed (Li et al. 2017) by hormones and stress stimuli at various developmental stages (Owji et al. 2017). The binding preferences of AP2/ERFs are also conserved. C-Repeat Element, which is also known as Dehydration-Responsive (DRE/CRT), is recognized by DREB to confer abiotic resistances, and ERFs bind to the Ethylene-Response Element (ERE) through a GCC box, participating in biotic stress tolerance (Guo and Ecker 2004; Franco-Zorrilla et al. 2014). Though this binding is flexible as in many *Arabidopsis* ERFs and DREBs, they have been reported to interchangeably bind in cooperation with ERE and DRE/CRT elements, leading to the conclusion that there is a role in different types of stresses, be it biotic or abiotic (Xie et al. 2019). Similar to *Oryza sativa* (Wan et al. 2011), *Glycine max* (Zhang et al. 2009), *Triticum aestivum* (Gao et al. 2018), *Zea mays* (Liu et al. 2013), and *Nicotiana tabacum*, the AP2/ERF conserved DNA-binding abilities have been extended to additional classes (Park et al. 2001). The structure and mode of action are described in Fig. 5.1.

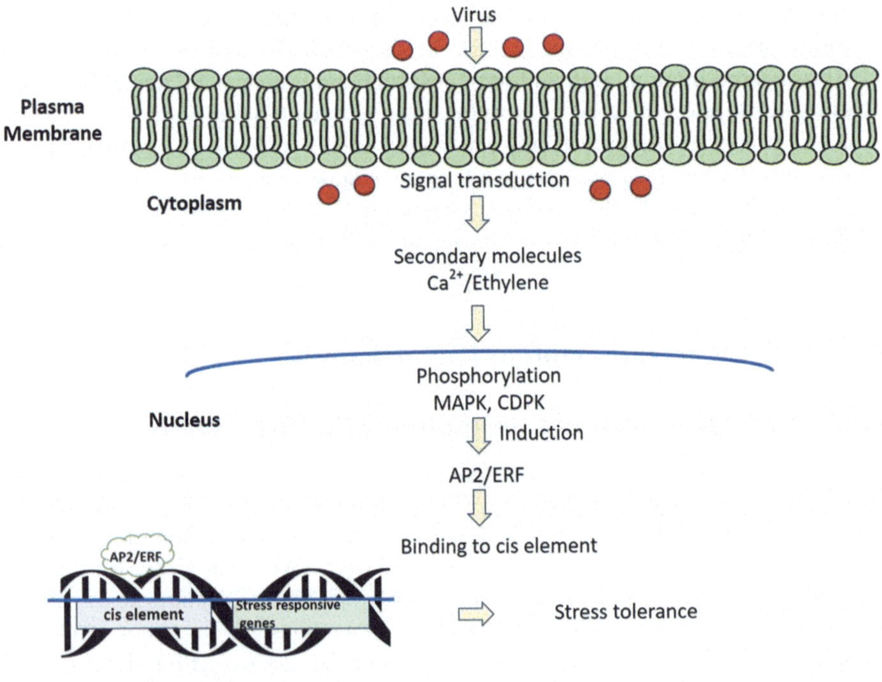

Fig. 5.1 On virus infection, induction of AP2/ERF TFs takes place, which in turn regulates the expression of many defense-related genes like pathogenesis-related (PR) genes. This induction also triggers small molecules like calcium ions and hormones like ethylene to activate phosphorylation and subsequent defense response regulation

5.3.2 Functional Characterization of AP2/ERF Family in Viral Defense Response

AP2/ERF family comprises transcription proteins and has a recognised role in biotic disease response patterns in plant life. These can function either through transcriptional activation or repression. Both N-DNA-binding and C-activation terminal domains participate in the DNA binding to modulate gene expression (Nakano et al. 2006). Activation domains in AP2-TF are generally acidic; for example, in tobacco protoplast, the N-terminal and/or C-terminal acidic portions of tobacco ERF2 and ERF4 function as activation domains (Ohta et al. 2000). ERF TFs regulate both the basal levels of transcription of target genes and other related genes to ensure biotic stress tolerance and hence are potential candidates for manipulation of host disease resistance pathways (Phukan et al. 2017). There have been studies showing the roles of AP2/ERF TFs participating in viral disease resistance, and they usually act as factors conferring tolerance to genes. Among various viruses, infiltration and consequent infection of *tomato yellow leaf curl virus* (TYLCV) in tomato modulated about 22 ERFs, and 5 ERF-B3 TFs were recognized in tomato cultivars Hongbeibei, Zheza-301/Zhefen-702, and Jinpeng-1/Xianke-6 (highly resistant, resistant, and susceptible, respectively). Consequent to the TYLCV infection in the five cultivars of tomato, gene expression patterns of five genes of ERF-B3, namely Soly19, Soly36, Soly66, Soly67, and Soly106, were identified using quantitative real-time polymerase chain reaction (qPCR). In five tomato cultivars, Soly106 manifestation was augmented. Soly19, Soly67, and Soly36 have their expressions elevated in Zhe-za-301 and Zhefen-702. In Hongbei-bei and the Xianke-6, expression of Soly66 and Soly36 was downregulated. According to a yeast one-hybrid investigation, the ERF-B3 TFs binding with the GCC box differed between sensitive and resistant tomato cultivars. The ability of SlERF TFs to bind to the GCC-box was connected to expression patterns in sensitive and tolerant tomato cultivars (Huang et al. 2016). Potato aphid resistance in tomatoes was amplified by the ERF Pti5 (Pto-interacting protein 5) through virus-induced gene silencing (VIGS) (Wu et al. 2015). In fact, multiple types of ERFs have been identified in tomatoes that could be potentially responsible for plant defense regulation under TYLCV stress (Huang et al. 2016). The *Nicotiana benthamiana* ERF5 was also identified as a factor whose overexpression led to weaker viral accumulation. NtERF5 was isolated using yeast one hybrid, and its recombinant form interacted with the GCC box cis-elements weakly, indicating potential regulation of PR genes. The ERF5 overexpressing plants under the constitutive promoter CaMV-35S demonstrated a higher level of resistance with a reduced HR response and impaired systemic virus spread (Fischer and Dröge-Laser 2004). P2/ERFs were found to be linked to several other tomato genes (such as MAPK) that were involved in the activation of plant resistance to fungal and virus pathogens. The AP2/ERF proteins may respond during TYLCV infection by interacting with other genes and altering the signaling pathways for JA, SA, ethylene, and H_2O_2.

5.4 bZIP Transcription Factor Family

5.4.1 Classification and Constitution of bZIP Family

In eukaryotes, it is one of the largest families of plant transcription factors. These factors take part in various pathways pertaining to growth, floral development, and abiotic and biotic stress response (Jakoby et al. 2002). The bZIP dimerization domain comprises two parts, i.e., a leucine-rich zipper domain, which is less conserved, and a basic region, which consists of a nuclear localization signal. bZIP proteins have a specificity of binding to the A-box (TACGTA), C-box (GACGTC), G-box (CACGTG), GLM (GTGAGTCAT), and PB-like (TGAAAA) sequences, which are components within the cis-elements of stress-related genes (Ali et al. 2016). The bZIP TF family of *Arabidopsis* is organized into ten categories, i.e., (A-S, alphabetically) with each group having comparable sequence similarities in the basic region, size of proteins as well as the leucine domain size (Jakoby et al. 2002). Multiple bZIP TFs have been identified in many different plants species, including *Arabidopsis* (Jakoby et al. 2002), *Malus sieversii* L. (Zhao et al. 2016), *Brassica oleracea* (Bai et al. 2016), *Hordeum vulgare* L. (Pourabed et al. 2015), *Brassica rapa* (Hwang et al. 2014), *Brachypodium distachyon* (Liu and Chu 2015), *Manihot esculenta* (Hu et al. 2016), *Ricinus communis* L. (Jin et al. 2014), *Cucumis sativus* (Baloglu et al. 2014), *Zea mays* L. (Wei et al. 2012), *Capsicum annum* (Hwang et al. 2005), *Phaseolus vulgaris* (Astudillo et al. 2013), *Lablab purpureus* L. (Wang et al. 2015), *Populus deltoides* (Ji et al. 2013), *Oryza sativa* L. (Nijhawan et al. 2008), *Solanum lycopersicum* L. (Li et al. 2015), and *Vitis vinifera* (Liu et al. 2014). Plant bZIPs are extensively involved in abiotic stress (Inaba et al. 2015; Moon et al. 2015; Banerjee and Roychoudhury 2017, Sornaraj et al. 2016; Zong et al. 2016; Xu et al. 2016), metabolic rate (Hartmann et al. 2015; Zhang et al. 2015; Sagor et al. 2016), and biotic stress (Kim and Delaney 2002; Alves et al. 2013, 2015; Shearer et al. 2012; Lim et al. 2015; Unel et al. 2019).

5.4.2 Functional Characterization of bZIP Family in Viral Defense Response

The pathogen responsiveness of the bZIP family of TFs has been extensively studied. During pathogen infection, the TGA proteins associate with Ankyrin repeat protein members, especially NPR1 (non- expresser of PR-1), which are important members of the SA signaling pathway in the defense network activated upon biotic stress (Li et al. 2012; Kaltdorf and Naseem 2013; Pajerowska-Mukhtar et al. 2013). During a pathogen attack, SA synthesis triggers alterations in the redox state within the cell, which monomerizes the NPR1 and relocates to the nucleus by the nucleopore complex (Li et al. 2012; Pieterse et al. 2012). The monomers of NPR1 then attach to the SA-regulated gene promoter with the help of the TGA family proteins (bZIP). NPR1 is phosphorylated during transport and association, and then E3 ubiquitin ligase ubiquitinates the protein, which has a strong binding potential for the phosphorylated version of NPR1. The NPR1 degradation is a result of its ubiquitination via proteasome complex. NPR1 homologs (NPR3 and NPR4) function similarly to SA receptors in the process of deterioration. NPR3/NPR4 are used here as Cullin-3-based E3 ubiquitin ligase adapter, and finally NPR1 is ubiquitinated and then degraded and are regulated by SA (Li et al. 2012; Kaltdorf and Naseem 2013; Pajerowska-Mukhtar et al. 2013; Pieterse et al. 2012) (Fig. 5.2).

A full-length gene called *PP1* (pepper PMMV interaction 1), which encodes a basic bZIP DNA-binding protein region, was identified and isolated from *Capsicum chinense*, which was infected with the *Pepper mild mottle virus* (PMMV). The encoded protein shared a similar amino acid sequence as well as activities with the ACGT- interacting domains of the bZIP TF family. This gene was induced on an incompatible interaction test using PMMV, *Pseudomonas syringae* 61, and *Xanthomonas campestris* pv. *Vesicatoria* race-3, but was not induced by any abiotic stressors. This was the first evidence that provided an indication of bZIP protein regulation under pathogen attack, with PP1 having a specific role in biotic

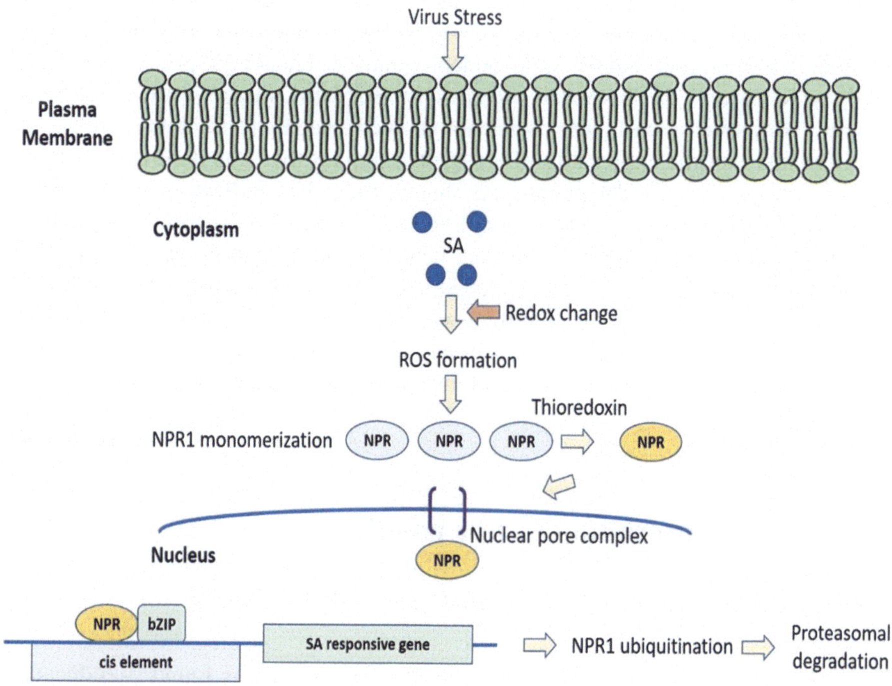

Fig. 5.2 Representation of bZIP TF mechanism of action under viral stress. Salicylic acid is an important regulator triggered during the infection process, which then leads to the change in cell redox potential. This leads to ROX formation and non-expressor of PR1 monomerization. The monomers are then engulfed by nuclear pore complexes where attachment takes place in the promoter region of SA-responsive genes along with the involvement of bZIP TFs

stress-responsive pathways (Lee et al. 2002). In another report, HAT1 protein (homeodomain-leucine zipper protein 1) of *Arabidopsis thaliana*, belonging to the hdZIP family of TFs, showed an important role in plant viral defense. Triple knockout mutant lines, hat1hat2hat3, and overexpressing lines (HAT1OX) were infected with *cucumber mosaic virus* (CMV), where the knockout mutants exhibited more tolerance to infection than the overexpressing lines. The triple mutant hat1hat2hat3 showed improved viral resistance relative to the hat1 and hat1hat3 mutants, indicating that HAT1 and its near homologs HAT2 and HAT3 act superfluously. Furthermore, as compared to Col 0 after CMV infection, the antioxidant system (functions and production of antioxidative enzymes) and the transcription of genes associated with defense response were downregulated in HAT1OX while upregulated in its expression in hat1hat2hat3. According to a new study, HAT1's participation in the anti-CMV defensive response appears to be salicylic acid dependent (SA), but jasmonic acid independent (JA). After CMV infection, the expression level of SA synthesizing genes was lower in HAT1OX but higher in hat1hat2hat3 triple mutant compared to Col-0, while there was no change in the level of JA or expression of JA synthesis-linked genes between HAT1OX and deficient plants. Therefore, HAT1 expression is reliant on the buildup of SA. HAT1 appears to be a downregulator of plant defensive responses to CMV, according to the findings (Zou et al. 2016). *Rice tungro bacilliform virus* (RTBV) is another persistent pathogen found in rice in predominantly South Asian countries. The RTBV promoter is regulated by RF2a and RF2b, which are host transcription factors critical for plant growth. Overexpression of these two genes showed a negative effect on morphogenesis of plants, but the expression of a negative mutant that was dominant of these genes in transgenic rice led to phenotypes that mimicked the indicators of RTD. After RTBV inoculation with *Agrobacterium*, lines with strong RF2a or RF2b expression showed minimal or no signs of infection, whereas control plants showed considerable shrinkage and leaf browning. Transgenic plants also had lower levels of RTBV RNA and viral DNA accumulation than wild type (Dai et al. 2008). These experiments established that the bZIP TFs, in association with SA signaling pathways, could be a capable strategy for virus defense plant genotypes to be created (Pandey et al. 2018).

5.5 MYB Transcription Factor Family

5.5.1 Classification and Constitution of MYB Family

The first plant MYB gene, *Zea mays* COLORED1 (C1), was found two decades ago, and its expressing MYB domain protein is necessary for anthocyanin synthesis in the maize aleurone (Paz-Ares et al. 1987). The existence of MYB domains that are very conserved and participate in DNA binding is used to characterize MYB TFs. Multiple repetitions (R) are seen in these domains; each repeat has 52 amino acids along with three α helix, the second and third of which create a helix–turn–helix (HTH) structure (Dubos et al. 2010). MYB TFs are divided into four groups based on the number of the adjacent repeats in the MYB domains: 1R-MYB (one repeat), R2R3-MYB (two repeats), 3R-MYB (three repeats), and 4R-MYB (four repeats) (Baillo et al. 2019). Each of the four incomplete amino acid sequence repetitions (R) in the MYB domain has 52 amino acids and three helices. The second and third helices of the repeats, as well as the second and third helices of the repeats with three consistently different tryptophan or hydrophobic residues, form a hydrophobic core component in the three-dimensional HTH structure (Pandey et al. 2018; Ogata et al. 1996).

5.5.2 *Functional Characterization of MYB Family in Viral Defense Response*

The behavior of MYB TFs in different plant processes, involving abiotic and biotic stress adaptation, has been widely examined in numerous plant species, with a full evaluation of the actions of MYB TFs in various plant processes. Considering the amount of studies in this field, it is worthwhile to note how MYB TFs' roles aid in improving stress tolerance in economically significant crops. A visual reported system was generated to track plant virus movement and infection in plants, which was centered on the activity of a MYB-linked Ros1 gene from *Antirrhinium majus*. This system allowed the activation of biosynthetic genes related to anthocyanin accumulation. Two different clones of *tobacco etch potyvirus*, gene labeled with Ros1, were constructed, and the infiltrated tobacco plants (*Nicotiana tabacum*) turned bright red upon pigment accumulation. This marker system also helped in reporting the viral load quantitatively and qualitatively through a simple extraction process. This was a very stable system that led to the establishment of an accurate tracking system in different plants as well, like tracking of *turnip mosaic potyvirus* and *tobacco* or *potato X virus* infecting *Arabidopsis* and *Nicotiana benthamiana*, respectively (Bedoya et al. 2012).

Plant PR genes are activated by overexpression of certain R2R3-MYB TFs, resulting in systemic acquired resistance (SAR). Phytohormones, particularly JA and SA, control this response, which defends the plant from bacterial, fungal, and viral diseases (Bostock 2005; Durrant and Dong 2004). Infection with biotic stressors and subsequent management with defense-related phytohormones increase the expression of AtMYB44. When transgenic plants overexpressing AtMYB44 are infected with *P. syringae*, they have an advanced amount of PR gene manifestation and improved resistance (Zou et al. 2012). In transgenic *Arabidopsis*, overexpression of AtMYB96 leads to increased disease tolerance. A subgroup of PR genes is also upregulated as a result of overexpression (Seo et al. 2009). OsMYB4 overexpressing *Solanum lycopersicum* plants have a better drought tolerance level and can successfully guard plants in the case of a viral infection (Vannini et al. 2007).

According to scientists, OsMYB4 is a critical knot in the interplay of stress signaling pathways because it activates many components, but its activity is dependent on the host's genetic makeup (Erpen et al. 2018). Overexpression of *Thinopyrum intermedium* MYB TF (TiMYB2R1) in wheat demonstrated that MYB TF is involved in disease resistance, and the progressive elevation of defense-correlated genes within a transgenic plant was demonstrated (Liu et al. 2013). One of the primary methods for plant antiviral resistance in virus–plant interplay is RNA silencing, which is frequently repressed by co-evolving virus suppressors, boosting viral pathogenicity in vulnerable hosts (Pumplin and Voinnet 2013). In *A. thaliana*, MYB TF acts as a negative controller to lower viral load and promote viral immunity (Zorzatto et al. 2015). A leucine-rich receptor-like kinase LRR-RLK called NIK1 interacted with LIMYB, leading to translational suppression. Overexpression of LIMYB causes transcriptional repression of ribosomal protein genes, leading to the synthesis of protein restriction, reduced viral messenger RNA interaction with polysome fractions, and increased begomovirus tolerance. Loss of LIMYB function, on the other hand, causes the repression of translation-related genes to be released, increasing viral infection vulnerability (Zorzatto et al. 2015) (Fig. 5.3). In a recent study,

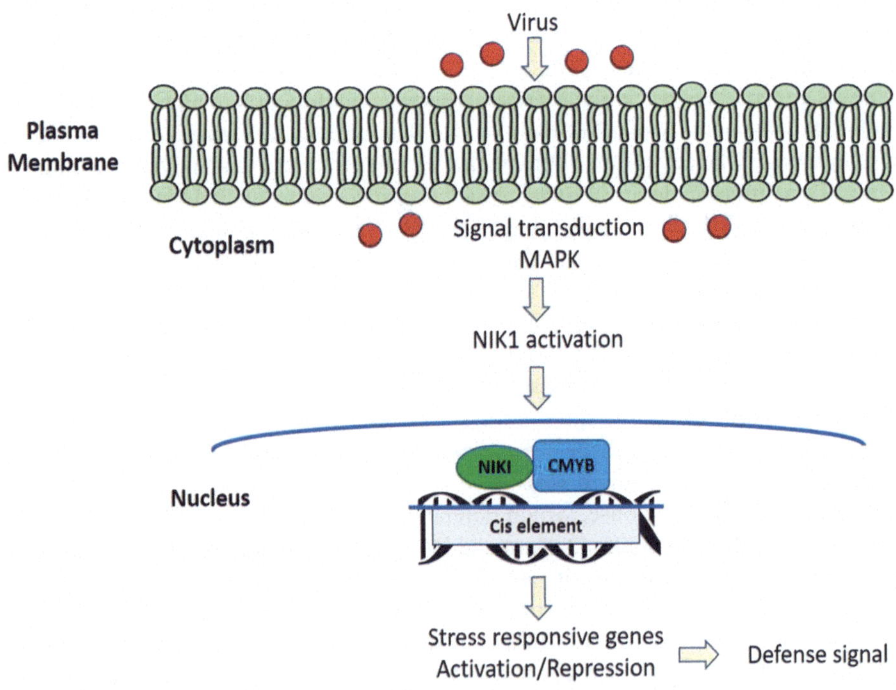

Fig. 5.3 Mechanism of MYB TFs in viral stress. On virus infection, modulation of very long fatty acid chains takes place, mounting an HR response within the plant cell. NIK1, which is an immune receptor, interacts with MYB and binds to the promoter elements, which then leads to a cascade of signaling mechanisms leading to plant defense responses in the case of infection by the begomovirus family

cucumber green mottle mosaic virus (CGMMV) was infected, and transcriptomic analysis was done for the susceptible infected *Cucumis sativus* inbred line 229 where six differentially regulated genes encoding MYB TFs were discovered expressing at 3 dpi, with just two DEGs seen in 20 dpi. In cucumber plants infected with CGMMV, two DEGs having lower transcript levels were upregulated early on, whereas four DEGs with lower or medium expression levels remained downregulated. Two weakly expressing MYB factors showed marginal upregulation toward the end (Slavokhotova et al. 2021). In *Nicotiana benthamiana*, MYB4-like TFs coupled with the ethylene pathway participates in viral resistance. NbMYB4L transcription was upregulated during TMV infection, while silencing the gene led to increased susceptibility to TMV replication. Also, on the treatment of the plants with 1-aminocyclopropanecarboxylic acid (ACC), which is an ethylene precursor, the ethylene signaling pathway was obstructed. The expression of NbMYB4L considerably got repressed in Ethylene Insensitive 3-like 1 (EIL1)-downregulated plants, according to gene expression analysis. NbEIL1 may directly bind to two particular areas of the NbMYB4L promoter, according to the results of EMSA (electrophoretic mobility shift assay) and also chromatin immunoprecipitation-quantitative PCR (ChIP-qPCR) tests. The regulation of NbMYB4L was considerably repressed in EIL1 silenced plants, according to gene expression analysis. NbEIL1 also dramatically increased the activity of the reporter of the promoter of MYB4L in *N. benthamiana*, based on a luciferase test. These findings suggest that NbEIL1 acts as a promoting regulator of NbMYB4L expression in *N. benthamiana* to protect it against TMV infection (Zhu et al. 2021).

5.6 NAC Transcription Factor Family

5.6.1 Classification and Constitution of NAC Family

NAC TF family is widespread in plants and comprises the NAC conserved region and was initially discovered in the following transcription factors: No Apical Meristem (NAM), *Arabidopsis thaliana* transcription activation factor (ATAF1/2), and cup-shaped cotyledon proteins and is similar in all members of this family (CUC2) (Aida et al. 1997). A total of 150 amino acid residues make up the conserved DNA-binding NAC domain N-terminal region of NAC TFs. Their C-terminal domains, which regulate transcription (TR), on the other hand, are all different. TR domains of NAC proteins have a transmembrane domain, which could upregulate or downregulate transcription (Puranik et al. 2012). Typically, functioning in an abscisic acid-dependent method, NAC TFs may also act in an ABA-independent manner to influence both biotic and abiotic stress mechanisms (Nakashima et al. 2007). Regulated NAC genes have promoters that comprising *cis*-regions that contain stress-related elements, which can bind to upstream TFs such as ABREs (ABA-responsive element-binding protein), DREBs (dehydration-responsive element-binding protein), and LTREs (low temperature-responsive elements), and

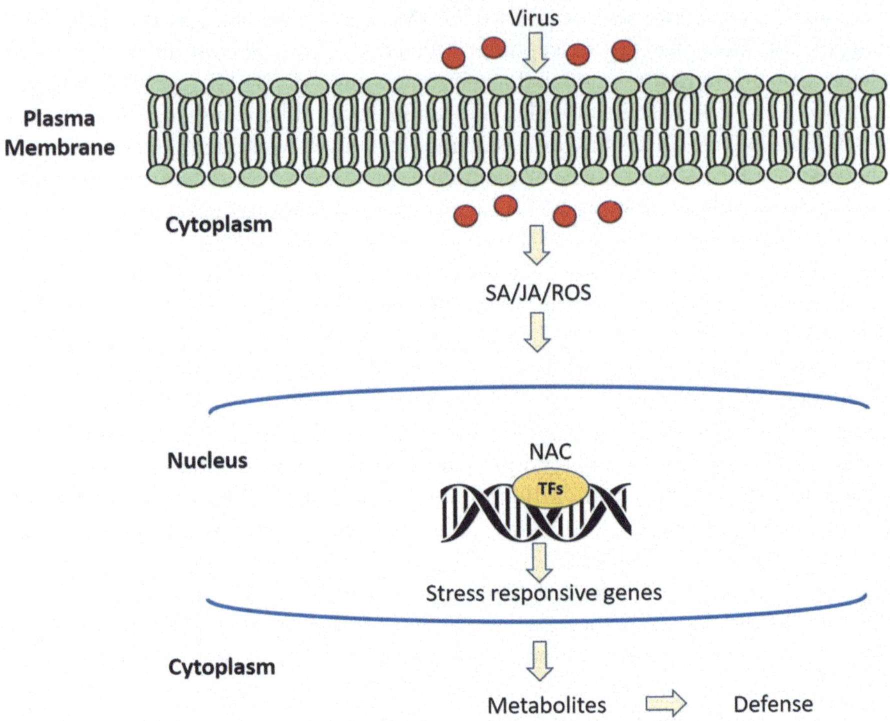

Fig. 5.4 A simplistic representation of NAC TFs mechanism under viral stress. On infection by virus and viral particles, a hypersensitive reaction is induced in the cell. Salicylic acid pathways get induced after cell redox potential balance collapses, leading to activation of resistance genes within plants

affect the transcription of these genic elements. Signaling pathways controlled by ABA are mediated by ABREs, which are typical *cis-acting* elements. The DRE/CRT element, on the other hand, acts in *cis*, which facilitates signaling pathways not dependent on ABA (Puranik et al. 2012). According to expression profiling studies on a whole-genome in *Arabidopsis*, most NAC TFs are engaged in the control of gene expression in regard to a minimum of one type of environmental stress signal (salinity, drought, cold, or heat); however, most NAC genes are susceptible to stress caused by salt and drought. However, many have been essential in viral stress down-regulation factors. A basic mode of action is explained in Fig. 5.4.

5.6.2 Functional Characterization of NAC Family in Viral Defense Response

NAC genes have been discovered to decrease the expression of defense genes, acting as negative regulators of disease resistance (Wang et al. 2009a, b). On direct interaction with viral proteins, virus replication can also be stunted or accelerated

(Ren et al. 2000; Selth et al. 2005; Yoshii et al. 2009). In a particular study with rice, five types of viruses were used for infection, namely *rice ragged stunt virus* (RRSV), *rice dwarf virus* (RDV), *rice grassy stunt virus* (RGSV), *rice black-streaked dwarf virus* (RBSDV), and *rice transitory yellowing virus* (RTYV). When compared to the control, a microarray study revealed that 75 (68%) OsNAC genic elements expressed differently following infection by RDV, RBSDV, RGSV, and RRSV. The most OsNAC genes were heightened in expression during RGSV infection, whereas the least were upregulated during RTYV infection. For all viral infections, the NAC subset genes NAC22, SND, ONAC2, ANAC34, and ONAC3 had the expression downregulated. These genes may be linked to the host's health upkeep. Interestingly, during RBSDV and RGSV infections, the majority of the genes belonging to TIP and SNAC groups were more substantial in expression. These results showed that OsNAC genes might be regulated by the immune reactions elicited by viral contagion (Nuruzzaman et al. 2015). Rice resistance to the fungus *Magnaporthe* is favorably regulated by OsNAC6, OsNAC58, OsNAC60, ONAC066, OsNAC111, ONAC122, and ONAC131, whereas rice tolerance to the virus RDV is negatively regulated by RIM1 (Yoshii et al. 2009). The helicase domain of the *tobacco mosaic virus* (TMV) 126-/183-kDa replicase protein(s) and the NAC domain transcription factor ATAF2 of *Arabidopsis thaliana* were found using yeast two-hybrid and immunoprecipitation assays *in planta* and their relationship was established. Transcriptional activation of ATAF2 takes place in response to TMV infection, and its overexpression significantly reduces viral buildup. It appears to be targeted for a breakdown during viral infection, according to proteasome inhibition experiments. Transgenic plants overexpressing ATAF2 had higher gene transcription of identified defense-related indicator genes PR1, PR2, and PDF1.2, whereas ATAF2 knockout or repressor plant lines have lower transcript levels. In a reaction to TMV infections, transcript accumulations of ATAF2 along with PR1 increase in damaged tissue but not in tissues infected systemically. The levels of ATAF2 along with PR1 transcripts rise in response to salicylic acid stimulation. The treatment of systemically infected hosts with salicylic acid did not result in a comparable rise in ATAF2/PR1 transcripts, indicating suppressed host defense mechanism during the systemic viral invasion. When treated with salicylic acid, uncompromised ATAF2 knockout or ATAF2 repressed lines show lower PR1 transcripts. These data collectively show that the connection between replicase and ATAF2 reduces basal host defenses to enhance systemic viral growth (Wang et al. 2019).

An *Arabidopsis thaliana* NAC transcription factor, TIP, and the coat protein (CP) of *turnip crinkle virus* (TCV) were found to directly interact in the Di-17 ecotype, which is resistant toward TCV infection (Donze et al. 2014). In the TCV-susceptible ecotype Col-0, the mutated CP fails to engage with TIP causing the associated R6A lines, a mutant for TCV, to induce added acute symptoms. Therefore, it was postulated that TCV modulated the basal viral immunity through TIP–CP interaction. In resistant plants, a robust HR-mediated resistance response may be identified, in which these plants express a hypersensitive response protein (termed HRT) that might potentially protect the TIP protein by sensing an alteration in TIP induced by the TIP–CP interface (Ren et al. 2000). Pathogen infection causes SlNAC1 to be expressed in tomatoes, although it has two functions in pathogen resistance. The replication enhancer (REn) of the *tomato leaf curl virus* (TLCV) selectively induces SlNAC1 regulation and expression in tomatoes. Its overexpression enhances virus DNA accumulation in tissues affected by the virus, indicating that SlNAC1 plays a negative role in TLCV tolerance (Selth et al. 2005). In tomatoes, six NAC TFs were discovered to react to TYLCV accumulation. Once there was infection spread in a resistant variety of tomatoes, transcripts of four NAC genes (SlNAC20, SlNAC24, SlNAC47, and SlNAC61) were found to be upregulated. Tomato NAC TFs were shown to be engaged in defense regulation and the progression of development and stress. Protein phosphatase (PP) and mitogen-activated protein kinase (MAPK) were among the proteins these NAC TFs interacted with. Some defensive response transcription factors, such as MYB, NAC, TGA, and WRKY, may interact with NAC proteins by binding cis-elements in NAC TF promoters. The newly discovered tomato NAC transcription factors interacted with other proteins as well as transcription factors, showing that the disclosed NAC TFs implicated in TYLCV infection have a complicated response mechanism (Huang et al. 2017). Viroids are uncoded RNAs that may infect plants and cause severe illness despite not coding for any known protein. As a novel feature of viroid pathogenesis, the effects of the *citrus exocortis viroid* (CEVd) was investigated on the translation mechanism in tomatoes. Tomato plants infected with the viroid had changes in ribosomal biogenesis, which hampered the development of 18S rRNA. The ribosomal stress mediator NAC082 was seen as overexpressed in tomato leaves infected by CEVd. The degree of viroid manifestations is associated with modifications in the rRNA processing and the initiation of expression of NAC082 (Cottilli et al. 2019). In an independent study, *cucumber green mottle mosaic virus* (CGMMV) was infected in the susceptible *Cucumis sativus* inbred line 229, and the differentially expressed genes were mined in 3dpi vs. 20 dpi samples. In response to CGMMV infection, 13 DEGs encoding NAC TFs stood revealed, most of which remained weakly expressed. Six genes were elevated early in infection; two (Csa 4G361820 and Csa 6G382950) exhibited higher levels of expression and the virus upregulated them. In the meantime, four weakly expressed genes (Csa 6G127320) and one gene that was highly expressed (Csa 6G127320) were downregulated simultaneously. DEGs, about seven in number, encoding NAC TFs were found at 20 dpi in which three showed marginal upregulation in response to CGMMV, whereas downregulation was observed in four others, comprising Csa 3G101810, which had a very high expression level and was robustly expressed (Csa 4G011770) (Slavokhotova et al. 2021). *LrNAC35*, a WRKY TF found upregulated in a microarray analysis of *cucumber mosaic virus* (CMV)-infected *Lilium regale*, was observed to be upregulated not only against CMV, but also *lily mottle virus* and *lily symptomless virus*. In petunia (*Petunia hybrida*), ectopic overexpression of LrNAC35 lowered susceptibility to CMV along with *tobacco mosaic virus* infection with increased lignin buildup in cell walls. PhC4H, Ph4CL, PhHCT, and PhCCR, four lignin biosynthesis genes, were discovered to have increased relative expression in CMV-infected LrNAC35 overexpressing petunia lines. That LrNAC35 controlled Ph4CL expression selectively was confirmed through in vivo promoter-binding assays. Therefore, transcriptome-derived LrNAC35 has a favorable role in the transcriptional regulation of innate immunity against viral infection (Sun et al. 2019). *Wheat dwarf geminivirus* (WDV) Rep A protein interacted with AtGRAB1 (Geminivirus Rep A-Binding) and AtGRAB2 (*Triticum aestivum* NAC TFs) and reduced WDV DNA replication. In *Arabidopsis thaliana*, AtAF2 binds with the *tobacco mosaic viral* (TMV) helicase domain, and its overexpression prevents viral infection.

5.7 WRKY Transcription Factor Family

5.7.1 Classification and Constitution of the WRKY Family

WRKY transcription factors come under the most significant families of regulators of transcriptional response in plants since their substantial involvement in a variety of signaling cascades that govern a variety of plant responses and functions (Rushton et al. 2010). The DNA-binding domain of the WRKY TFs is a 60-amino-acid expanded region with a Zn finger-like motif and a conserved N-terminal region of WRKYGQK. Based on the amount of WRKY domains, WRKY proteins are divided into three categories: group I having a double domain (2WRKY DBD) while groups II (DBD + C2H2 variable zinc finger) and III (DBD + different C2H2 zinc finger) have single domains. They have WRKY DNA-binding domain(s) (DBD) and Zinc-finger motifs, which are 60 amino acids long and four-stranded sheets. Based on the main amino acid sequence, Group II, which is non-monophyletic, is classified into IIa, IIb, IIc, IId, and IIe (Rushton et al. 2010). Furthermore, they comprise a basic nuclear localization domain, leucine zippers, a serine-threonine-rich region, a glutamine-rich region, a proline-rich region, a kinase domain, and a TIR-NBS-LRR domain (Chen et al. 2012). WRKY TFs interface with W-box (with core motif TTGACC/T) and grouped W-boxes situated within the promoters of downstream genes to modulate a flexible grid of communication through kinase and additional phosphorylation cascades (Phukan et al. 2016). WRKY TFs are essentially widespread in plants and participate in many metabolic pathways.

5.7.2 Functional Characterization of the WRKY Family in Viral Defense Response

WRKY-TF has also been shown to have a function in plant–virus interactions. This transcription factor functions at a distinct level of the defense system, and its interaction with defense-associated genes has been demonstrated in tobacco plants for N-mediated *tobacco mosaic virus* resistance. Silencing of WRKY 1–3 altered N-mediated resistance as well as MYB downregulation (Liu et al. 2004). In another study, a WRKY gene was isolated through a domain-specific differential procedure, which was highly induced through an incompatible contact between *tobacco mosaic virus* and hot pepper. This CaWRKY expression could be induced by both salicylic acid (SA) and wounding treatment, signifying that this gene could play an imperative role in defense-related processes (Park et al. 2006). In another study, a *Capsicum annuum* WRKY TF d (CaWRKYd) was identified through a microarray analysis in (TMV)-P0-injected hot pepper plants. Silencing of this gene enhanced cell death and accumulation of viral particles in both local and systemically inoculated leaves, which also led to the downregulation of some PR- and HR-related gene expression. This established that WRKY seemed to be involved in TMV viral pathogenesis (Huh et al. 2012). Continuing the TMV studies, it was recently reported that a GTPase called NtRHO1 (RHO type) was upregulated on TMV infection. The overexpression and silencing experiments showed contrasting results with TMV production acceleration in the former and reduction of sensitivity in the latter. Yeast one hybrid and EMSA studies brought forward the knowledge that NtWRKY50 binds to the WK box of the PR1 promoter, and the strength of interaction being essential is the susceptibility or resistance of the plants. Therefore, this study proved the negative role of GTPase in plant resistance (Han et al. 2021). The *Rsv1* gene is a resistance gene against *soybean mosaic virus* (SMV), which participates in conferring huge resistance against many SMV strains reported. Two WRKY TFs were identified in soybean, which compromised Rsv1-mediated resistance on being silent. This was an important study that gave insight into the signaling network in soybean under SMV stress (Zhang et al. 2012). NbWRKY40, a newly identified TF, was investigated for its role in *Nicotiana benthamiana* resistance to *tomato mosaic virus* (ToMV). ToMV infection causes NbWRKY40 to be dramatically downregulated, and subcellular localization studies show that NbWRKY40 is localized to the nucleus. NbWRKY40 also displays transcriptional activation in yeast cells and promotes W-box-dependent transcription in plants. ToMV infection is inhibited by overexpressing NbWRKY40 (OEWRKY40), but NbWRKY40 suppression enhances susceptibility. The plasmodesmata is reduced in OEWRKY40 plants but enhanced in NbWRKY40-silenced plants, according to callose staining, while exogenous administration of SA decreases viral accumulation in ToMV-infected NbWRKY40-silenced plants. Anti-ToMV resistance is likely regulated by NbWRKY40, which regulates the production of SA, leading to the deposition of callose at the plasmodesmata neck, which restricts viral migration (Jiang et al. 2021). RNAi-suppressed transgenic plants were used to validate NtWRKY4 in biotic stress tolerance. The leaves of *tobacco mosaic virus* (TMV)-infected transgenic plants were more deformed and exhibited a more obvious mosaic pattern than those of vector transgenic plants. In vector-transformed plants, a lesser amount of TMV viral RNA accrued than in transgenic plants. Therefore, NtWRKY4 was found to be involved in leaf growth and development and antiviral defense (Ren et al. 2010). *Zucchini yellow mosaic virus* (ZYMV) is an agriculturally significant virus causing huge damage every year. qPCR analysis of WRKY genes in melon plants treated with salicylic acid at 3 mM and jasmonic acid at 0.5 mM, as well as the control plant following ZYMV inoculation, was performed. Following inoculation of the virus into the plant, treatment with these two hormones, notably salicylic acid, resulted in significant alterations in the expression of these genes. The study looked at the induction of PR1, NPR1, PAL, and WRKY genes in melon resistance to ZYMV (Mirhosseini et al. 2021). *Arabidopsis thaliana* was used to functionally validate the role of a versatile TF WRKY 6 against TMV. TMV was injected into protoplasts produced from WRKY6-overexpressed or -knockout plants, and the accumulation of TMV was unaffected 24 hours later. In comparison to wild-type inoculation plants, TMV accumulation was decreased in both WRKY6-overexpressed and WRKY6-knockout plants. As a result, WRKY6 may not be required for a viral infection to begin, but it may have a dual function in virus infection support and inhibition (Chen and Yeh 2010). Another *Nicotiana* WRKY gene, NbWRKY1, was found to alleviate the induction of cell death of RepA protein of *mulberry mosaic dwarf-associated virus* (MMDaV) in *Nicotiana benthamiana* upon its downregulation. This study shows that MMDaV causes an antiviral defense response in *Nicotiana benthamiana* plants through the hypersensitive response (HR). The MMDaV RepA protein was involved in HR-type cell death induction, according to this study. It was discovered that RepA mutants with impaired nuclear localization abilities are less capable of inducing cell death. On downregulating the NbWRKY1, which is nucleus targeting, alleviation of RepA's cell death-inducing activity takes place, according to virus-induced gene silencing of essential constituents of the R protein-mediated signaling cascade. We also found that RepA increases the amount of NbWRKY1 transcripts. Furthermore, RepA expression provides plant resistance to two begomoviruses in *N. benthamiana.* So, plant tolerance to RepA might be utilized to increase plant immunity to geminiviral attacks in crops (Sun et al. 2021). The MtWRKY gene of *Medicago truncatula* gives tolerance to the *tobacco mosaic virus* in *Nicotiana tabacum*. The GhWRKY15 gene from *Gossypium hirsutum* was shown to be effective against the *tobacco mosaic virus* when it was introduced into *Nicotiana tabacum* (Erpen et al. 2018). Therefore, the WRKY family plays an essential role in viral stress repression.

5.8 Concluding Remarks

Transcription factors play critical roles at the transcriptional level, inhibiting or activating genes in response to various stressors. TFs regulate genes at the transcriptional level, accounting for around 7% of the coding capacity of the vascular plant genome. Thousands of transcription factors have been discovered in plants, with the most important ones being distinct signal transduction pathways that are mediated by different TF families (WRKY, NAC, MYB, AP2/ERF, etc.). Several crops have used methods to cope with abiotic and biotic challenges during the last two decades, and TFs have definitely contributed to a huge amount in that. However, further in-depth field research is needed to determine the applicability of TF. As demonstrated by the overexpression and silencing of these TFs, they are useful targets for biotechnological manipulation. TF may play a role in regulating several genes linked to defense mechanisms, and these have been worked out in many plants. We can boost biotic stress resistance in plants by manipulating these transcription factors. As we have seen from the instances of various TF families, there was a lot of cross talk among the various genes involved. Several signal transduction pathways were discussed, which on regulation and modulation lead to viral stress inhibition. The use of TFs allows the plant to make adjustments to its environment, defending against a variety of insects and pathogens. As complicated as virus infections are and also because pest sprays do not work against them, TF modulation can be a key to defend plants against plant viral stress. Defense mechanisms in plants are intricate under bacterial, fungal, and viral stressors, and transcriptional reprogramming and cross talk among pathways are key to understanding these mechanisms. From decades, genetically modified crops have been under trial for a variety of improved traits, and TF modification is an essential aspect to it. These present as tools against combating viral stress in particular and biotic stress in general. With time, more and more factors are being discovered, which could lead to more tolerant plants against viral stress.

References

Aida M, Ishida T, Fukaki H, Fujisawa H, Tasaka M (1997) Genes involved in organ separation in Arabidopsis: an analysis of the cup-shaped cotyledon mutant. Plant Cell 9(6):841–857

Ali Z, Sarwat SS, Karim I, Faridi R, Jaskani MJ, Khan AA (2016) Functions of plant's bZIP transcription factors. Pak J Agric Sci 53(2)

Alves MS, Dadalto SP, Gonçalves AB, De Souza GB, Barros VA, Fietto LG (2013) Plant bZIP transcription factors responsive to pathogens: a review. Int J Mol Sci 14(4):7815–7828

Alves MS, Soares ZG, Vidigal PM, Barros EG, Poddanosqui AM, Aoyagi LN, Abdelnoor RV, Marcelino-Guimarães FC, Fietto LG (2015) Differential expression of four soybean bZIP genes during Phakopsora pachyrhizi infection. Funct Integr Genomics 15(6):685–696

Astudillo C, Fernandez A, Blair MW, Cichy KA (2013) The Phaseolus vulgaris ZIP gene family: identification, characterization, mapping, and gene expression. Front Plant Sci 4:286

Bai Y, Zhu W, Hu X, Sun C, Li Y, Wang D, Wang Q, Pei G, Zhang Y, Guo A, Zhao H (2016) Genome-wide analysis of the bZIP gene family identifies two ABI5-like bZIP transcription factors, BrABI5a and BrABI5b, as positive modulators of ABA signalling in Chinese cabbage. PLoS One 11(7):e0158966

Baillo EH, Kimotho RN, Zhang Z, Xu P (2019) Transcription factors associated with abiotic and biotic stress tolerance and their potential for crops improvement. Genes 10(10):771

Baloglu MC, Eldem V, Hajyzadeh M, Unver T (2014) Genome-wide analysis of the bZIP transcription factors in cucumber. PLoS One 9(4):e96014

Banerjee A, Roychoudhury A (2017) Abscisic-acid-dependent basic leucine zipper (bZIP) transcription factors in plant abiotic stress. Protoplasma 254(1):3–16

Bedoya LC, Martínez F, Orzáez D, Daròs JA (2012) Visual tracking of plant virus infection and movement using a reporter MYB transcription factor that activates anthocyanin biosynthesis. Plant Physiol 158(3):1130–1138

Bostock RM (2005) Signal crosstalk and induced resistance: straddling the line between cost and benefit. Annu Rev Phytopathol 43:545–580

Boualem A, Dogimont C, Bendahmane A (2016) The battle for survival between viruses and their host plants. Curr Opin Virol 17:32–38

Calil IP, Fontes EP (2017) Plant immunity against viruses: antiviral immune receptors in focus. Ann Bot 119(5):711–723

Chen CE, Yeh HH (2010) Plant defense-related transcription factor WRKY 6 plays both supportive and inhibitory roles in tobacco mosaic virus infection. Plant Pathol Bull 19(1):31–40

Chen L, Song Y, Li S, Zhang L, Zou C, Yu D (2012) The role of WRKY transcription factors in plant abiotic stresses. Biochim Biophys Acta (BBA)-Gene Regulatory Mechanisms 1819(2):120–128

Coll NS, Epple P, Dangl JL (2011) Programmed cell death in the plant immune system. Cell Death Different 18(8):1247–1256

Cottilli P, Belda-Palazón B, Adkar-Purushothama CR, Perreault JP, Schleiff E, Rodrigo I, Ferrando A, Lisón P (2019) Citrus exocortis viroid causes ribosomal stress in tomato plants. Nucleic Acids Res 47(16):8649–8661

Dai S, Wei X, Alfonso AA, Pei L, Duque UG, Zhang Z, Babb GM, Beachy RN (2008) Transgenic rice plants that overexpress transcription factors RF2a and RF2b are tolerant to rice tungro virus replication and disease. Proc Natl Acad Sci 105(52):21012–21016

Donze T, Qu F, Twigg P, Morris TJ (2014) Turnip crinkle virus coat protein inhibits the basal immune response to virus invasion in Arabidopsis by binding to the NAC transcription factor TIP. Virology 449:207–214

Dubos C, Stracke R, Grotewold E, Weisshaar B, Martin C, Lepiniec L (2010) MYB transcription factors in Arabidopsis. Trends Plant Sci 15(10):573–581

Durrant WE, Dong X (2004) Systemic acquired resistance. Annu Rev Phytopathol 42:185–209

Erpen L, Devi HS, Grosser JW, Dutt M (2018) Potential use of the DREB/ERF, MYB, NAC and WRKY transcription factors to improve abiotic and biotic stress in transgenic plants. Plant Cell Tissue and Organ Culture (PCTOC) 132(1):1–25

Eulgem T (2005) Regulation of the Arabidopsis defense transcriptome. Trends Plant Sci 10(2):71–78

Fan M, Bai MY, Kim JG, Wang T, Oh E, Chen L, Park CH, Son SH, Kim SK, Mudgett MB, Wang ZY (2014) The bHLH transcription factor HBI1 mediates the trade-off between growth and pathogen-associated molecular pattern–triggered immunity in Arabidopsis. Plant Cell 26(2):828–841

Feng JX, Liu DI, Pan YI, Gong W, Ma LG, Luo JC, Deng XW, Zhu YX (2005) An annotation update via cDNA sequence analysis and comprehensive profiling of developmental, hormonal or environmental responsiveness of the Arabidopsis AP2/EREBP transcription factor gene family. Plant Mol Biol 59(6):853–868

Fischer U, Dröge-Laser W (2004) Overexpression of NtERF5, a new member of the tobacco ethylene response transcription factor family enhances resistance to tobacco mosaic virus. Mol Plant-Microbe Interact 17(10):1162–1171

Franco-Zorrilla JM, López-Vidriero I, Carrasco JL, Godoy M, Vera P, Solano R (2014) DNA-binding specificities of plant transcription factors and their potential to define target genes. Proc Natl Acad Sci 111(6):2367–2372

Gao T, Li GZ, Wang CR, Dong J, Yuan SS, Wang YH, Kang GZ (2018) Function of the ERFL1a transcription factor in wheat responses to water deficiency. Int J Mol Sci 19(5):1465

Guo H, Ecker JR (2004) The ethylene signaling pathway: new insights. Curr Opin Plant Biol 7(1):40–49

Han H, Zou J, Zhou J, Zeng M, Zheng D, Yuan X, Xi D (2021) The small GTPase NtRHO1 negatively regulates tobacco defense response to tobacco mosaic virus by interacting with NtWRKY50. J Exp Bot

Hartmann L, Pedrotti L, Weiste C, Fekete A, Schierstaedt J, Göttler J, Kempa S, Krischke M, Dietrich K, Mueller MJ, Vicente-Carbajosa J (2015) Crosstalk between two bZIP signaling pathways orchestrates salt-induced metabolic reprogramming in Arabidopsis roots. Plant Cell 27(8):2244–2260

Hu W, Yang H, Yan Y, Wei Y, Tie W, Ding Z, Zuo J, Peng M, Li K (2016) Genome-wide characterization and analysis of bZIP transcription factor gene family related to abiotic stress in cassava. Sci Rep 6(1):1–12

Huang Y, Zhang BL, Sun S, Xing GM, Wang F, Li MY, Tian YS, Xiong AS (2016) AP2/ERF transcription factors involved in response to tomato yellow leaf curly virus in tomato. Plant Genome 9(2).:plantgenome2015-09

Huang Y, Li T, Xu ZS, Wang F, Xiong AS (2017) Six NAC transcription factors involved in response to TYLCV infection in resistant and susceptible tomato cultivars. Plant Physiol Biochem 120:61–74

Huh SU, Lee GJ, Kim YJ, Paek KH (2012) Capsicum annuum WRKY transcription factor d (CaWRKYd) regulates hypersensitive response and defense response upon Tobacco mosaic virus infection. Plant Sci 197:50–58

Hwang EW, Kim KA, Park SC, Jeong MJ, Byun MO, Kwon HB (2005) Expression profiles of hot pepper (Capsicum annuum) genes under cold stress conditions. J Biosci 30(5):657–667

Hwang I, Jung HJ, Park JI, Yang TJ, Nou IS (2014) Transcriptome analysis of newly classified bZIP transcription factors of Brassica rapa in cold stress response. Genomics 104(3):194–202

Ikeda M, Ohme-Takagi M (2009) A novel group of transcriptional repressors in Arabidopsis. Plant Cell Physiol 50(5):970–975

Inaba S, Kurata R, Kobayashi M, Yamagishi Y, Mori I, Ogata Y, Fukao Y (2015) Identification of putative target genes of bZIP19, a transcription factor essential for Arabidopsis adaptation to Zn deficiency in roots. Plant J 84(2):323–334

Jakoby M, Weisshaar B, Dröge-Laser W, Vicente-Carbajosa J, Tiedemann J, Kroj T, Parcy F (2002) bZIP transcription factors in Arabidopsis. Trends Plant Sci 7(3):106–111

Ji L, Wang J, Ye M, Li Y, Guo B, Chen Z, Li H, An X (2013) Identification and characterization of the Populus AREB/ABF subfamily. J Integr Plant Biol 55(2):177–186

Jiang Y, Zheng W, Li J, Liu P, Zhong K, Jin P, Xu M, Yang J, Chen J (2021) NbWRKY40 positively regulates the response of Nicotiana benthamiana to tomato mosaic virus via salicylic acid signaling. Front Plant Sci, p 2090

Jin Z, Xu W, Liu A (2014) Genomic surveys and expression analysis of bZIP gene family in castor bean (Ricinus communis L.). Planta 239(2):299–312

Kaltdorf M, Naseem M (2013) How many salicylic acid receptors does a plant cell need? Sci Signal 6(279):jc3-jc3

Kim HS, Delaney TP (2002) Over-expression of TGA5, which encodes a bZIP transcription factor that interacts with NIM1/NPR1, confers SAR-independent resistance in Arabidopsis thaliana to Peronospora parasitica. Plant J 32(2):151–163

Lee SJ, Lee MY, Yi SY, Oh SK, Choi SH, Her NH, Choi D, Min BW, Yang SG, Harn CH (2002) PPI1: a novel pathogen-induced basic region-leucine zipper (bZIP) transcription factor from pepper. Mol Plant-Microbe Interact 15(6):540–548

Lefkowitz EJ, Dempsey DM, Hendrickson RC, Orton RJ, Siddell SG, Smith DB (2018) Virus taxonomy: the database of the International Committee on Taxonomy of Viruses (ICTV). Nucleic Acids Res 46(D1):D708–D717

Lei R, Jiang H, Hu F, Yan J, Zhu S (2017) Chlorophyll fluorescence lifetime imaging provides new insight into the chlorosis induced by plant virus infection. Plant Cell Rep 36(2):327–341

Li X, Chen L, Hong M, Zhang Y, Zu F, Wen J, Yi B, Ma C, Shen J, Tu J, Fu, T (2012) A large insertion in bHLH transcription factor BrTT8 resulting in yellow seed coat in Brassica rapa

Li D, Fu F, Zhang H, Song F (2015) Genome-wide systematic characterization of the bZIP transcriptional factor family in tomato (Solanum lycopersicum L.). BMC Genomics 16(1):1–18

Li H, Wang Y, Wu M, Li L, Li C, Han Z, Yuan J, Chen C, Song W, Wang C (2017) Genome-wide identification of AP2/ERF transcription factors in cauliflower and expression profiling of the ERF family under salt and drought stresses. Front Plant Sci 8:946

Lim CW, Baek W, Lim S, Han SW, Lee SC (2015) Expression and functional roles of the pepper pathogen–induced bZIP transcription factor CabZIP2 in enhanced disease resistance to bacterial pathogen infection. Mol Plant-Microbe Interact 28(7):825–833

Liu X, Chu Z (2015) Genome-wide evolutionary characterization and analysis of bZIP transcription factors and their expression profiles in response to multiple abiotic stresses in Brachypodium distachyon. BMC Genomics 16(1):1–15

Liu Y, Schiff M, Dinesh-Kumar SP (2004) Involvement of MEK1 MAPKK, NTF6 MAPK, WRKY/MYB transcription factors, COI1 and CTR1 in N-mediated resistance to tobacco mosaic virus. Plant J 38(5):800–809

Liu S, Wang X, Wang H, Xin H, Yang X, Yan J, Li J, Tran LSP, Shinozaki K, Yamaguchi-Shinozaki K, Qin F (2013) Genome-wide analysis of ZmDREB genes and their association with natural variation in drought tolerance at seedling stage of Zea mays L. PLoS Genet 9(9):e1003790

Liu J, Chen N, Chen F, Cai B, Dal Santo S, Tornielli GB, Pezzotti M, Cheng ZMM (2014) Genome-wide analysis and expression profile of the bZIP transcription factor gene family in grapevine (Vitis vinifera). BMC Genomics 15(1):1–18

Lozano-Durán R, Macho AP, Boutrot F, Segonzac C, Somssich IE, Zipfel C (2013) The transcriptional regulator BZR1 mediates trade-off between plant innate immunity and growth. elife 2:e00983

Maisonneuve B, Pitrat M, Gognalons P, Moury B (2018) Growth stage-dependent resistance to the potyviruses lettuce Italian necrotic virus and Lettuce mosaic virus displayed by Lactuca sativa introgression lines carrying the Mo3 locus from L. virosa. Plant Pathol 67(9):2013–2018

Malinovsky FG, Batoux M, Schwessinger B, Youn JH, Stransfeld L, Win J, Kim SK, Zipfel C (2014) Antagonistic regulation of growth and immunity by the Arabidopsis basic helix-loop-helix transcription factor homolog of brassinosteroid enhanced expression2 interacting with increased leaf inclination1 binding bHLH1. Plant Physiol 164(3):1443–1455

Marwal A, Gaur RK (2020) Host plant strategies to combat against viruses effector proteins. Curr Genomics 21(6):401–410

Meng X, Zhang S (2013) MAPK cascades in plant disease resistance signaling. Annu Rev Phytopathol 51:245–266

Mirhosseini HA, Nasrollahnejad S, Babaeizad V (2021) Examining the effect of salicylic acid and jasmonic acid on the expression of PR1, NPR1, PAL and WRKY genes of melon against zucchini yellow mosaic virus. Int J Mod Agri 10(2):4375–4384

Moon SJ, Han SY, Kim DY, Yoon IS, Shin D, Byun MO, Kwon HB, Kim BG (2015) Ectopic expression of a hot pepper bZIP-like transcription factor in potato enhances drought tolerance without decreasing tuber yield. Plant Mol Biol 89(4):421–431

Nakano T, Suzuki K, Fujimura T, Shinshi H (2006) Genome-wide analysis of the ERF gene family in Arabidopsis and rice. Plant Physiol 140(2):411–432

Nakashima K, Tran LSP, Van Nguyen D, Fujita M, Maruyama K, Todaka D, Ito Y, Hayashi N, Shinozaki K, Yamaguchi-Shinozaki K (2007) Functional analysis of a NAC-type transcription factor OsNAC6 involved in abiotic and biotic stress-responsive gene expression in rice. Plant J 51(4):617–630

Nijhawan A, Jain M, Tyagi AK, Khurana JP (2008) Genomic survey and gene expression analysis of the basic leucine zipper transcription factor family in rice. Plant Physiol 146(2):333–350

Nuruzzaman M, Sharoni AM, Satoh K, Karim MR, Harikrishna JA, Shimizu T, Sasaya T, Oomura T, Haque MA, Kikuchi S (2015) NAC transcription factor family genes are differentially expressed in rice during infections with Rice dwarf virus, Rice black-streaked dwarf virus, Rice grassy stunt virus, Rice ragged stunt virus, and Rice transitory yellowing virus. Front Plant Sci 6:676

Ogata K, Kanei-Ishii C, Sasaki M, Hatanaka H, Nagadoi A, Enari M, Nakamura H, Nishimura Y, Ishii S, Sarai A (1996) The cavity in the hydrophobic core of Myb DNA-binding domain is reserved for DNA recognition and trans-activation. Nat Struct Biol 3(2):178–187

Ohta M, Ohme-Takagi M, Shinshi H (2000) Three ethylene-responsive transcription factors in tobacco with distinct transactivation functions. Plant J 22(1):29–38

Owji H, Hajiebrahimi A, Seradj H, Hemmati S (2017) Identification and functional prediction of stress responsive AP2/ERF transcription factors in Brassica napus by genome-wide analysis. Comput Biol Chem 71:32–56

Pajerowska-Mukhtar KM, Emerine DK, Mukhtar MS (2013) Tell me more: roles of NPRs in plant immunity. Trends Plant Sci 18(7):402–411

Pandey S, Sahu PP, Kulshreshtha R, Prasad M (2018) Role of host transcription factors in modulating defense response during plant-virus interaction. Caister Academic Press

Park JM, Park CJ, Lee SB, Ham BK, Shin R, Paek KH (2001) Overexpression of the tobacco Tsi1 gene encoding an EREBP/AP2–type transcription factor enhances resistance against pathogen attack and osmotic stress in tobacco. Plant Cell 13(5):1035–1046

Park CJ, Shin YC, Lee BJ, Kim KJ, Kim JK, Paek KH (2006) A hot pepper gene encoding WRKY transcription factor is induced during hypersensitive response to Tobacco mosaic virus and Xanthomonas campestris. Planta 223(2):168–179

Paz-Ares J, Ghosal D, Wienand U, Peterson PA, Saedler H (1987) The regulatory c1 locus of Zea mays encodes a protein with homology to myb proto-oncogene products and with structural similarities to transcriptional activators. EMBO J 6(12):3553–3558

Phukan UJ, Jeena GS, Shukla RK (2016) WRKY transcription factors: molecular regulation and stress responses in plants. Front Plant Sci 7:760

Phukan UJ, Jeena GS, Tripathi V, Shukla RK (2017) Regulation of Apetala2/Ethylene response factors in plants. Front Plant Sci 8:150

Pieterse CM, Van der Does D, Zamioudis C, Leon-Reyes A, Van Wees SC (2012) Hormonal modulation of plant immunity. Annu Rev Cell Dev Biol 28:489–521

Pourabed E, Golmohamadi FG, Monfared PS, Razavi SM, Shobbar ZS (2015) Basic leucine zipper family in barley: genome-wide characterization of members and expression analysis. Mol Biotechnol 57(1):12–26

Pumplin N, Voinnet O (2013) RNA silencing suppression by plant pathogens: defence, counter-defence and counter-counter-defence. Nat Rev Microbiol 11(11):745–760

Puranik S, Sahu PP, Srivastava PS, Prasad M (2012) NAC proteins: regulation and role in stress tolerance. Trends Plant Sci 17(6):369–381

Ren T, Qu F, Morris TJ (2000) HRT gene function requires interaction between a NAC protein and viral capsid protein to confer resistance to turnip crinkle virus. Plant Cell 12(10):1917–1925

Ren XJ, Huang WD, Li WZ, Yu DQ (2010) Tobacco transcription factor WRKY4 is a modulator of leaf development and disease resistance. Biol Plant 54(4):684–690

Rushton PJ, Somssich IE, Ringler P, Shen QJ (2010) WRKY transcription factors. Trends in plant science 15(5):247–258

Sagor GHM, Berberich T, Tanaka S, Nishiyama M, Kanayama Y, Kojima S, Muramoto K, Kusano T (2016) A novel strategy to produce sweeter tomato fruits with high sugar contents by fruit-specific expression of a single bZIP transcription factor gene. Plant Biotechnol J 14(4):1116–1126

Sakuma Y, Liu Q, Dubouzet JG, Abe H, Shinozaki K, Yamaguchi-Shinozaki K (2002) DNA-binding specificity of the ERF/AP2 domain of Arabidopsis DREBs, transcription factors involved in dehydration-and cold-inducible gene expression. Biochem Biophys Res Commun 290(3):998–1009

Selth LA, Dogra SC, Rasheed MS, Healy H, Randles JW, Rezaian MA (2005) A NAC domain protein interacts with tomato leaf curl virus replication accessory protein and enhances viral replication. Plant Cell 17(1):311–325

Seo PJ, Xiang F, Qiao M, Park JY, Lee YN, Kim SG, Lee YH, Park WJ, Park CM (2009) The MYB96 transcription factor mediates abscisic acid signaling during drought stress response in Arabidopsis. Plant Physiol 151(1):275–289

Shearer HL, Cheng YT, Wang L, Liu J, Boyle P, Després C, Zhang Y, Li X, Fobert PR (2012) Arabidopsis clade I TGA transcription factors regulate plant defenses in an NPR1-independent fashion. Mol Plant-Microbe Interact 25(11):1459–1468

Slavokhotova A, Korostyleva T, Shelenkov A, Pukhalskiy V, Korottseva I, Slezina M, Istomina E, Odintsova T (2021) Transcriptomic analysis of genes involved in plant defense response to the cucumber green mottle mosaic virus infection. Life 11(10):1064

Sornaraj P, Luang S, Lopato S, Hrmova M (2016) Basic leucine zipper (bZIP) transcription factors involved in abiotic stresses: a molecular model of a wheat bZIP factor and implications of its structure in function. Biochim Biophys Acta (BBA)-General Subjects 1860(1):46–56

Sun D, Zhang X, Zhang Q, Ji X, Jia Y, Wang H, Niu L, Zhang Y (2019) Comparative transcriptome profiling uncovers a Lilium regale NAC transcription factor, LrNAC35, contributing to defence response against cucumber mosaic virus and tobacco mosaic virus. Mol Plant Pathol 20(12):1662–1681

Sun S, Ren Y, Wang D, Farooq T, He Z, Zhang C, Li S, Yang X, Zhou X (2021) A group I WRKY transcription factor regulates mulberry mosaic dwarf-associated virus-triggered cell death in Nicotiana benthamiana. Mol Plant Pathol

Tolosa LN, Zhang Z (2020) The role of major transcription factors in Solanaceous food crops under different stress conditions: current and future perspectives. Plan Theory 9(1):56

Unel NM, Cetin F, Karaca Y, Altunoglu YC, Baloglu MC (2019) Comparative identification, characterization, and expression analysis of bZIP gene family members in watermelon and melon genomes. Plant Growth Regul 87(2):227–243

Van Verk MC, Gatz C, Linthorst HJ (2009) Transcriptional regulation of plant defense responses. Adv Bot Res 51:397–438

Vannini C, Campa M, Iriti M, Genga A, Faoro F, Carravieri S, Rotino GL, Rossoni M, Spinardi A, Bracale M (2007) Evaluation of transgenic tomato plants ectopically expressing the rice Osmyb4 gene. Plant Sci 173(2):231–239

Wan L, Zhang J, Zhang H, Zhang Z, Quan R, Zhou S, Huang R (2011) Transcriptional activation of OsDERF1 in OsERF3 and OsAP2-39 negatively modulates ethylene synthesis and drought tolerance in rice. PLoS One 6(9):e25216

Wang X, Goregaoker SP, Culver JN (2009a) Interaction of the Tobacco mosaic virus replicase protein with a NAC domain transcription factor is associated with the suppression of systemic host defenses. J Virol 83(19):9720–9730

Wang Y, Bao Z, Zhu Y, Hua J (2009b) Analysis of temperature modulation of plant defense against biotrophic microbes. Mol Plant-Microbe Interact 22(5):498–506

Wang Z, Cheng K, Wan L, Yan L, Jiang H, Liu S, Lei Y, Liao B (2015) Genome-wide analysis of the basic leucine zipper (bZIP) transcription factor gene family in six legume genomes. BMC Genomics 16(1):1–15

Wang Z, Yan L, Wan L, Huai D, Kang Y, Shi L, Jiang H, Lei Y, Liao B (2019) Genome-wide systematic characterization of bZIP transcription factors and their expression profiles during seed development and in response to salt stress in peanut. BMC Genomics 20(1):1–14

Wei KAIFA, Chen JUAN, Wang Y, Chen Y, Chen S, Lin Y, Pan S, Zhong X, Xie D (2012) Genome-wide analysis of bZIP-encoding genes in maize. DNA Res 19(6):463–476

Wu C, Avila CA, Goggin FL (2015) The ethylene response factor Pti5 contributes to potato aphid resistance in tomato independent of ethylene signalling. J Exp Bot 66(2):559–570

Xie Z, Nolan TM, Jiang H, Yin Y (2019) AP2/ERF transcription factor regulatory networks in hormone and abiotic stress responses in Arabidopsis. Front Plant Sci 10:228

Xu D, Jiang Y, Li J, Lin F, Holm M, Deng XW (2016) BBX21, an Arabidopsis B-box protein, directly activates HY5 and is targeted by COP1 for 26S proteasome-mediated degradation. Proc Natl Acad Sci 113(27):7655–7660

Yoshii M, Shimizu T, Yamazaki M, Higashi T, Miyao A, Hirochika H, Omura T (2009) Disruption of a novel gene for a NAC-domain protein in rice confers resistance to Rice dwarf virus. Plant J 57(4):615–625

Zhang G, Chen M, Li L, Xu Z, Chen X, Guo J, Ma Y (2009) Overexpression of the soybean GmERF3 gene, an AP2/ERF type transcription factor for increased tolerances to salt, drought, and diseases in transgenic tobacco. J Exp Bot 60(13):3781–3796

Zhang C, Grosic S, Whitham SA, Hill JH (2012) The requirement of multiple defense genes in soybean Rsv1–mediated extreme resistance to Soybean mosaic virus. Mol Plant-Microbe Interact 25(10):1307–1313

Zhang F, Fu X, Lv Z, Lu X, Shen Q, Zhang L, Zhu M, Wang G, Sun X, Liao Z, Tang K (2015) A basic leucine zipper transcription factor, AabZIP1, connects abscisic acid signaling with artemisinin biosynthesis in Artemisia annua. Mol Plant 8(1):163–175

Zhao J, Guo R, Guo C, Hou H, Wang X, Gao H (2016) Evolutionary and expression analyses of the apple basic leucine zipper transcription factor family. Front Plant Sci 7:376

Zhu T, Zhou X, Zhang JL, Zhang WH, Zhang LP, You CX, Jameson PE, Ma PT, Guo SL (2021) Ethylene-induced NbMYB4L is involved in resistance against tobacco mosaic virus in Nicotiana benthamiana. Mol Plant Pathol

Zong W, Tang N, Yang J, Peng L, Ma S, Xu Y, Li G, Xiong L (2016) Feedback regulation of ABA signaling and biosynthesis by a bZIP transcription factor targets drought-resistance-related genes. Plant Physiol 171(4):2810–2825

Zorzatto C, Machado JPB, Lopes KV, Nascimento KJ, Pereira WA, Brustolini OJ, Reis PA, Calil IP, Deguchi M, Sachetto-Martins G, Gouveia BC (2015) NIK1-mediated translation suppression functions as a plant antiviral immunity mechanism. Nature 520(7549):679–682

Zou B, Jia Z, Tian S, Wang X, Gou Z, Lü B, Dong H (2012) AtMYB44 positively modulates disease resistance to Pseudomonas syringae through the salicylic acid signalling pathway in Arabidopsis. Funct Plant Biol 40(3):304–313

Zou LJ, Deng XG, Han XY, Tan WR, Zhu LJ, Xi DH, Zhang DW, Lin HH (2016) Role of transcription factor HAT1 in modulating Arabidopsis thaliana response to Cucumber mosaic virus. Plant Cell Physiol 57(9):1879–1889

Chapter 6
Role of Microbial Bioagents as Elicitors in Plant Defense Regulation

Mukesh Meena , Garima Yadav, Priyankaraj Sonigra, Adhishree Nagda, Tushar Mehta, Andleeb Zehra, and Prashant Swapnil

Abstract Plants are constantly challenged by an array of potential pathogens like fungi, bacteria, viruses, insects, nematodes, etc., which lead to a significant loss to plant yield. Plants commonly overcome these phytopathogens by showing resistance through plant defense mechanisms. Several general microbe elicitors allow plants to mitigate the harmful effects of pathogenic microbes by enhancing the capability of plants to identify anonymous pathogenic agents and act as surveillance systems for plants. Elicitors are small drug-like compounds released by pathogens that are composed of molecules like oligosaccharides, lipids, peptides, and proteins, and they activate various kinds of defense responses in plants. They deliver information to plants through perception and identification of signaling molecules by cell surface-localized receptors, which is followed by the triggering of signal transmission pathways that commonly induces the synthesis of active oxygen species (AOS), phytoalexin production, production of defense enzymes, and the aggregation of pathogenesis-related (PR) proteins. This article chiefly highlights the role of microbial elicitors in improving plant defense mechanisms as well as their modes of action that have been used to boost up the plant immune system.

Keywords Microbial bioagents · Elicitors · Plant defense regulation · Systemic acquired resistance · Phytopathogens

---

M. Meena (✉) · G. Yadav · P. Sonigra · A. Nagda · T. Mehta
Laboratory of Phytopathology and Microbial Biotechnology, Department of Botany, Mohanlal Sukhadia University, Udaipur, Rajasthan, India

A. Zehra
Centre of Advanced Study in Botany, Institute of Science, Banaras Hindu University, Varanasi, India

P. Swapnil
Department of Botany, University of Delhi, New Delhi, India

School of Basic Sciences, Department of Botany, Central University of Punjab, Bhatinda, Punjab, India

© The Author(s), under exclusive license to Springer Nature Switzerland AG 2022

S. H. Wani et al. (eds.), *Transcription Factors for Biotic Stress Tolerance in Plants*, https://doi.org/10.1007/978-3-031-12990-2_6

6.1 Introduction

In the course of development, plants are systematically challenged by a broad range of biotic stresses in their natural habitat, such as fungi, bacteria, viruses, insects, nematodes, etc. There are numerous choices available for the plants to protect themselves from the disease (Abdul Malik et al. 2020). Plants usually overcome these biotic stresses by activating their separate defense pathways according to perceived signals from potential pathogens (Sarma et al. 2015; Barupal et al. 2020). There is an intricate type of defense mechanism employed by plants to detect microorganisms based on elicitor molecules produced during plant–pathogen interaction. Numerous elicitors of microbial origin belonging to distinct chemical groups have been identified, i.e., glycopeptides, carbohydrate polymers, glycoproteins, and lipids. This elicitor perception is followed by the stimulation of signal transmission pathways that commonly induces the synthesis of active oxygen species, production of phytoalexin, accumulation of pathogenesis-related proteins, deposition of callose, strengthening of the cell wall of plant cell related to phenyl propanoic compounds, and production of defense enzymes (Van Loon and Van Strien 1999; Patel et al. 2019). Active oxygen species (AOS) induce localized or fast death of limited cells at the site of infection, which induces a hypersensitive response in host plants to restrict the growth of invading pathogens. Activation of hypersensitive response (HR) results in the development of resistance in uninfected distal parts of the host plant to upcoming infection, which is called systemic acquired resistance (SAR) (Thakur and Sohal 2013). Systemic acquired resistance is mainly relying up on salicylic acid, where the first set of reactions brings on a complex modification in gene expression, enzymatic action, and metabolic changes (Garcia-Brugger et al. 2006; Barupal et al. 2019). Salicylic acid-dependent reaction is stimulated by biotrophic pathogens and distinct types of elicitors. Several microbial elicitors allow plants to mitigate the harmful effects of pathogenic microbes by enhancing the capability of plants to identify anonymous pathogenic agents and act as surveillance systems for plants (Newman et al. 2013). Elicitors are small drug-like compounds composed of molecules like oligosaccharides, lipids, peptides, and proteins, which activate various kinds of defense responses in plants. They are either secreted by pathogens or plants or pathogen cell walls by hydrolytic enzymes. Elicitor-activated signal transduction pathways bring on a hypersensitive response and systemic acquired resistance type of defense responses against a broad range of pathogens (Garcia-Brugger et al. 2006). Microbial biocontrol agents suppress the growth of phytopathogens through a wide array of distinct modes of actions. The most important advantage of using microbial biocontrol agents is that they display specificity for a particular pathogen and are expected to be harmless to nontarget species (Hussain et al. 2020a, b). In the last few decades, many studies have been done on the broad range of applications of microbial biocontrol agents in the plant disease management data given in Table 6.1 (Kokalis-burelle et al. 2002; Mavrodi et al. 2012; Singh et al. 2020). Environmentally friendly and sustainable attributes of biocontrol agents have driven profound investigation into the promising microbial

Table 6.1 Biocontrol agents and their target phytopathogens

Biocontrol agents	Crop	Pathogen	References
Bacillus polymyxa	Rice (*Oryza sativa*)	*Rhizoctonia solani*, *Pyricularia grisea*	Kavitha et al. (2005)
Trichoderma viride, *Trichoderma harzianum*, *Pseudomonas fluorescens*	Groundnut (*Arachis hypogaea*)	*Macrophomina phaseolina*	Karthikeyan et al. (2006)
Acremonium strictum, *Trichoderma harzianum*	Tomato (*Solanum lycopersicum*)	*Meloidogyne incognita*	Goswami et al. (2008)
Trichoderma harzianum	Tomato (*Solanum lycopersicum*)	*Meloidogyne javanica*	Sahebani and Hadavi (2008)
Trichoderma viride	Soybean (*Glycine max*)	*Fusarium oxysporum* f. sp. *adzuki*, *Pythium arrenomanes*	John et al. (2010)
Trichoderma harzianum, *Pseudomonas fluorescens*, *Bacillus subtilis*	Safflower (*Carthamus tinctorius*)	*Macrophomina phaseolina* (root rot disease)	Govindappa et al. (2010)
Paecilomyces lilacinus	Tomato (*Solanum lycopersicum*)	*Meloidogyne incognita*	Oclarit and Cumagun (2009)
Trichoderma asperellum	Cocoyam (*Xanthosoma sagittifolium*)	*Pythium myriotylum*	Mbarga et al. (2012)
Verticillium chlamydosporium, *Photorhabdus luminescens*	Cucumber (*Cucumis sativus*)	*Meloidogyne incognita*	Zakaria et al. (2013)
Bacillus amyloliquefaciens	Wheat (*Triticum aestivum*)	*Fusarium graminearum* (*Gibberella zeae*)	Dunlap et al. (2013)
Bacillus spp.	Ginseng (*Panax ginseng*)	*Fusarium* c.f. *incarnatum*	Song et al. (2014)
Bacillus cereus	Thale cress (*Arabidopsis thaliana*)	*Pseudomonas syringae*	Chowdhury et al. (2015)

candidates for the production of elicitors. In this chapter, we address the role of microbial elicitors in improving plant defense mechanisms as well as their modes of action that have been used to boost up the plant immune system.

6.2 Elicitors

Elicitors are small drug-like compounds composed of molecules like oligosaccharides, lipids, peptides, and proteins, which activate various kinds of defense responses in plants. Elicitors produced by pathogenic agents can be classified into two groups: general elicitors and specific elicitors (Montesano et al. 2003). General elicitors are engaged in the conventional resistance, which has the capacity to trigger defense reactions in both host and nonhost plants, whereas race-specific elicitors are released by specialized pathogens involved in *R* gene-mediated signal transduction (Gowthami 2018). General elicitors have the capacity to trigger defense in both nonhost and host plants through the realized incidence of potential pathogens (Onaga and Wydra 2016). Commonly, general elicitors are found in the cell walls of pathogens as structural constituents, for example, glucan, flagellin, chitin, and lipopolysaccharides (LPS) (Abdul Malik et al. 2020). Elicitor molecules act as ligands and generally bind to the specific receptor proteins located on the surface of plant cell membranes. According to the molecular pattern of elicitors recognized by receptors, an intracellular defense signaling has been triggered, which is echoed by the synthesis of secondary metabolites (Gowthami 2018; Zehra et al. 2021). It has long been recognized that microbial elicitors can induce many cellular defense responses in plants. Currently, elicitors of microbial origin have also been stated as microbe- or pathogen-associated molecular patterns (MAMPs or PAMPs). Following MAMP recognition, production of reactive nitrogen species, ion fluxes across the membrane, medium alkalinization, reactive oxygen species, and ethylene synthesis lead to activate plant pattern-triggered immunity (PTI) against broad range of microbial attack (Wu et al. 2014). Newman et al. (2013) stated that N-acetyl-chito-oligosaccharides, i.e., chitin oligomers, a fungal cell wall-derived elicitor molecule, can activate several defense responses in monocot as well as dicot plants. In recent years, numerous MAMPs and their corresponding PRRs have been recognized, such as flagellin, peptidoglycan, elongation factor (Tu), lipopolysaccharides, β-glucans from oomycetes and Ax21, fungal chitin, etc. (Newman et al. 2013).

6.2.1 Microbial Agents as a Source of Elicitors

Induction of plant defense response is a crucial step during plant–pathogen interaction via several factors. The first step of inducible response is carried out by the plant by the perception of molecules derived by microbes known as elicitors. While the plant percept these molecules, it results in a plant response that provides effective resistance toward pathogens; hence, they can be described as "defense elicitors" (Wiesel et al. 2014). These elicitors may be of proteinaceous, polysaccharide, laminarin, and other chemical nature. Apart from the pathogenic role of microbes, there are some beneficial microbes that live in plant tissue as endophytes, plant growth promoting rhizobacteria (PGPR), arbuscular mycorrhizal fungi (AMF), and some algae, oomycetes, and viruses also play a significant role in inducing resistance in plants (Siah et al. 2018; Yadav and Meena 2021). The primary work of these elicitors is to induce production of reactive oxygen species (ROS) or oxidative burst, which ultimately evokes plant defense responses like cross-linking of plant cell wall proteins, upregulation of defense-related genes, stimulation of synthesis of phenolic compounds (phytoalexins), and induction of hypersensitive response (Low and Merida 1996). The biological agents evoke plant defense via several modes like production of siderophores, antibiotic secretion, lytic enzyme production, hyperparasitism, and inducing systemic response (ISR); all mechanisms are induced by secretion of elicitor molecules (Pieterse et al. 2014; Navarro et al. 2019; Singh et al. 2020).

6.2.2 Some Potent Elicitor-Producing Microbial Agents

6.2.2.1 Fungi

Fungal groups possess some cell wall breakdown products like chitin, β-glucans, and mannoproteins that act as potent elicitors and can evoke defense response; for example, yeast extract can be used widely for the study of defense response in plants via closing of stomata and peroxidase-mediated ROS production (Khokon et al. 2010). *Sclerotinia* culture filtrate elicitor1 (SCFE1) is a proteinaceous elicitor secreted by *Sclerotinia sclerotiorum* that induces BAK1-dependent PTI responses in *A. thaliana* (Zhang et al. 2013). Among fungal-derived elicitors, chitin and chitosan (a deacetylated derivative of chitin) are potent elicitors that increase resistance in plants toward several fungal and bacterial pathogens (Hadrami et al. 2010). Fungal cell wall polysaccharides, especially chitin and carboxymethyl cellulose, are active elicitors that stimulate synthesis and accumulation of a secondary metabolite tolytoxin (phytoalexin) in a cyanobacterium *Scytonema ocellatum*, which provides chemical defense against fungal pathogens (Patterson and Bolis 1997; Meena and Samal 2019). Transcription of retrotransposons is also carried out by some fungal genera to increase host defense; for example, application of crude extracts of *Trichoderma viride* induces transcription of *Tnt1* gene, which accumulates capsidiol (a phytoalexin) in tobacco plants (Pouteau et al. 1994; Meena and Swapnil 2019). Some other examples also suggest transcription activation by fungal elicitors as it has been seen in *Phaseolus vulgaris* where plant cells show upregulation of genes related to phytoalexin metabolism such as phenylalanine ammonium lyase (*PAL*) and chalcone synthase (*CHS*) (Lawton and Lamb 1987). Some other species of *Trichoderma* like *T. virens* induce plant defense response by producing an elicitor named Sm1 (small protein 1), which triggers an increased production of reactive oxygen species in seedlings of monocot and dicot plants and proves as a potent elicitor in defense against foliar pathogen *Colletotrichum* sp. (Djonović et al. 2006). *Trichoderma harzianum* is also reported as an inducer of antioxidant defense system in tomatoes against Fusarium wilt disease (Zehra et al. 2017a, b). It is reported that oxidative burst during plant defense is dependent on external calcium (Ca^{+2}) and protein kinase activity (Schwacke and Hager 1992). Hypersensitive response is also stimulated by the same fungus in *Vitis vinifera* by increasing the level of endogenous H_2O_2, which ultimately activates oxidative phenolic metabolism in respective plants (Calderón et al. 1993). A proteinaceous elicitor PeaT1 produced by *Alternaria tenuissima* enhances plant defense response against tomato aphid (*Myzus persicae*), which is evidenced by accumulation of defense-related substances such as jasmonic acid (JA), salicylic acid (SA), and ethylene (ET) (Meena et al. 2017a, b; Basit et al.

2021). PeaT1 is also responsible for systemic acquired resistance (SAR) in tobacco plants (Mao et al. 2010). Other than the above described fungal genera, there are several fungi that are sources of potent elicitors and regulate plant defense responses as given in Table 6.2.

6.2.2.2 Bacteria

In addition to fungal-derived elicitors, bacteria-derived elicitors have also been shown to regulate plant defense mechanisms and reduce pathogen infections in plants. There are several pieces of evidence that justify this statement, for example, *Ralstonia solanacearum* produce extracellular polysaccharides (EPS), which trigger a defense response in tomato plants in the case of bacterial wilt (Milling et al. 2011). Gram-negative bacteria-derived lipopolysaccharides (LPS)-mediated induction resistance is also shown in many crop plants (Erbs and Newman 2012). At concentrations of 1 g/ml, lipopolysaccharides from *Xanthomonas campestris* induce transcription of genes of β-1,3-glucanase, which ultimately shows defense responses in turnip (Newman et al. 1995). Cold shock protein (Csp)-related elicitor activity has been detected in bacterial extracts; there are many aromatic and basic side chains of csp domains that are necessary for elicitor activity; hence, RNA-binding motif RNP-1 of bacterial cold shock proteins that are highly conserved is recognized as an elicitor signal in *Nicotiana sylvestris* plant (Felix and Boller 2003). The two bacterial microbe-associated molecular patterns (MAMPs) are flagellin and the elongation factor Tu (EF-Tu), which are recognized by a variety of plant species (Deslandes and Rivas 2012). *Botrytis cinerea* and *Erwinia carotovora* produce a wide array of elicitors that enhance the expression of conserved plant defense-associated genes such as *HrpN* gene and show responses like shrinkage of cytoplasm, programmed cell death (PCD), etc. in *Physcomitrella patens* (de León et al. 2007). *Hrp* genes are crucial for HR response in plants; Wei et al. (1992) reported that *hrp* genes *(hrpN)* of *Erwinia amylovora* encode harpin, a proteinaceous elicitor, which shows HR necrosis in respective plants. Surfactin lipopeptide is secreted by *Bacillus* sp., which triggers induced systemic response in host plants and defense responses like oxidative burst, etc. (Cawoy et al. 2014). In elicitation, not only free-living or plant-associated bacteria but also animal-associated bacteria are also involved; for example, it is observed that insects named *Helicoverpa zea*, gut-associated bacteria, induce defenses in tomatoes indirectly by secreting a salivary elicitor that induces expression of genes of defense-related enzymes like polyphenol oxidase and jasmonic acid (JA) and suppression of pathogenesis-related genes of salicylic acid (SA) response (Wang et al. 2017). Twenty-three bacteria isolated from gut segments of *Spodoptera exigua*, *Agrotis segetum*, and *Mamestra brassicae* produce surfactants such as *N*-acylglutamine, which is recognized as a potent elicitor for plant defense response (Spiteller et al. 2000). There are several other bacterial groups identified as sources of elicitors, which are mentioned in Table 6.2.

Table 6.2 Table showing elicitor producing microbial agents, host, and their mode of plant defense regulation

Microbial agent		Host	Mechanism of host defense regulation	References
Fungi	*Trichoderma harzianum*	Sunflower	Induce resistance by increasing phenolics as well as stress enzymes	Singh et al. (2014), Swapnil et al. (2021)
	Trichoderma harzianum T3	Grapevine	Enhance expression of defense-related genes	Banani et al. 2015
	Trichoderma viride	Potato	Increase total phenol content	Rosyidah et al. (2014), Meena et al. (2020)
	Trichoderma viride	Black gram	Induction of defense enzymes and total phenolic content	Surekha et al. (2014)
	Trichoderma asperellum	Onion	Increase of glucanase, chitinase, and peroxidase activity	Guzmán-Valle et al. (2014)
	Trichoderma asperelloides	*Arabidopsis*	Suppress nitric oxide generation, elicited by pathogen	Gupta et al. 2014
	Fusarium oxysporum Fo47	Pepper	Production of caffeic, ferulic, and chlorogenic acids	Veloso et al. (2016)
	Penicillium oxalicum	Pearl millet	Increase peroxidase and chitinase activity	Murali and Amruthesh (2015)
	Clonostachys rosea	Canola	Upregulation of host genes involved in biosynthesis of jasmonic acid, ethylene, and auxin	Lahlali et al. (2014)
Arbuscular mycorrhizal fungi (AMF)	*Glomus fasciculatum*	Tomato	Higher expression of genes involved in jasmonic acid biosynthesis	Nair et al. (2015)
	Funneliformis mosseae, *Rhizophagus irregularis*	Wheat	Accumulation of polyphenolic compounds and reduction of pathogen conidia	Mustafa et al. (2016)

(continued)

Table 6.2 (continued)

Microbial agent		Host	Mechanism of host defense regulation	References
Bacteria	*Bacillus subtilis* QST 713	Tomato	Increase expression of *Pin2* gene in host	Fousia et al. (2016)
	Bacillus cereus AR156	*Arabidopsis*	Activation of PAMP-triggered immunity and ISR through NPR1- and SA-dependent signaling pathway in host	Niu et al. (2016), Meena et al. (2019)
	Bacillus amyloliquefaciens S13-3	Tomato	Induction of ISR through antibiotic production	Yamamoto et al. (2015)
	Bacillus oryzicola	Rice	Induced systemic response in host	Chung et al. (2015)
	Paenibacillus polymyxa CF05	Tomato	Induction of defense-related enzymes (PAL, SOD, and PPO) and accumulation of H_2O_2 and phenolics in host plant	Mei et al. (2014)
	Pseudomonas sp. LBUM223	Potato	Induction of defense-related genes like *LOX*, *PIN2*, *PAL-2*, *ERF3*, *ChtA*, *PR-1b*, *PR-2*, and *PR-5*	Arseneault et al. (2014)
	Streptomyces rochei A-1	Apple	Increased activities of POD, CAT, SOD, PAL, β-1,3-glucanase, and chitinase, promoted H_2O_2 generation, decreased lipid peroxidation, and upregulation of related genes	Zhang et al. (2016)
	Brevibacterium iodinum KUDC1716	Pepper	Elicit systemic acquired resistance (SAR)	Son et al. (2014)
	Carnobacterium sp. SJ-5	Soybean	Higher expression of defense-related proteins	Jain and Choudhary (2014)

(continued)

Table 6.2 (continued)

Microbial agent		Host	Mechanism of host defense regulation	References
Oomycetes	*Pythium oligandrum*	Grapevine	Induction of genes related to phenylpropanoid pathways, PR proteins, oxylipins, and oxydo-reduction systems	Yacou et al. (2016)
	Phytophthora parasitica	Tobacco	Formation of physical barriers like phloem proteins, impregnation of pectin, etc. in the host plant	Lherminier et al. (2003)

6.2.2.3 Oomycetes

Oomycetes are taxonomically and structurally different from plants and fungi. There are several plant pathogenic oomycetes known, but genera *Phytophthora* and *Pythium* show superiority in causing disease of crop plants. The cell walls of these groups consist of several elicitor factors such as cellulose, glycan, and hydroxyproline-rich proteins. Some potent elicitors reported from oomycetes are CBEL, cryptogein, eicosapentaenoic acid, Pep-13, and INF1 (Wiesel et al. 2014). Necrosis and ethylene-inducing peptide 1 (Nep1)-like proteins (NLP) has been identified in dicot plants, which are associated with defense response in *Arabidopsis thaliana* (Qutob et al. 2006). In *Nicotiana benthamiana*, HR response is induced by INF1 elicitin of *Phytophthora infestans* (Kamoun et al. 1998). These responses are dependent on the receptor-like kinase SERK3/BAK1, required for multiple resistance responses in plants (Heese et al. 2007). Pathogenic species of *Phytophthora* release some extracellular and intracellular effectors into plants encoding protease or glucanase inhibitors to suppress pattern-triggered immunity in plants (Hein et al. 2009; Schornack et al. 2009). RXLR effector Avrblb2 of *P. infestans* prevents secretion of an immune-associated protease (Bozkurt et al. 2011). An intracellular RXLR effector named Avr3a of *P. infestans* interacts with potato E3 ubiquitin ligase CMPG1 and stabilizes it, which results in perturbation in cell death response induced by INF1 (Bos et al. 2010). The other examples of oomycete elicitors and plant defense regulations are mentioned in Table 6.2.

6.2.2.4 Virus

Among well-known elicitor-producing microbes like fungi, bacteria, and oomycetes, some viruses are also known that immunize plants and regulate their defense response. Plant virus coat proteins (CPs) can act as elicitors that triggers R-gene-mediated HR response (Moffett 2009). Several viral silencing suppressors misregulate AUXIN RESPONSE FACTOR 8, which finally causes chlorotic symptoms in plants (García and Pallás 2015). Strain-specific P3 of *Soybean Mosaic Virus* G7 is identified as an elicitor for *Rsv1* (a single dominant resistance gene)-mediated HR response (Hajimorad et al. 2005). It is also observed that TMV replicase sequence of 126/183 kDa activates N-gene mediated hypersensitive response in tobacco plants (Padgett et al. 1997). BV1 protein of bean dwarf mosaic virus is also recognized as a determinant factor for the hypersensitive response and avirulence in French bean (*Phaseolus vulgaris*) (Garrido-Ramirez et al. 2000).

6.2.3 Mode of Action by Which Microbial Bioagents Bring About Plant Defense

The microbial bioagents show antagonism, competition, and parasitism against different pathogenic microbes. These activities of microbial bioagents provide defense to plants directly or indirectly, such as plant defense response stimulation. These mechanisms include antimicrobial compound production, competition for niches and nutrients, elicitation of plant defenses, etc. (Jamalizadeh et al. 2011; Compant et al. 2013; Hussain et al. 2020a). Different mechanisms of biocontrol agents, which have been shown in Fig. 6.1, are described in the following sections.

6.2.3.1 Antagonisms

In antagonism, actions of one organism inhibit or obstruct the normal growth and development of other organisms appearing in its near vicinity. If these types of organisms inhibit phytopathogens, they can be used as biocontrol agents against pests and pathogens (Heydari and Pessarakli 2010). According to Shoda (2000), microorganisms that have capability to multiply in the rhizospheres are regarded as ideal biocontrol agents. Microorganisms colonize in the root of the host, produce some metabolites, and secrete into the root system, which are toxic to pathogens and directly suppress the pathogen growth. These metabolites directly offer protection to the host or sometimes trigger defense in the host plant (Nihorimbere et al. 2012; Chandran et al. 2020). The elicitation of the host plant defense system by microbial bioagents is known as direct antagonism (Ab Rahman et al. 2018).

6.2.3.2 Parasitism

In parasitism, one microorganism is ubiquitous for another. The microbial bioagents produce lytic enzymes like glucosaminidases and chitinases, which lead to the degradation of the cell wall of phytopathogens (Guigón-López et al. 2015). Urbina et al. (2016) investigated the role of enzymes synthesized by *Candida oleophila*

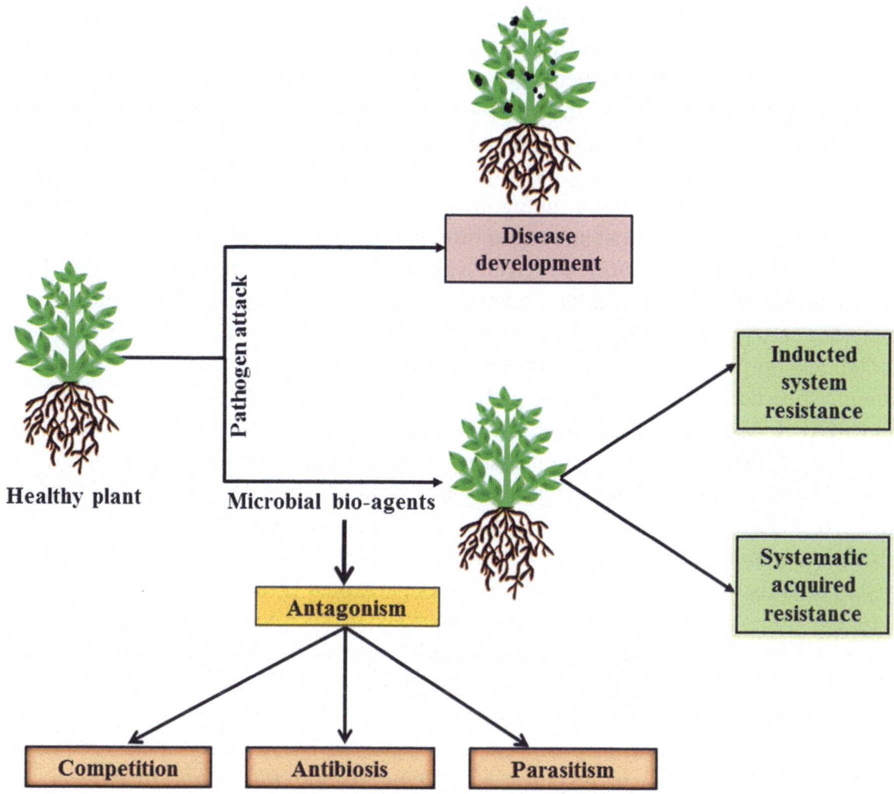

Fig. 6.1 Diagrammatic representation of different antagonistic action and plant defense inducted by microbial bio-agents

(extracellular 3-glucanase exo-β-1) in biocontrol of *Penicillium expansum* causing apple spoilage. Moreover, many researchers reported different antagonistic microbial species (Meena et al. 2017c, d). Jeffries (1995) reported that *Rhizoctonia solani* could be controlled using 30 different hyperparasitic species belonging to 16 genera. Powdery mildew, which is caused by an obligate biotrophic pathogen, was controlled with eight hyperparasites by Hijmegen and Buchenauer (Hijwegen and Buchenauer 1984). In a study, it was observed that *Pseudomonas flocculosa* release some cell wall lytic enzymes, which cause cell collapse in powdery mildew cells (Bélanger et al. 2012). Some fungi release protease enzymes such as *Pochonia chlamydosporia*, which causes infection in eggs of the nematodes (Escudero et al. 2016). Rust pathogens *Puccinia violae* and *Puccinia striiformis* f. sp. *tritici* were tested with more than 30 hyperparasitic fungal species, including *Cladosporium uredinicola* and *Alternaria alternata*, respectively, and positive results were obtained (Zheng et al. 2017). Köhl et al. (2019) reviewed that *Alternaria alternata* had the capability to penetrate the urediniospore of wheat rust fungus by germ tubes; the urediniospores were completely collapsed and lost their ability to germinate. In an experiment, it was reported that the urediniospores treated with *A. alternata* pustules had reduced ability to germinate up to 25% as compared to untreated urediniospores pustules (80%) (Zheng et al. 2017). In *Trichoderma*, parasitism was observed most frequently against *Pythium myriotylum* and *Macrophomina phaseolina* (Kubicek et al. 2001). *Trichoderma* and *Clonostachy* are the most studied mycoparasites, and members belonging to these genera have a wide and varied plant pathogenic host range. These antagonistic isolates form different structures by which they attach to the host and cause infection and death of their hosts by producing cell wall degrading enzymes (Karlsson et al. 2017; Nygren et al. 2018). The synthesis of cell wall degrading substances is not constitutive. The synthesis of enzymes is triggered upon host recognition. Host contain some specific types of molecules on their surface (lectins or secondary metabolites), and these molecules trigger specific types of signaling pathways (G-protein signaling cAMP pathway, and MAPK cascades) (Zhai et al. 2017; Karlsson et al. 2017; Zehra et al. 2015; Meena et al. 2017e, f). Signaling pathways lead to upregulation and transcription of certain genes known as "molecular weapons" including lytic enzymes, which attack and cause lysis of the host. In *Trichoderma*, there are two types of mycoparasitism-related gene families, namely *ech42* and *prb1*, which are overexpressed throughout mycoparasitism (Barbara et al. 2011). Mycoparasitics (*Trichoderma*) first release lytic enzymes; as a result, some oligosaccharides are secreted from the host that are identified by receptors and trigger increased synthesis of lytic enzymes (Karlsson et al. 2017; Meena et al. 2016a, b). This increased level of lytic enzymes results in increased permeability, degradation, and death of the host plant. These types of collaborative transcriptional results were also reported by Reithner et al. (2011) in *Trichoderma atroviride* in response to *B. cinerea* and *Phytophthora capsici*. In *Metschnikowia fructicola*, induced chitinase activity was observed to be regulated by *MfChi* gene due to close contact with the cell wall of yeast *Monilinia fructicola* (Banani et al. 2015). The same type of result was also observed with *Pichia pastoris* when used against *Monilinia fructicola* and *Monilinia laxa* that cause postharvest disease in peach fruits (Dukare et al. 2019).

6.2.3.3 Competition

Competition is a mechanism in which two or more organisms utilize the same type of nutrition or space or both for their survival; therefore, the interaction becomes competitive. The microbial bioagents exploit the nutrients, prevent the pathogen growth and proliferation, and reduce the virulence of the pathogen. For a microbe to thrive in the phyllosphere or rhizosphere, it must be able to make use of accessible nutrients in the form of leachates and exudates or senescent tissue. In rhizosphere, plants release different photosynthates, which are a great source of nutrients (specific sugars, organic acids, and amino acids) for microbes; therefore, the rhizosphere works as a niche. High availability of carbon (40%) around the root surface attracts different microbes. Microbial bioagents compete with pathogens for nutrients and protect the host from disease occurrence (Degenhardt et al. 2003). This type of approach has been observed in different pathogens such as *Pythium* and *Fusarium*; these are soil-borne pathogens and cause infection by mycelial penetration. *Enterobacter cloacae* act as a microbial bioagent against *Pythium ultimum* by increasing catabolism of nutrients (van Dijk and Nelson 2000; Kageyama and Nelson 2003). Some microbial bioagents, namely *Pseudomonas fluorescence*, chelate iron, which is essential for *Fusarium oxysporum*, whereas *Chryseobacterium* sp. WR21 exploits root exudates and competes with *Ralstonia solanacearum* (Huang et al. 2017). Moreover, antagonistic fungus *Pichia guilliermondii* was found to show competition against certain known pathogenic fungi isolated from wounds of fruit such as apple, namely *Penicillium expansum*, *Penicillium digitatum*, *Colletotrichum* spp. or *B. cinerea*, and *Aureobasidium pullulans* (Spadaro and Droby 2016). It was reported that microbes compete for nitrogen sources in a carbohydrate-rich environment. Besides nitrogen, they also compete for iron because it is a limiting factor for microbial growth and also has low solubility, thus playing a vital role in antagonistic activity such as competition (Spadaro and Droby 2016). Microorganisms have the ability to produce a variety of siderophores, which are low-molecular-weight chelating compounds with a great affinity for iron (van Loon 2000). Pathogenic strains use the chelating compounds to accumulate the ions, and they can be used as microbial bioagents for disease suppression through competition with pathogenic strains that produce siderophores but with low affinity (van Loon 2000; Lugtenberg and Kamilova 2009). *Pseudomonas* spp. have shown siderophore-facilitated iron competition with pathogenic populations present in rhizospheres and reduced their number in soil (Raaijmakers et al. 1995). Fungal antagonists such as *Trichoderma asperellum* and *Metschnikowia pulcherrima* produce iron-binding siderophores and control the growth of *Fusarium* and *A. alternata*, *B. cinerea*, and *P. expansum*, respectively (Saravanakumar et al. 2008; Segarra et al. 2010).

6.2.3.4 Production of Antimicrobial Compounds

Active microbes and the produced allelochemicals as secondary metabolites are potent options for treating plant diseases (Puopolo et al. 2018; Zhao et al. 2021). The most common mechanism associated with biocontrol activity is the production of antibiotics. Besides that, many biocontrol strains produce antifungal enzymes like β-1,3-glucanases, chitinases, proteases, or lipases that are involved in fungal cell wall lysis, produce siderophores, and chelate iron in the rhizosphere, thus inhibiting the proliferation of pathogens (Bais et al. 2004; Latz et al. 2018; Köhl et al. 2019; Pirttilä et al. 2021).

6.2.3.5 Antibiotics

Antibiotics are small, heterogenous molecular compounds, which can inhibit the growth of pathogens at low concentrations (Huang et al. 2021). The general mechanism of antibiotic action is cell wall synthesis inhibition, disruption of cell membrane structure and function, nucleic acids structure and function inhibition, and blocking of key metabolic pathways (Wu et al. 2021). Some antibiotic-producing strains among rhizobacteria are *Bacillus* sp. producing surfactin and iturin A, *Pseudomonas* spp. producing phenazine derivatives, *Erwinia* sp. producing herbicolin A, *Agrobacterium* sp. producing agrocin 84, etc. (Viswanathan and Samiyappan 1999; Compant et al. 2005a, b; Sonigra and Meena 2021).

6.2.3.6 Siderophores

Iron is a trace element that affects the growth, germination, and virulence of a pathogen and hence the development of the pathogen (Spadaro and Droby 2016; Chen et al. 2020; Huang et al. 2021). The bacterial siderophores compete for zinc, copper, manganese, and most importantly iron. These BCA limit the availability of iron in the soil by solubilization and the competitive acquisition of Fe^{3+} and subsequently inhibit the plant pathogen by limiting their growth (Leong 1986; Loper and Henkels 1997; Chin-A-Woeng et al. 2003; Haas and Défago 2005; Ab Rahman et al. 2018). Bacteria produce many types of siderophores, for example, catecholate, carboxylate, hydroxamate, and salicylate (Rajkumar et al. 2010; Kumari et al. 2018a, b). Dual inoculation of *Pseudomonas koreensis* and *B. subtilis* strains have been proved to have antagonistic activity and produce siderophore in controlling *Cephalosporium maydis* in maize plants (Ghazy and El-Nahrawy 2021). *Paenibacillus polymyxa*, a siderophore producer, has been proved as a growth promoter of *Lilium lancifolium* and showed antifungal activity against *Botryosphaeria dothidea*, *F. oxysporum*, *Fusarium fujikuroi*, and *B. cinerea* (Khan et al. 2020).

6.2.3.7 Volatile Organic Compounds (VOCs)

VOCs are low-molecular-weight compounds that, under low normal atmospheric temperature and pressure, can evaporate below 300 Da (Vespermann et al. 2007). The main composition of VOC mixture is alcohols, esters, aldehydes, terpenes, aliphatic and aromatic hydrocarbons, nitrides, and sulfides, which exhibit strong antimicrobial effects (Strobel 2011; Lemfack et al. 2018; Huang et al. 2021). Delgado et al. (2021) developed a new consortium PUCV-VBL, composed of *Hanseniaspora osmophila* and *Gluconobacter cerinus*, to control fungal rots in the grapes. The VOCs produced by this consortium showed 86% mycelial inhibition against *B. cinerea*.

6.2.3.8 Lytic Enzymes

Microbial enzymes assist microbes in reproducing in a particular niche and function as biocatalysts for key biochemical reactions (Chaudhari and Patel 2021). The microbes extracellularly produce hydrolytic enzymes to prevent potential plant pathogens (Umer et al. 2021). The antagonists release various enzymes, such as lipase, cellulases, chitinases, xylanases, mannanases, laminarinase, chitosanase, glucose oxidase, protease, and betaglucosidases for biocontrol activity (Picard et al. 2000). Two novel *Bacillus* strains (*simplex* and *subtilis* species) have been found to produce lytic enzymes (protease and β-glucanase), which aided in the biofungicidal activity against *Zymoseptoria tritici* causing *Septoria tritici* blotch of wheat (Allioui et al. 2021).

6.3 Conclusion

During the past few years, beneficial plant microbes have received attention as a substitute for chemical fertilizers because of their sustainable plant protection property. The microbial bioagents produce different elicitors and MAMPs, which trigger induced systemic resistance. A distinctive feature of ISR-eliciting microbial bioagents is local suppression of root immune response in a cell-type specific manner. The studies of root cell-type-specific metabolome and transcriptome profiles in response to microbial bioagents will aid in providing information to develop consistent and reliable methods of crop production. Agrochemicals pose a danger to the health of living beings and the environment due to their toxicity, while elicitors have no adverse effects and leave no residues. The isolation of novel microbial bioagents with high effectiveness against plant pathogens is important and essential. The microbial bioagents with synergistic action against plant pathogens may provide desirable results.

Acknowledgments The authors are thankful to their respective departments and universities for providing the necessary facilities during the course of study.

Author Contributions MM: Provided the general concept, conceived, and drafted part of the manuscript; writing – original draft preparation; prepared the figures and tables; conceptualization; investigation; resources; supervision; validation; visualization; writing – review & editing.

GY, PR, AN, and TM: Provided the general concept, conceived, and drafted part of the manuscript; writing – original draft preparation; prepared the figures and tables.

AZ and PS: Provided the general concept; validation; writing – review & editing.

All authors have read and agreed to the published version of the manuscript.

Data Availability Statement No associated data marked.

Conflicts of Interest The authors declare no conflict of interest. The funders had no role in the design of the study; in the collection, analyses, or interpretation of data; in the writing of the manuscript; or in the decision to publish the results.

Funding The authors are thankful to the University Grant Commission (UGC) under Startup Research Grant (UGC Faculty Research Promotion Scheme; FRPS), New Delhi, India, for its financial assistance (No.F. 30-476/2019 (BSR) FD Diary No. 5662).

References

Ab Rahman SFS, Singh E, Pieterse CM, Schenk PM (2018) Emerging microbial biocontrol strategies for plant pathogens. Plant Sci 267:102–111. https://doi.org/10.1016/j.plantsci.2017.11.012

Abdul Malik NA, Kumar IS, Nadarajah K (2020) Elicitor and receptor molecules: orchestrators of plant defense and immunity. Int J Mol Sci 21(3):963. https://doi.org/10.3390/ijms21030963

Allioui N, Driss F, Dhouib H, Jlail L, Tounsi S, Frikha-Gargouri O (2021) Two novel *Bacillus* strains (subtilis and simplex species) with promising potential for the biocontrol of *Zymoseptoria tritici*, the causal agent of *Septoria Tritici* blotch of wheat. Biomed Res Int. https://doi.org/10.1155/2021/6611657

Arseneault T, Pieterse CM, Gérin-Ouellet M, Goyer C, Filion M (2014) Long-term induction of defense gene expression in potato by *Pseudomonas* sp. LBUM223 and *Streptomyces scabies*. Phytopathology 104(9):926–932. https://doi.org/10.1094/PHYTO-11-13-0321-R

Bais HP, Park SW, Weir TL, Callaway RM, Vivanco JM (2004) How plants communicate using the underground information superhighway. Trends Plant Sci 9(1):26–32. https://doi.org/10.1016/j.tplants.2003.11.008

Banani H, Spadaro D, Zhang D, Matic S, Garibaldi A, Gullino ML (2015) Postharvest application of a novel chitinase cloned from *Metschnikowia fructicola* and overexpressed in *Pichia pastoris* to control brown rot of peaches. Int J Food Microbiol 199:54–61. https://doi.org/10.1016/j.ijfoodmicro.2015.01.002

Barbara R, Enrique IL, Robert LM, Alfredo HE (2011) Identification of mycoparasitism-related genes in *Trichoderma atroviride*. Appl Environ Microbiol 70:4361–4370. https://doi.org/10.1128/AEM.00129-11

Barupal T, Meena M, Sharma K (2019) Inhibitory effects of leaf extract of *Lawsonia inermis* on *Curvularia lunata* and characterization of novel inhibitory compounds by GC–MS analysis. Biotechnol Rep 23:e00335. https://doi.org/10.1016/j.btre.2019.e00335

Barupal T, Meena M, Sharma K (2020) A study on preventive effects of *Lawsonia inermis* L. bioformulations against leaf spot disease of maize. Biocatal Agric Biotechnol 23:101473. https://doi.org/10.1016/j.bcab.2019.101473

Basit A, Farhan M, Abbas M, Wang Y, Zhao D.G, Mridha A.U, Al-Tawaha ARMS, Bashir, MA, Arif M, Ahmed S, Alajmi RA (2021) Do microbial protein elicitors PeaT1 obtained from *Alternaria tenuissima* and PeBL1 from *Brevibacillus laterosporus* enhance defense response against tomato aphid (*Myzus persicae*)?. Saudi J Biol Sci 28(6):3242–3248. https://doi.org/10.1016/j.sjbs.2021.02.063

Bélanger RR, Labbé C, Lefebvre F, Teichmann B (2012) Mode of action of biocontrol agents: all that glitters is not gold. Can J Plant Pathol 34(4):469–478. https://doi.org/10.1080/07060661.2012.726649

Bos JI, Armstron MR, Gilroy EM, Boevink PC, Hein I, Taylor RM, Zhendong T, Engelhardt S, Vetukuri RR, Harrower B, Dixelius C (2010) *Phytophthora infestans* effector AVR3a is essential for virulence and manipulates plant immunity by stabilizing host E3 ligase CMPG1. Proc Natl Acad Sci U S A 107(21):9909–9914. https://doi.org/10.1073/pnas.0914408107

Bozkurt TO, Schornack S, Win J, Shindo T, Ilyas M, Oliva R (2011) *Phytophthora infestans* effector Avrblb2 prevents secretion of a plant immune protease at the haustorial interface. Proc Natl Acad Sci U S A 108:20832–20837. https://doi.org/10.1073/pnas.1112708109

Calderón AA, Zapata JM, Muñoz R, Pedreño MA, Barceló AR (1993) Resveratrol production as a part of the hypersensitive-like response of grapevine cells to an elicitor from *Trichoderma viride*. New Phytol 124(3):455–463. https://doi.org/10.1111/j.1469-8137.1993.tb03836.x

Cawoy H, Mariutto M, Henry G, Fisher C, Vasilyeva N, Thonart P, Dommes J, Ongena M (2014) Plant defense stimulation by natural isolates of *Bacillus* depends on efficient surfactin production. Mol Plant-Microbe Interact 27(2):87–100. https://doi.org/10.1094/MPMI-09-13-0262-R

Chandran H, Meena M, Barupal T, Sharma K (2020) Plant tissue culture as a perpetual source for production of industrially important bioactive compounds. Biotechnol Rep 26:e00450. https://doi.org/10.1016/j.btre.2020.e00450

Chaudhari A, Patel J (2021) Current status and future perspective on enzyme involving in biocontrol of plant pathogen. Int J Appl Sci Biotechnol 8(4):49–55. https://doi.org/10.31033/ijrasb.8.4.8

Chen T, Dong G, Zhang S, Zhang X, Zhao Y, Cao J, Zhou TWQ (2020) Effects of iron on the growth, biofilm formation and virulence of *Klebsiella pneumoniae* causing liver abscess. BMC Microbiol 20(1):1–7. https://doi.org/10.1186/s12866-020-01727-5

Chin-A-Woeng TF, Bloemberg GV, Lugtenberg BJ (2003) Phenazines and their role in biocontrol by *Pseudomonas* bacteria. New Phytol 157(3):503–523. https://doi.org/10.1046/j.1469-8137.2003.00686.x

Chowdhury SP, Hartmann A, Gao X, Borriss R (2015) Biocontrol mechanism by root-associated *Bacillus amyloliquefaciens* FZB42 – a review. Front Microbiol 6:780. https://doi.org/10.3389/fmicb.2015.00780

Chung EJ, Hossain MT, Khan A, Kim KH, Jeon CO, Chung YR (2015) *Bacillus oryzicola* sp. nov., an endophytic bacterium isolated from the roots of rice with antimicrobial, plant growth promoting, and systemic resistance inducing activities in rice. Plant Pathol J 31(2):152. https://doi.org/10.5423/PPJ.OA.12.2014.0136

Compant S, Duffy B, Nowak J, Clément C, Barka EA (2005a) Use of plant growth-promoting bacteria for biocontrol of plant diseases: principles, mechanisms of action, and future prospects. Appl Environ Microbiol 71(9):4951–4959. https://doi.org/10.1128/AEM.71.9.4951-4959.2005

Compant S, Reiter B, Sessitsch A, Nowak J, Clément C, Ait Barka E (2005b) Endophytic colonization of *Vitis vinifera* L. by plant growth-promoting bacterium *Burkholderia* sp. strain PsJN. Appl Environ Microbiol 71(4):1685–1693. https://doi.org/10.1128/AEM.71.4.1685-1693.2005

Compant S, Duffy B, Nowak J, Clement C, Barka EA (2013) Use of plant growth promoting bacteria for biocontrol of plant diseases: principles, mechanisms of action, and future prospects. Appl Environ Microbiol 71(9):4951–4959. https://doi.org/10.1128/AEM.71.9.4951-4959.2005

de León IP, Oliver JP, Castro A, Gaggero C, Bentancor M, Vidal S (2007) *Erwinia carotovora* elicitors and *Botrytis cinerea* activate defense responses in *Physcomitrella patens*. BMC Plant Biol 7(1):1–11. https://doi.org/10.1186/1471-2229-7-52

Degenhardt J, Gershenzon J, Baldwin IT, Kessler A (2003) Attracting friends to feast on foes: engineering terpene emission to make crop plants more attractive to herbivore enemies. Curr Opin Biotech 14(2):169–176. https://doi.org/10.1016/s0958-1669(03)00025-9

Delgado N, Olivera M, Cádiz F, Bravo G, Montenegro I, Madrid A, Besoain X (2021) Volatile organic compounds (VOCs) produced by *Gluconobacter cerinus* and *Hanseniaspora osmophila* displaying control effect against table grape-rot pathogens. Antibiotics 10(6):663. https://doi.org/10.3390/antibiotics10060663

Deslandes L, Rivas S (2012) Catch me if you can: bacterial effectors and plant targets. Trends Plant Sci 17(11):644–655. https://doi.org/10.1016/j.tplants.2012.06.011

Djonović S, Pozo MJ, Dangott LJ, Howell CR, Kenerley CM (2006) Sm1, a proteinaceous elicitor secreted by the biocontrol fungus *Trichoderma virens* induces plant defense responses and systemic resistance. Mol Plant-Microbe Interact 19(8):838–853. https://doi.org/10.1094/MPMI-19-0838

Dukare AS, Paul S, Nambi VE, Gupta RK, Singh R, Sharma K, Vishwakarma RK (2019) Exploitation of microbial antagonists for the control of postharvest diseases of fruits: a review. Crit Rev Food Sci Nutr 59(9):1498–1513. https://doi.org/10.1080/10408398.2017.1417235

Dunlap CA, Bowman MJ, Schisler DA (2013) Genomic analysis and secondary metabolite production in *Bacillus amyloliquefaciens* AS 43.3: a biocontrol antagonist of *Fusarium* head blight. Biol Control 64(2):166–175. https://doi.org/10.1016/j.biocontrol.2012.11.002

Erbs G, Newman MA (2012) The role of lipopolysaccharide and peptidoglycan, two glycosylated bacterial microbe–associated molecular patterns (MAMPs), in plant innate immunity. Mol Plant Pathol 13(1):95–104. https://doi.org/10.1111/j.1364-3703.2011.00730.x

Escudero N, Ferreira SR, Lopez-Moya F, Naranjo-Ortiz MA, Marin-Ortiz AI, Thornton CR, Lopez-Llorca LV (2016) Chitosan enhances parasitism of *Meloidogyne javanica* eggs by the nematophagous fungus *Pochonia chlamydosporia*. Fungal Biol 120(4):572–585. https://doi.org/10.1016/j.funbio.2015.12.005

Felix G, Boller T (2003) Molecular sensing of bacteria in plants: the highly conserved RNA-binding motif RNP-1 of bacterial cold shock proteins is recognized as an elicitor signal in tobacco. J Biol Chem 278(8):6201–6208. https://doi.org/10.1074/jbc.M209880200

Fousia S, Paplomatas EJ, Tjamos SE (2016) *Bacillus subtilis* QST 713 confers protection to tomato plants against *Pseudomonas syringae* pv. *tomato* and induces plant defence-related genes. Phytopathology 164(4):264–270. https://doi.org/10.1111/jph.12455

García JA, Pallás V (2015) Viral factors involved in plant pathogenesis. Curr Opin Virol 11:21–30. https://doi.org/10.1016/j.coviro.2015.01.001

Garcia-Brugger A, Lamotte O, Vandelle E, Bourque S, Lecourieux D, Poinssot B, Wendehenne D, Pugin A (2006) Early signaling events induced by elicitors of plant defenses. Mol Plant-Microbe Interact 19(7):711–724. https://doi.org/10.1094/MPMI-19-0711

Garrido-Ramirez ER, Sudarshana MR, Lucas WJ, Gilbertson RL (2000) Bean dwarf mosaic virus BV1 protein is a determinant of the hypersensitive response and avirulence in *Phaseolus vulgaris*. Mol Plant-Microbe Interact 13(11):1184–1194. https://doi.org/10.1094/MPMI.2000.13.11.1184

Ghazy N, El-Nahrawy S (2021) Siderophore production by *Bacillus subtilis* MF497446 and *Pseudomonas koreensis* MG209738 and their efficacy in controlling *Cephalosporium maydis* in maize plant. Arch Microbiol 203(3):1195–1209. https://doi.org/10.1007/s00203-020-02113-5

Goswami J, Pandey RK, Tewari JP, Goswami BK (2008) Management of root knot nematode on tomato through application of fungal antagonists, *Acremonium strictum* and *Trichoderma harzianum*. J Environ Sci Health B 43(3):237–240. https://doi.org/10.1080/03601230701771164

Govindappa M, Lokesh S, Rai VR, Naik VR, Raju SG (2010) Induction of systemic resistance and management of safflower *Macrophomina phaseolina* root-rot disease by biocontrol agents. Arch Phytopathol Prot 43(1):26–40. https://doi.org/10.1080/03235400701652227

Gowthami L (2018) Role of elicitors in plant defense mechanism. Int J Pharmacogn Phytochem Res 7(6):2806–2812

Guigón-López C, Vargas-Albores F, Guerrero-Prieto V, Ruocco M, Lorito M (2015) Changes in *Trichoderma asperellum* enzyme expression during parasitism of the cotton root rot pathogen *Phymatotrichopsis omnivora*. Fungal Biol 119(4):264–273. https://doi.org/10.1016/j.funbio.2014.12.013

Gupta KJ, Mur LA, Brotman Y (2014) *Trichoderma asperelloides* suppresses nitric oxide generation elicited by *Fusarium oxysporum* in *Arabidopsis* roots. Mol Plant-Microbe Interact 27(4):307–314. https://doi.org/10.1094/MPMI-06-13-0160-R

Guzmán-Valle P, Bravo-Luna L, Montes-Belmont R, Guigón-López C, Sepúlveda-Jiménez G (2014) Induction of resistance to *Sclerotium rolfsii* in different varieties of onion by inoculation with *Trichoderma asperellum*. Eur J Plant Pathol 138(2):223–229. https://doi.org/10.1007/s10658-013-0336-y

Haas D, Défago G (2005) Biological control of soil-borne pathogens by fluorescent pseudomonads. Nat Rev Microb 3(4):307–319. https://doi.org/10.1038/nrmicro1129

Hadrami AE, Adam LR, Hadrami IE, Daayf F (2010) Chitosan in plant protection. Mar Drugs 8(4):968–987. https://doi.org/10.3390/md8040968

Hajimorad MR, Eggenberger AL, Hill JH (2005) Loss and gain of elicitor function of Soybean mosaic virus G7 provoking Rsv1-mediated lethal systemic hypersensitive response maps to P3. J Virol 79(2):1215–1222. https://doi.org/10.1128/JVI.79.2.1215-1222.2005

Heese A, Hann DR, Gimenez-Ibanez S, Jones AM, He K, Li J (2007) The receptor-like kinase SERK3/BAK1 is a central regulator of innate immunity in plants. Proc Natl Acad Sci U S A 104:12217–12222. https://doi.org/10.1073/pnas.0705306104

Hein I, Gilroy EM, Armstrong MR, Birch PRJ (2009) The zig-zag-zig in oomycete? Plant interactions. Mol Plant Pathol 10:547–562. https://doi.org/10.1111/j.1364-3703.2009.00547.x

Heydari A, Pessarakli M (2010) A review on biological control of fungal plant pathogens using microbial antagonists. J Biol Sci 10(4):273–290

Hijwegen T, Buchenauer H (1984) Isolation and identification of hyperparasitic fungi associated with Erysiphaceae. Neth J Plant Pathol 90(2):79–83. https://doi.org/10.1007/BF01999956

Huang J, Wei Z, Hu J, Yang C, Mei X, Shen Q, Riaz (2017) *Chryseobacterium nankingense* sp. nov. WR21 effectively suppresses *Ralstonia solanacearum* growth via intensive root exudates competition. Biol Control 62(4):567–577. https://doi.org/10.1007/s10526-017-9812-1

Huang X, Ren J, Li P, Feng S, Dong P, Ren M (2021) Potential of microbial endophytes to enhance the resistance to postharvest diseases of fruit and vegetables. J Sci Food Agric 101(5):1744–1757. https://doi.org/10.1002/jsfa.10829

Hussain T, Akthar N, Aminedi R, Danish M, Nishat Y, Patel S (2020a) Role of the potent microbial based bioagents and their emerging strategies for the ecofriendly management of agricultural phytopathogens. In: Singh J, Yadav A (eds) Natural bioactive products in sustainable agriculture. Springer, Singapore, pp 45–66. https://doi.org/10.1007/978-981-15-3024-1_4

Hussain T, Singh S, Danish M, Pervez R, Hussain K, Husain R (2020b) Natural metabolites: an eco-friendly approach to manage plant diseases and natural bioactive products. J Sustain Agric 2020:1. https://doi.org/10.1007/978-981-15-3024-1_1

Jain S, Choudhary DK (2014) Induced defense-related proteins in soybean (*Glycine max* L. Merrill) plants by *Carnobacterium* sp. SJ-5 upon challenge inoculation of *Fusarium oxysporum*. Planta 239(5):1027–1040. https://doi.org/10.1007/s00425-014-2032-3

Jamalizadeh M, Etebarian HR, Aminian H, Alizadeh A (2011) A review of mechanisms of action of biological control organisms against post-harvest fruit spoilage. Bull OEPP 41(1):65–71. https://doi.org/10.1111/j.1365-2338.2011.02438.x

Jeffries P (1995) Biology and ecology of mycoparasitism. Can J Bot 73(S1):1284–1290. https://doi.org/10.1139/b95-389

John RP, Tyagi RD, Prévost D, Brar SK, Pouleur S, Surampalli RY (2010) Mycoparasitic *Trichoderma viride* as a biocontrol agent against *Fusarium oxysporum* f. sp. *adzuki* and *Pythium arrhenomanes* and as a growth promoter of soybean. Crop Prot 29(12):1452–1459. https://doi.org/10.1016/j.cropro.2010.08.004

Kageyama K, Nelson EB (2003) Differential inactivation of seed exudate stimulation of *Pythium ultimum* sporangium germination by *Enterobacter cloacae* influences biological control efficacy on different plant species. Appl Environ Microbiol 69(2):1114–1120. https://doi.org/10.1128/AEM.69.2.1114-1120.2003

Kamoun S, Van West P, Vleeshouwers VGAA, de Groot KE, Govers F (1998) Resistance of *Nicotiana benthamiana* to *Phytophthora infestans* is mediated by the recognition of the elicitor protein INF1. Plant Cell 10:1413–1425. https://doi.org/10.1105/tpc.10.9.1413

Karlsson M, Atanasova L, Jensen DF, Zeilinger S (2017) Necrotrophic mycoparasites and their genomes. Microbiol Spectr 5(2). https://doi.org/10.1128/microbiolspec.FUNK-0016-2016

Karthikeyan V, Sankaralingam A, Nakkeeran S (2006) Biological control of groundnut stem rot caused by *Sclerotium rolfsii* (Sacc.). Arch Phytopathol Plant Prot 39(3):239–246. https://doi.org/10.1080/03235400500094688

Kavitha S, Senthilkumar S, Gnanamanickam S, Inayathullah M, Jayakumar R (2005) Isolation and partial characterization of antifungal protein from *Bacillus polymyxa* strain VLB16. Process Biochem 40(10):3236–3243. https://doi.org/10.1016/j.procbio.2005.03.060

Khan MS, Gao J, Chen X, Zhang M, Yang F, Du Y, Zhang X (2020) Isolation and characterization of plant growth-promoting endophytic bacteria *Paenibacillus polymyxa* SK1 from *Lilium lancifolium*. Biomed Res Int 1:1–7. https://doi.org/10.1155/2020/8650957

Khokon MAR, Hossain MA, Munemasa S, Uraji M, Nakamura Y, Mori IC, Murata Y (2010) Yeast elicitor-induced stomatal closure and peroxidase-mediated ROS production in *Arabidopsis*. Plant Cell Physiol 51(11):1915–1921. https://doi.org/10.1093/pcp/pcq145

Köhl J, Kolnaar R, Ravensberg WJ (2019) Mode of action of microbial biological control agents against plant diseases: relevance beyond efficacy. Front Plant Sci 10:845. https://doi.org/10.3389/fpls.2019.00845

Kokalis-Burelle N, Vavrina CS, Rosskopf EN, Shelby RA (2002) Field evaluation of plant growth-promoting rhizobacteria amended transplant mixes and soil solarization for tomato and pepper production in Florida. Plant Soil 238(2):257–266. https://doi.org/10.1023/A:1014464716261

Kubicek CP, Mach RL, Peterbauer CK, Lorito M (2001) *Trichoderma*: From genes to biocontrol. J Plant Pathol 83(2):11–23. https://www.jstor.org/stable/41998018

Kumari P, Meena M, Gupta P, Dubey MK, Nath G, Upadhyay RS (2018a) Plant growth promoting rhizobacteria and their biopriming for growth promotion in mung bean (*Vigna radiata* (L.) R. Wilczek). Biocatal Agric Biotechnol 16:163–171. https://doi.org/10.1016/j.bcab.2018.07.030

Kumari P, Meena M, Upadhyay RS (2018b) Characterization of plant growth promoting rhizobacteria (PGPR) isolated from the rhizosphere of *Vigna radiata* (mung bean). Biocatal Agric Biotechnol 16:155–162. https://doi.org/10.1016/j.bcab.2018.07.029

Lahlali R, McGregor L, Song T, Gossen BD, Narisawa K, Peng G (2014) *Heteroconium chaetospira* induces resistance to clubroot via upregulation of host genes involved in jasmonic acid, ethylene, and auxin biosynthesis. PLoS One 9(4):94144. https://doi.org/10.1371/journal.pone.0094144

Latz MA, Jensen B, Collinge DB, Jørgensen HJ (2018) Endophytic fungi as biocontrol agents: elucidating mechanisms in disease suppression. Plant Ecol Divers 11(5–6):555–567. https://doi.org/10.1080/17550874.2018.1534146

Lawton MA, Lamb CJ (1987) Transcriptional activation of plant defense genes by fungal elicitor wounding and infection. Mol Cell Biol 7(1):335–341. https://doi.org/10.1128/mcb.7.1.335-341.1987

Lemfack MC, Gohlke BO, Toguem SMT, Preissner S, Piechulla B, Preissner R (2018) mVOC 2.0: a database of microbial volatiles. Nucleic Acids Res 46(D1):D1261–D1265. https://doi.org/10.1093/nar/gkx1016

Leong J (1986) Siderophores: their biochemistry and possible role in the biocontrol of plant pathogens. Annu Rev Phytopathol 24(1):187–209

Lherminier J, Benhamou N, Larrue J, Milat ML, Boudon-Padieu E, Nicole M, Blein JP (2003) Cytological characterization of elicitin-induced protection in tobacco plants infected by *Phytophthora parasitica* or phytoplasma. Phytopathology 93(10):1308–1319. https://doi.org/10.1094/PHYTO.2003.93.10.1308

Loper JE, Henkels MD (1997) Availability of iron to *Pseudomonas fluorescens* in rhizosphere and bulk soil evaluated with an ice nucleation reporter gene. Appl Environ Microb 63(1):99–105. https://doi.org/10.1128/aem.63.1.99-105.1997

Low PS, Merida JR (1996) The oxidative burst in plant defense: function and signal transduction. Physiol Plant 96(3):533–542. https://doi.org/10.1111/j.1399-3054.1996.tb00469.x

Lugtenberg B, Kamilova F (2009) Plant-growth-promoting rhizobacteria. Annu Rev Microbiol 63:541–556. https://doi.org/10.1146/annurev.micro.62.081307

Mao J, Liu Q, Yang X, Long C, Zhao M, Zeng H, Liu H, Yuan J, Qiu D (2010) Purification and expression of a protein elicitor from *Alternaria tenuissima* and elicitor–mediated defence responses in tobacco. Ann Appl Biol 156(3):411–420. https://doi.org/10.1111/j.1744-7348.2010.00398.x

Mavrodi OV, Walter N, Elateek S, Taylor CG, Okubara PA (2012) Suppression of *Rhizoctonia* and *Pythium* root rot of wheat by new strains of *Pseudomonas*. Biol Control 62(2):93–102. https://doi.org/10.1016/j.biocontrol.2012.03.013

Mbarga JB, Ten Hoopen GM, KuatÚ J, Adiobo A, Ngonkeu MEL, Ambang Z, Akoa A, Tondje PR, Begoude BAD (2012) *Trichoderma asperellum*: A potential biocontrol agent for *Pythium myriotylum*, causal agent of cocoyam (*Xanthosoma sagittifolium*) root rot disease in Cameroon. Crop Prot 36:18–22. https://doi.org/10.1016/j.cropro.2012.02.004

Meena M, Samal S (2019) *Alternaria* host-specific (HSTs) toxins: an overview of chemical characterization, target sites, regulation and their toxic effects. Toxicol Rep 6:745–758. https://doi.org/10.1016/j.toxrep.2019.06.021

Meena M, Swapnil P (2019) Regulation of *WRKY* genes in plant defense with beneficial fungus *Trichoderma*: current perspectives and future prospects. Arch Phytopathol Plant Protect 52(1–2):1–17. https://doi.org/10.1080/03235408.2019.1606490

Meena M, Prasad V, Upadhyay RS (2016a) Assessment of the bioweedicidal effects of *Alternaria alternata* metabolites against *Parthenium* species. Bull Environ Sci Res 5(1):1–7

Meena M, Zehra A, Dubey MK, Aamir M, Gupta VK, Upadhyay RS (2016b) Comparative evaluation of biochemical changes in tomato (*Lycopersicon esculentum* Mill.) infected by *Alternaria alternata* and its toxic metabolites (TeA, AOH, and AME). Front Plant Sci 7:1408. https://doi.org/10.3389/fpls.2016.01408

Meena M, Prasad V, Upadhyay RS (2017a) Evaluation of biochemical changes in leaves of tomato infected with *Alternaria alternata* and its metabolites. Vegetos 30:2. https://doi.org/10.5958/2229-4473.2017.00020.9

Meena M, Swapnil P, Upadhyay RS (2017b) Isolation, characterization and toxicological potential of tenuazonic acid, alternariol and alternariol monomethyl ether produced by *Alternaria* species phytopathogenic on plants. Sci Rep 7:8777. https://doi.org/10.1038/s41598-017-09138-9

Meena M, Swapnil P, Zehra A, Aamir M, Dubey MK, Upadhyay RS (2017c) Beneficial microbes for disease suppression and plant growth promotion. In: Singh DP, Singh HB, Prabha R (eds) Plant-microbe interactions in agro-ecological perspectives. Springer, Singapore, pp 395–432. https://doi.org/10.1007/978-981-10-6593-4_16

Meena M, Swapnil P, Zehra A, Dubey MK, Upadhyay RS (2017d) Antagonistic assessment of *Trichoderma* spp. by producing volatile and non-volatile compounds against different fungal pathogens. Arch Phytopathol Plant Protect 50(13–14):629–648. https://doi.org/10.1080/03235408.2017.1357360

Meena M, Gupta SK, Swapnil P, Zehra A, Dubey MK, Upadhyay RS (2017e) *Alternaria* toxins: potential virulence factors and genes related to pathogenesis. Front Microbiol 8:1451. https://doi.org/10.3389/fmicb.2017.01451

Meena M, Prasad V, Upadhyay RS (2017f) Evaluation of *Alternaria alternata* isolates for metabolite production isolated from different sites of Varanasi, India. J Agric Res 2(1):00012

Meena M, Swapnil P, Zehra A, Dubey MK, Aamir M, Patel CB, Upadhyay RS (2019) Virulence factors and their associated genes in microbes. In: Singh HB, Gupta VK, Jogaiah S (eds) New and future developments in microbial biotechnology and bioengineering. Elsevier. https://doi.org/10.1016/B978-0-444-63503-7.00011-5

Meena M, Swapnil P, Divyanshu K, Kumar S, Harish TYN, Zehra A, Marwal A, Upadhyay RS (2020) PGPR-mediated induction of systemic resistance and physiochemical alterations in plants against the pathogens: current perspectives. J Basic Microbiol 60(8):1–34. https://doi.org/10.1002/jobm.202000370

Mei L, Liang Y, Zhang L, Wang Y, Guo Y (2014) Induced systemic resistance and growth promotion in tomato by an indole-3-acetic acid-producing strain of *Paenibacillus polymyxa*. Ann Appl Biol 165(2):270–279. https://doi.org/10.1111/aab.12135

Milling A, Babujee L, Allen C (2011) *Ralstonia solanacearum* extracellular polysaccharide is a specific elicitor of defense responses in wilt-resistant tomato plants. PLoS One 6(1):5853. https://doi.org/10.1371/journal.pone.0015853

Moffett P (2009) Mechanisms of recognition in dominant *R* gene mediated resistance. Adv Virus Res 75:1–229. https://doi.org/10.1016/S0065-3527(09)07501-0

Montesano M, Brader G, Palva ET (2003) Pathogen derived elicitors: searching for receptors in plants. Mol Plant Pathol 4(1):73–79. https://doi.org/10.1046/j.1364-3703.2003.00150.x

Murali M, Amruthesh KN (2015) Plant growth-promoting fungus *Penicillium oxalicum* enhances plant growth and induces resistance in pearl millet against downy mildew disease. Phytopathology 163(9):743–754. https://doi.org/10.1111/jph.12371

Mustafa G, Randoux B, Tisserant B, Fontaine J, Magnin-Robert M, Sahraoui ALH, Reignault P (2016) Phosphorus supply, arbuscular mycorrhizal fungal species, and plant genotype impact on the protective efficacy of mycorrhizal inoculation against wheat powdery mildew. Mycorrhiza 26(7):685–697. https://doi.org/10.1007/s00572-016-0698-z

Nair A, Kolet SP, Thulasiram HV, Bhargava S (2015) Systemic jasmonic acid modulation in mycorrhizal tomato plants and its role in induced resistance against *Alternaria alternata*. Plant Biol 17(3):625–631. https://doi.org/10.1111/plb.12277

Navarro MO, Piva AC, Simionato AS, Spago FR, Modolon F, Emiliano J, Azul AM, Chryssafidis AL, Andrade G (2019) Bioactive compounds produced by biocontrol agents driving plant health. In: Kumar V, Prasad R, Kumar M, Choudhary DK (eds) Microbiome in plant health and disease. Springer, Singapore, pp 337–374. https://doi.org/10.1007/978-981-13-8495-015

Newman MA, Daniels MJ, Dow JM (1995) Lipopolysaccharide from *Xanthomonas campestris* induces defense-related gene expression in *Brassica campestris*. Mol Plant-Microbe Interact 8(5):778–780

Newman MA, Sundelin T, Nielsen JT, Erbs G (2013) MAMP (microbe-associated molecular pattern) triggered immunity in plants. Front Plant Sci 4:139. https://doi.org/10.3389/fpls.2013.00139

Nihorimbere V, Cawoy H, Seyer A, Brunelle A, Thonart P, Ongena M (2012) Impact of rhizosphere factors on cyclic lipopeptide signature from the plant beneficial strain *Bacillus amyloliquefaciens* S499. FEMS Microbiol Ecol 79(1):176–191. https://doi.org/10.1111/j.1574-6941.2011.01208.x

Niu D, Wang X, Wang Y, Song X, Wang J, Guo J, Zhao H (2016) *Bacillus cereus AR156* activates PAMP-triggered immunity and induces a systemic acquired resistance through a NPR1-and SA-dependent signaling pathway. Biochem Biophys Res Commun 469(1):120–125. https://doi.org/10.1016/j.bbrc.2015.11.081

Nygren K, Dubey M, Zapparata A, Iqbal M, Tzelepis GD, Durling MB, Karlsson M (2018) The mycoparasitic fungus *Clonostachys rosea* responds with both common and specific gene expression during interspecific interactions with fungal prey. Evol Appl 11(6):931–949. https://doi.org/10.1111/eva.12609

Oclarit E, Cumagun C (2009) Evaluation of efficacy of *Paecilomyces lilacinus* as biological control agent of *Meloidogyne incognita* attacking tomato. J Plant Prot Res 49(4). https://doi.org/10.2478/v10045-009-0053-x

Onaga G, Wydra K (2016) Advances in plant tolerance to abiotic stresses. Plant Genome 10:229–272

Padgett HS, Watanabe Y, Beachy RN (1997) Identification of the TMV replicase sequence that activates the N gene-mediated hypersensitive response. Mol Plant-Microbe Interact 10(6):709–715. https://doi.org/10.1094/MPMI.1997.10.6.709

Patel CB, Singh VK, Singh AP, Meena M, Upadhyay RS (2019) Microbial genes involved in interaction with plants. In: Singh HB, Gupta VK, Jogaiah S (eds) New and future developments in microbial biotechnology and bioengineering. Elsevier, pp 171–180. https://doi.org/10.1016/B978-0-444-63503-7.00011-5

Patterson GM, Bolis CM (1997) Fungal cell wall polysaccharides elicit an antifungal secondary metabolite (phytoalexin) in the cyanobacterium *Scytonema ocelutum*[2]. J Phycol 33(1):54–60. https://doi.org/10.1111/j.0022-3646.1997.00054.x

Picard K, Ponchet M, Belin JP, Rey P, Tirily Y, Benhamou N (2000) Oligandrin. A proteinaceous molecule produced by the mycoparasite *Pythium oligandrum* induces resistance to *Phytophthora parasitica* infection in tomato plants. Plant Physiol 124:379–395. https://doi.org/10.1104/pp.124.1.379

Pieterse CM, Zamioudis C, Berendsen RL, Weller DM, Van Wees SC, Bakker PA (2014) Induced systemic resistance by beneficial microbes. Annu Rev Phytopathol 52:347–375. https://doi.org/10.1146/annurev-phyto-082712-102340

Pirttilä AM, Mohammad Parast Tabas H, Baruah N, Koskimäki JJ (2021) Biofertilizers and biocontrol agents for agriculture: how to identify and develop new potent microbial strains and traits. Microorganisms 9(4):817. https://doi.org/10.3390/microorganisms9040817

Pouteau S, Grandbastien MA, Boccara M (1994) Microbial elicitors of plant defence responses activate transcription of a retrotransposon. Plant J 5(4):535–542. https://doi.org/10.1046/j.1365-313X.1994.05040535.x

Puopolo G, Tomada S, Pertot I (2018) The impact of the omics era on the knowledge and use of *Lysobacter* species to control phytopathogenic micro-organisms. J Appl Microbiol 124(1):15–27. https://doi.org/10.1111/jam.13607

Qutob D, Kemmerling B, Brunner F, Kufner I, Engelhardt S, Gust AA, Luberacki B, Seitz HU, Stahl D, Rauhut T, Glawischnig E (2006) Phytotoxicity and innate immune responses induced by Nep1-like proteins. Plant Cell 18(12):3721–3744. https://doi.org/10.1105/tpc.106.044180

Raaijmakers JM, Sluis LVD, Bakker PA, Schippers B, Koster M, Weisbeek PJ (1995) Utilization of heterologous siderophores and rhizosphere competence of fluorescent *Pseudomonas* spp. Can J Microbiol 41(2):126–135. https://doi.org/10.1139/m95-017

Rajkumar M, Ae N, Prasad MNV, Freitas H (2010) Potential of siderophore-producing bacteria for improving heavy metal phytoextraction. Trends Biotechnol 28(3):142–149. https://doi.org/10.1016/j.tibtech.2009.12.002

Reithner B, Ibarra-Laclette E, Mach RL, Herrera-Estrella A (2011) Identification of mycoparasitism-related genes in *Trichoderma atroviride*. Appl Environ Microbiol 77(13):4361–4370. https://doi.org/10.1128/AEM.00129-11

Rosyidah A, Wardiyati T, Abadi AL, Maghfoer MD, Aini LQ (2014) Induced resistance of potato (*Solanum tuberosum L.*) toward *Ralstonia solanacearum* disease with combination of several bio-control microbes. J Biol Agri Healthc 4(2):90–98

Sahebani N, Hadavi N (2008) Biological control of the root-knot nematode *Meloidogyne javanica* by *Trichoderma harzianum*. Soil Biol Biochem 40(8):2016–2020. https://doi.org/10.1016/j.soilbio.2008.03.011

Saravanakumar D, Ciavorella A, Spadaro D, Garibaldi A, Gullino ML (2008) *Metschnikowia pulcherrima* strain MACH1 outcompetes *Botrytis cinerea*, *Alternaria alternata* and *Penicillium expansum* in apples through iron depletion. Postharvest Biol Tec 49(1):121–128. https://doi.org/10.1016/j.postharvbio.2007.11.006

Sarma BK, Yadav SK, Singh S, Singh HB (2015) Microbial consortium-mediated plant defense against phytopathogens: readdressing for enhancing efficacy. Soil Biol Biochem 87:25–33. https://doi.org/10.1016/j.soilbio.2015.04.001

Schornack S, Huitema E, Cano LM, Bozkurt TO, Oliva R, Van Damme M, Schwizer S, Raffaele S, CHAPARRO-GARCIA ANGELA, Farrer R, Segretin ME (2009) Ten things to know about oomycete effectors. Mol Plant Pathol 10(6):795–803. https://doi.org/10.1111/j.1364-3703.2009.00593.x

Schwacke R, Hager A (1992) Fungal elicitors induce a transient release of active oxygen species from cultured spruce cells that is dependent on Ca^{2+} and protein-kinase activity. Planta 87(1):136–141. https://doi.org/10.1007/BF00201635

Segarra G, Casanova E, Avilés M, Trillas I (2010) *Trichoderma asperellum* strain T34 controls fusarium wilt disease in tomato plants in soilless culture through competition for iron. Microb Ecol 59(1):141–149. https://doi.org/10.1007/s00248-009-9545-5

Shoda M (2000) Bacterial control of plant diseases. J Biosci Bioeng 89(6):515–521. https://doi.org/10.1016/S1389-1723(00)80049-3

Siah A, Magnin-Robert M, Randoux B, Choma C, Rivière C, Halama P, Reignault P (2018) Natural agents inducing plant resistance against pests and diseases. Int J Antimicrob Agents. Springer, Cham:121–159. https://doi.org/10.1007/978-3-319-67045-46

Singh BN, Singh A, Singh BR, Singh HB (2014) *Trichoderma harzianum* elicits induced resistance in sunflower challenged by *Rhizoctonia solani*. J Appl Microbiol 116(3):654–666. https://doi.org/10.1111/jam.12387

Singh S, Kumar V, Dhanjal DS, Singh J (2020) Biological control agents: diversity, ecological significances, and biotechnological applications. In: Singh J, Ajar Nath Y (eds) Natural bioactive products in sustainable agriculture. Springer, Singapore, pp 31–44. https://doi.org/10.1007/978-981-15-3024-1_3

Son JS, Sumayo M, Hwang YJ, Kim BS, Ghim SY (2014) Screening of plant growth-promoting rhizobacteria as elicitor of systemic resistance against gray leaf spot disease in pepper. Agric Ecosyst Environ Appl 73:1–8. https://doi.org/10.1016/j.apsoil.2013.07.016

Song M, Yun HY, Kim YH (2014) Antagonistic *Bacillus* species as a biological control of ginseng root rot caused by *Fusarium* cf. *incarnatum*. J Ginseng Res 38(2):136–145

Sonigra P, Meena M (2021) Metabolic profile, bioactivities, and variations in the chemical constituents of essential oils of the *Ferula* genus (Apiaceae). Front Pharmacol 11:608649. https://doi.org/10.3389/fphar.2020.608649

Spadaro D, Droby S (2016) Development of biocontrol products for postharvest diseases of fruit: the importance of elucidating the mechanisms of action of yeast antagonists. Trends Food Sci Technol 47:39–49. https://doi.org/10.1016/j.tifs.2015.11.003

Spiteller D, Dettner K, Bolan W (2000) Gut bacteria may be involved in interactions between plants, herbivores and their predators: microbial biosynthesis of N-acylglutamine surfactants as elicitors of plant volatiles. Biol Chem 381(8):755–762. https://doi.org/10.1515/BC.2000.096

Strobel G (2011) Muscodor species-endophytes with biological promise. Phytochem Rev 10(2):165–172. https://doi.org/10.1007/s11101-010-9163-3

Surekha CH, Neelapu NRR, Prasad BS, Ganesh PS (2014) Induction of defense enzymes and phenolic content by *Trichoderma viride* in *Vigna mungo* infested with *Fusarium oxysporum* and *Alternaria alternata*. Int J Agric Sci 4(4):31–40

Swapnil P, Meena M, Singh SK, Dhuldhaj UP, Harish MA (2021) Vital roles of carotenoids in plants and humans to deteriorate stress with its structure, biosynthesis, metabolic engineering and functional aspects. Curr Plant Biol 26:100203. https://doi.org/10.1016/j.cpb.2021.100203

Thakur M, Sohal BS (2013) Role of elicitors in inducing resistance in plants against pathogen infection: a review. International Scholarly Research Notices. https://doi.org/10.1155/2013/762412

Umer M, Mubeen M, Iftikhar Y, Shad MA, Usman HM, Sohail MA, Ateeq M (2021) Role of rhizobacteria on plants growth and biological control of plant diseases: a review. Plant Protect 5(1):59–73. https://doi.org/10.33804/pp.005.01.3565

Urbina CT, Prieto VG, Lopez CG, Albores FV, Reyes DB, Muniz CA, Barrios DO (2016) Purification and characterization of β-1,3-glucanase from *Candida oleophila* for the biocontrol of *Penicillium expansum*. Res Rev J Bot Sci 5(1):38–45

Van Dijk K, Nelson EB (2000) Fatty acid competition as a mechanism by which *Enterobacter cloacae* suppresses *Pythium ultimum* sporangium germination and damping-off. Appl Environ Microbiol 66(12):340–5347. https://doi.org/10.1128/AEM.66.12.5340-5347.2000

Van Loon LC (2000) Helping plants to defend themselves: biocontrol by disease-suppressing rhizobacteria. In: Developments in plant genetics and breeding, vol 6. Elsevier, pp 203–213. https://doi.org/10.1016/s0168-7972(00)80123-1

Van Loon LC, Van Strien EA (1999) The families of pathogenesis-related proteins, their activities, and comparative analysis of PR-1 type proteins. Physiol Mol Plant Pathol 55(2):85–97. https://doi.org/10.1006/pmpp.1999.0213

Veloso J, Alabouvette C, Olivain C, Flors V, Pastor V, García T, Díaz J (2016) Modes of action of the protective strain Fo47 in controlling verticillium wilt of pepper. Plant Pathol 65(6):997–1007. https://doi.org/10.1111/ppa.12477

Vespermann A, Kai M, Piechulla B (2007) Rhizobacterial volatiles affect the growth of fungi and *Arabidopsis thaliana*. Appl Environ Microb 73(17):5639–5641. https://doi.org/10.1128/AEM.01078-07

Viswanathan R, Samiyappan R (1999) Induction of systemic resistance by plant growth promoting rhizobacteria against red rot disease in sugarcane. Sugar Tech 1(3):67–76. https://doi.org/10.1016/S0261-2194(00)00056-9

Wang J, Peiffer M, Hoove K, Rosa C, Zeng R, Felton GW (2017) *Helicoverpa zea* gut-associated bacteria indirectly induce defenses in tomato by triggering a salivary elicitor. New Phytol 214(3):1294–1306. https://doi.org/10.1111/nph.14429

Wei ZM, Laby RJ, Zumoff CH, Bauer DW, He SY, Collmer A, Beer SV (1992) Harpin, elicitor of the hypersensitive response produced by the plant pathogen *Erwinia amylovora*. Science 257(5066):85–88. https://doi.org/10.1126/science.1621099

Wiesel L, Newton AC, Elliott I, Booty D, Gilroy EM, Birch PR, Hein I (2014) Molecular effects of resistance elicitors from biological origin and their potential for crop protection. Front Plant Sci 5:655. https://doi.org/10.3389/fpls.2014.00655

Wu S, Shan L, He P (2014) Microbial signature-triggered plant defense responses and early signaling mechanisms. Plant Sci 228:118–126. https://doi.org/10.1016/j.plantsci.2014.03.001

Wu ZH, Ma Q, Sun ZN, Cui HC, Liu HR (2021) Biocontrol mechanism of *Myxococcus fulvus* B25-I-3 against *Phytophthora infestans* and its control efficiency on potato late blight. Folia Microbiol (Praha) 66(4):555–567. https://doi.org/10.1007/s12223-021-00865-1

Yacou AJ, Gerbore N, Magnin P, Chambon MC, Dufour MF, Corio-Costet R, Guyoneaud P, Rey (2016) Ability of *Pythium oligandrum* strains to protect *Vitis vinifera* L., by inducing plant resistance against *Phaeomoniella chlamydospora*, a pathogen involved in Esca, a grapevine trunk disease. Biol Control 92:7–16. https://doi.org/10.1016/j.biocontrol.2015.08.005

Yadav G, Meena M (2021) Bioprospecting of endophytes in medicinal plants of Thar Desert: an attractive resource for biopharmaceuticals. Biotechnol Rep 30:e00629. https://doi.org/10.1016/j.btre.2021.e00629

Yamamoto S, Shiraishi S, Kawagoe Y, Mochizuki M, Suzuki S (2015) Impact of *Bacillus amyloliquefaciens S13-3* on control of bacterial wilt and powdery mildew in tomato. Pest Manag Sci 71:722–727. https://doi.org/10.1002/ps.3837

Zehra A, Dubey MK, Tiwari A, Meena M, Kumari P, Singh VK, Gupta VK, Upadhyay RS (2015) Fungal biomolecules and their implications. In: Gupta VK, Mach RL, Sreenivasaprasad S (eds) Fungal biomolecules: source applications and recent developments. Wiley Blackwell/John Wiley & Sons Ltd., USA, pp 363–375

Zehra A, Meena M, Dubey MK, Aamir M, Upadhyay RS (2017a) Synergistic effects of plant defense elicitors and *Trichoderma harzianum* on enhanced induction of antioxidant defense system in tomato against Fusarium wilt disease. Bot Stud 58(1):44. https://doi.org/10.1186/s40529-017-0198-2

Zehra A, Meena M, Dubey MK, Aamir M, Upadhyay RS (2017b) Activation of defense response in tomato against Fusarium wilt disease triggered by *Trichoderma harzianum* supplemented with exogenous chemical inducers (SA and MeJA). Braz J Bot 40(3):651–664. https://doi.org/10.1007/s40415-017-0382-3

Zehra A, Raytekar NA, Meena M, Swapnil P (2021) Efficiency of microbial bio-agents as elicitors in plant defense mechanism under biotic stress: a review. Curr Res Microb Sci 2:100054. https://doi.org/10.1016/j.crmicr.2021.100054

Zhai X, Jia M, Chen L, Zheng CJ, Rahman K, Han T, Qin LP (2017) The regulatory mechanism of fungal elicitor-induced secondary metabolite biosynthesis in medical plants. Crit Rev Microbiol 43(2):238–261. https://doi.org/10.1080/1040841X.2016.1201041

Zhang W, Fraiture M, Kolb D, Löffelhardt B, Desaki Y, Boutrot FF., Tör M, Zipfel C, Gust AA, Brunner F (2013) *Arabidopsis* receptor-like protein30 and receptor-like kinase suppressor of BIR1-1/EVERSHED mediate innate immunity to necrotrophic fungi. Plant Cell 25(10):4227–4241. https://doi.org/10.1105/tpc.113.117010

Zhang Q, Yong D, Zhang Y, Shi X, Li B, Li G, Liang W, Wang C (2016) *Streptomyces rochei* A-1 induces resistance and defense-related responses against *Botryosphaeria dothidea* in apple fruit during storage. Postharvest Biol Technol 115:30–37. https://doi.org/10.1016/j.postharvbio.2015.12.013

Zhao Y, Jiang T, Xu H, Xu G, Qian G, Liu F (2021) Characterization of *Lysobacter* spp. strains and their potential use as biocontrol agents against pear anthracnose. Microbiol Res 242:126624. https://doi.org/10.1016/j.micres.2020.126624

Zheng L, Zhao J, Liang X, Zhan G, Jiang S, Kang Z (2017) Identification of a novel *Alternaria alternata* strain able to hyperparasitize *Puccinia striiformis* f. sp. *tritici*, the causal agent of wheat stripe rust. Front Microbiol 87:1. https://doi.org/10.3389/fmicb.2017.00071

Zakaria H, Kassab AS, Shamseldean M, Oraby M, El-Mourshedy MMF (2013) Controlling the root-knot nematode, *Meloidogyne incognita* in cucumber plants using some soil bioagents and some amendments under simulated field conditions. Ann Agric Sci 58(1):77–82. https://doi.org/10.1016/j.aoas.2013.01.011

Chapter 7
Transcriptional Factors' Response Under Biotic Stress in Wheat

Fahad Masoud Wattoo, Rashid Mehmood Rana, and Sajid Fiaz

Abstract Wheat is a vital food crop both economically and socially. It is, however, susceptible to a variety of ailments caused primarily by fungus and bacteria. To combat these illnesses, wheat has developed a genetically regulated defence system. This defensive mechanism relies heavily on transcription factors. Several studies show that transcription factors play a regulatory function in disease resistance. As a result, these transcription factors could be used to improve disease resistance in wheat either directly (genetic transformation) or indirectly (marker-assisted selection). We have compiled a list of some of the most essential transcription factor families and discussed their roles in disease resistance.

Keywords Transcription factors (TFs) · Regulation · Metabolism · Marker-assisted selection · Vertical resistance · Quantitative resistance

7.1 Introduction

Bread wheat (*Triticum aestivum* L.) is a staple food of a huge population across the globe. It is widely consumed for flat bread and leavened purposes aside from many other baked confectionary products (Branlard et al. 2001). In nature, bread wheat is hexaploid ($2n = 42$, AABBDD genome) and is thought to have evolved from diploid progenitors such as *Triticum urartu* (AA), *Aegilops speltoides* (BB) and *Triticum tauschii* (DD) (El Baidouri et al. 2017). In hot and dry climatic

F. M. Wattoo
Department of Plant Breeding and Genetics, PMAS-Arid Agriculture University, Rawalpindi, Pakistan

National Center for Industrial Biotechnology, PMAS-Arid Agriculture University, Rawalpindi, Pakistan

R. M. Rana
Department of Plant Breeding and Genetics, PMAS-Arid Agriculture University, Rawalpindi, Pakistan

S. Fiaz (✉)
Department of Plant Breeding, The University of Haripur, Haripur, Pakistan
e-mail: sfiaz@uoh.edu.pk

© The Author(s), under exclusive license to Springer Nature Switzerland AG 2022

S. H. Wani et al. (eds.), *Transcription Factors for Biotic Stress Tolerance in Plants*, https://doi.org/10.1007/978-3-031-12990-2_7

zones, tetraploid wheat (*Triticum turgidum* L. var. *durum*, $2n = 28$, AABB genome) widely known as 'pasta wheat' contributes roughly 35–40 million tonnes of production (Pena 2002). Ancient wheat species einkorn (*Triticum monococcum*, $2n = 14$ AA genome), emmer (*Triticum turgidum* L. var. *dicoccum*, AABB genome, tetraploid) and spelt (*Triticum aestivum* subsp. *spelta*, AABBDD genome) differ little from modern wheat in terms of various bioactive components (Shewry and Hey 2015).

Wheat is the third most important crop after maize and rice; it has been forecasted to about 780 million tonnes from an area of 216 million hectare production in years 2017–2019 (http://www.fao.org/faostat/en/). Globally it is considered as the most important source of carbohydrates, high vegetable protein content (about 13%), minerals, vitamins and lipids compared to other cereals (Lafiandra et al. 2014). Wheat-based diet is higher in fibre content as compared to meat-based diet (Zhu et al. 2010). Balanced wheat production will ensure food security as demand for wheat is projected to rise by 60% by 2050. A huge challenge of massive population increase, abiotic (high temperature, salinity and drought) and biotic (casual organism; bacteria and fungi) stresses due to changing climate (Rind et al. 2019; Saira et al. 2019). Moreover, extreme change in climatic conditions has made crop vulnerable to pests and diseases. There are about 50 destructive diseases reported in wheat caused by various fungal, bacterial and viral pathogens which could lead up to 70% yield losses (Singh et al. 2016). Among these pathogens, fungi are considered as the most common agents of disease (Zhou 2011). Intensity and response of diseases are dependent on climatic conditions. Moreover, whole plant is susceptible to diseases; in severe cases many diseases can attack on single plant at the same time. Major diseases causing significant losses in wheat are listed in Table 7.1.

Unpredictable shift in climatic conditions worldwide due to increase in CO_2 concentration has changed the patterns of rainfall, temperature and humidity giving advantage to survival of several seasonal pathogens against standing crop in field; thus the term 'disease triangle' could help understand the interaction of pathogen, host and climate. In this situation crop plants are exposed to both abiotic (heat, drought, water logging, salinity, etc.) and biotic stresses (insect pests and pathogens) which enhances the potential effect of wheat crop (Miedaner and Juroszek 2021). Mitigation strategies for disease resistance in wheat are dependent on the integrated disease management procedures including the use of resistant varieties as well as adoption of cultural practices along with chemical control. Currently, it is mandatory for all the wheat breeding programs to focus on major diseases of wheat, and all improved varieties must have resistance. Mostly, these breeding efforts are dependent on phenotypic appearance of resistance. However, phenotype of a plant is the result of genetic makeup of plants. Disease resistance is obviously genetically controlled by relatively complex phenomena. Thus understanding genetics is very important to develop/incorporate resistance. Moreover, molecular markers, if developed, could help in precise and accurate selection of resistance (Goutam et al. 2015).

Table 7.1 Major disease of wheat

Disease	Causal organism	Disease transmission	Epidemic conditions
Brown rust	*Puccinia recondita* f. sp. *tritici*	Air	High humidity, low temperature
Stripe rust/yellow rust	*Puccinia striiformis* f. sp. *tritici*	Air	High humidity, low temperature
Stem rust (black rust)	*Puccinia graminis* f. sp. *tritici*	Air	High humidity, low temperature
Powdery mildew	*Blumeria graminis* f. sp. *tritici*	Air, soil, plant debris	High humidity, low temperature
Fusarium head blight	*Fusarium graminearum*	Plant debris, water, air	High temperature, high humidity
Leaf spot/ leaf blotch	*Zymoseptoria tritici*	Seed	High humidity, low temperature
Yellow spot/yellow leaf blotch	*Pyrenophora tritici*	Seed	High humidity, low temperature
Helminthosporium leaf blight	*Cochliobolus sativus*	Seed	High temperature, high humidity
Septoria glum blotch	*Parastagonospora nodorum*	Seed	Low temperature, high humidity
Wheat blast	*Magnaporthe oryzae*	Air, seed	High temperature, high humidity
Loose smut	*Ustilago tritici*	Seed	Low temperature, high humidity

7.2 Genetic Control of Disease Resilience

Disease resilience mechanism is regulated in two distinct patterns, viz. (i) monogenic/qualitative resistance and (ii) polygenic/quantitative resistance (Stuthman et al. 2007). Monogenic resistance primarily relies on single gene and is worked on based on host-pathogen interaction. This type of resistance is considered as absolute but limited to particular strains of pathogen. Thus, a separate gene is required to confer resistance against specific strain/race of pathogen. Accumulation of such genes in a single wheat genotype is essential to provide a broad spectrum of resistance. This accumulation is termed as gene pyramiding (Liu et al. 2020). Polygenic/quantitative resistance depends upon multiple genes. However, these genes do not interact with pathogen. Instead, they reduce infection by mobilizing cellular machinery to minimize the harmful effects caused by pathogen. Though resistance provided by quantitative inheritance is not absolute, they provide relatively durable resistance (Stuthman et al. 2007). Generally, working with monogenic resistance is easy as compared to polygenic; however, it breaks quite easily as pathogen strains mutate quite often (Lindhout 2002). Several scientists worked on the discovery of the phenomena of polygenic disease resistance and highlighted the role of transcription factors.

7.3 Transcription Factors

All biological functions of living organisms are controlled by genes, while expression of all the genes is controlled by regulatory proteins called as transcription factors (Johnson and McKnight 1989). The transcription factors (TFs) have a transmembrane motif which allows them to translocate into the nucleus from the cytoplasm. Another important feature is the presence of DNA-binding motif/domain which enables them to bind with a signature sequence of DNA (*cis*-acting site) (Hobert 2008). Based on DNA-binding domain and other motifs, the transcription factors are classified into different families. Generally, the TFs perform their regulatory function in networks and modulate various pathways. Thus, TFs hold a vital position in functional modulation of cellular pathways by regulating participating proteins (Century et al. 2008). Significant research has been conducted to elucidate the role of various transcription factors and indicated their involvement in various processes including biotic stress resistance in plants (Akio Amorim et al. 2017). Transcription factor families including NAC, MYB, WRKY, ERF/DREB and bZIP have been studied in model plants, and many TFs of these families showed their involvement in disease resistance (Baillo et al. 2019). Certain members of TF families have also been characterized in wheat and showed functional homology (Andersen et al. 2020). Here, we summarized the role of these TF families in biotic stress resistance of wheat.

7.3.1 NAC *Transcription Factors*

NAC (NAM, ATAF and CUC) transcription factors (TFs) belong to the largest TF superfamilies in plants. Bread wheat contains 359 *NAC* genes (Guérin et al. 2019). This number is quite high if compared with other plant species like *Arabidopsis* (117), rice (151) and soybean (152) (Singh et al. 2021). The probable reason of this high number of *NAC* genes in wheat is its hexaploid nature. These 359 *NAC* genes have been reported to locate on all 21 chromosomes (A, B and D genome). Moreover, the *NAC* genes are located in clusters and supposed to have functional homology. Typically, NAC protein consists of DNA-binding domains (BD) and regulatory regions (TR). The BD are subdivided in five sub-domains (A–E). The sub-domains including C and D are considered to be highly conserved, while sub-domains B and E are relatively divergent, thought to be providing functional divergence to the family members (Singh et al. 2021).

Functional studies revealed the involvement of several *NAC* genes in regulation (positive or negative) of disease resistance (Table 7.2, Fig. 7.1). For example, *TaNACL-D1* is induced by virulence factor (deoxynivalenol, DON) of *Fusarium* head blight (FHB) and interacts with *TaFROG*. The complex between TaNACL-D1 and TaFROG produces FHB resistance. The lines overexpressing *TaNACL-D1*

Table 7.2 Transcription factors' modulation disease resistance in wheat

Gene	Trans activity	Function	Possible mechanism	Target gene	Refs.
TaNAC8	Activator	Novel NAC transcription factor gene in wheat, counter stripe rust infection and abiotic stresses	Induced by wounding, pathogen attack, drought and ABA	ATAF1	Xia et al. (2010b)
TaFROG	Activator	*Fusarium* head blight resistance improved	Pleotropic effect	TaSnRKα1, ATQQS	Perochon et al. (2019)
TaNAC069	Activator	Regulating resistance to wheat leaf rust fungus	Fungal elicitor-responsive element	TaCAT, TaSOD, TaCP3	Zhang et al. (2021)
TaNAC30	Repressor	Negatively regulates resistance of wheat to stripe rust	Knockdown of TaNAC30 increases resistance to Pst		Wang et al. (2018)
TaNAC4	Activator	Higher expression of OsNAC4 in hypersensitive response (HR) cell death, against rust	SA, ethylene, MeJA and ABA pathway	TaNAC4, OsNAC19 and GmNAC1	Xia et al. (2010)
TaLHY (1R-MYB)	Activator	Overexpression cause resistance	The SA signalling pathway is different from the ABA and JA/Et treatment pathways		Zijin Zhang et al. (2015)
TaRIM1 (R2R3-MYB)	Activator	Participates in resistance response against the pathogen *Rhizoctonia cerealis* infection through regulating defence genes	Silencing of *TaRIM1* impairs wheat resistance	Transcript of BSMV coat protein	Tianlei Shan and Wei Rong et al. (2016)
TaPIMP1, *AtMYB108*	Activator	Higher levels of AtMYB30 expression in tobacco and *Arabidopsis* hasten the emergence of HR and improve plant resistance to infections from *Cercospora nicotianae* and *Pseudomonas syringae*	Defence gene activation		Liu et al. (2011)

(continued)

Table 7.2 (continued)

Gene	Trans activity	Function	Possible mechanism	Target gene	Refs.
TabZIP1	Activator	This gene is involved in defence response and stress tolerance against infection caused by stripe rust pathogen	Signal transduction pathways that are dependent on ET/MeJA. Wounding and environmental stressors, such as low temperature and high salt, also increased TabZIP1 transcription levels		Zhang, et al. (2008)
Fhb1	Activator	Retrotransposon and gene activation response to mycotoxigenic and non-mycotoxigenic-associated *Fusarium* stress	They may activate undiscovered genes that function to maintain cellular activity in senescing and cold-stressed guard cells by upregulating DON-responsive genes in wheat. Plant immunity is harmed by ABA signalling, which suppresses SA-dependent responses	Lip19	Ansari et al. (2007)
TabZIP74, *AtbZIP60* and *OsbZIP50/OsbZIP74*	Repressor	Silencing TabZIP74 with barley stripe mosaic virus-induced gene silencing (BSMV-VIGS) increased stripe rust susceptibility in wheat seedlings while lowering drought tolerance and lateral roots in silenced plants			Wang et al. (2019)
TaWRKY45	Activator	The TaWRKY45 transgene was overexpressed in a way that imparted increased resistance to *F. graminearum*		Fhb1	Insaf Bahrini et al. (2011)

(continued)

Table 7.2 (continued)

Gene	Trans activity	Function	Possible mechanism	Target gene	Refs.
HvWRKY6, HvWRKY70	Activator	BTH-induced resistance and NPR1-mediated acquired resistance enhance broad-spectrum disease resistance in wheat	Salicylic acid (SA) or its chemical analog benzothiadiazole (BTH) can also stimulate SAR	MLOC_66134 and MLOC_78461	Li et al. (2020)
TaWRKY49 and *TaWRKY62*	Activators or repressors	Confer differential high-temperature seedling-plant resistance to *Puccinia striiformis* f. sp. *tritici*		Lr22, Lr34, Lr37, Lr 24	Wang et al. (2017)
YrAS2388, YrU1	Activator	Stripe rust resistance in wheat is conferred through ankyrin-repeat and WRKY domain-containing immunological receptors	NLR proteins (such as RRS1 and RGA5) start the resistance response by recognising and binding a stripe rust effector via integrated decoy domains		Wang et al. (2020)

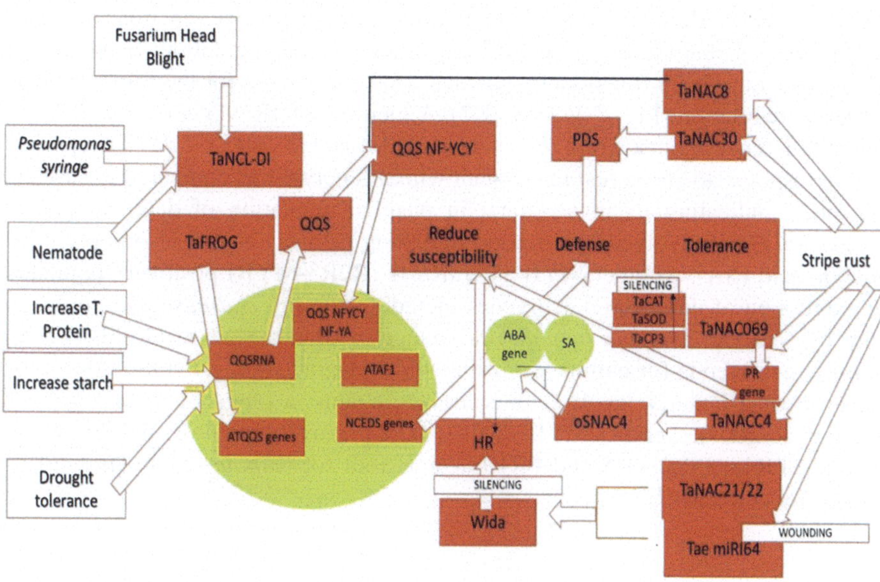

Fig. 7.1 Regulation of disease resistance by NAC transcription factors

showed considerable resistance to FHB (Perochon et al. 2019). Likewise, NAC genes *TaNAC4* and *TaNAC8* are also induced by stripe rust (*Puccinia striiformis* f. sp. *tritici*) infection and regulate the defence mechanism (Xia et al. 2010b). *TaNAC4* is also induced by methyl jasmonate (MeJ), ABA and ethylene, indicating its involvement in ABA-dependent hypersensitive response against stripe rust (Xia et al. 2010a). Another gene, *TaNAC30*, is induced by virulent *Puccinia striiformis* (*Pst*) strains and acts as a negative modulator of leaf rust resistance in wheat by inhibiting the expression of genes involved in pathogenesis and H_2O_2 accumulation (Wang et al. 2018). A NAC gene *TaNAC21/22* is targeted by microRNA (tae-miR164) and acts a negative regulator of leaf rust resistance (Feng et al. 2014). The NAC genes are not only involved in negative regulation of disease resistance. Some NAC genes, such as *TaNAC069*, enhance disease resistance in wheat. The *TaNAC069* is regulated by salicylic acid (SA) signalling and enhances resistance to *Pst* in wheat by inhabiting ROS-scavenging genes or activating pathogenesis-related (PR) genes.

7.3.2 MYB Transcription Factors

The myeloblastosis (MYB) transcription factors constitute a big TF family in plants. The MYB TF members are divided into four subfamilies based on repeat (R) segment of the domain, viz. R1/R2-MYB, R2R3-MYB, R1R2R3-MYB and R1R2R2R1/R2-MYB, containing 1R, 2R, 3R and 4R proteins, respectively (Dubos et al. 2010). Wheat contains 6 R1/R2-MYB, 393 R2R3-MYB and 12 R1R2R3-MYB genes, while little or no information available regarding R1R2R2R1/R2-MYB subfamily (Wei et al. 2020). Members of MYB TF families perform diverse functions in plants ranging from metabolic processes to biotic/abiotic stress regulation. Some wheat MYB genes have been functionally characterized and proved for their involvement in disease regulation (Fig. 7.2; Table 7.2). A member of 1R-MYB family, *TaLHY*, regulates defence against stripe rust in wheat (Zhang et al. 2015). *TaLHY* is upregulated in response to stripe rust in resistant wheat genotypes as compared to susceptible ones, indicating its involvement in positive regulation of defence against disease. *TaLHY* is involved in hormone signalling pathway as it is also induced by salicylic acid (SA). Another MYB TF *TaRIM1* (R2R3-MYB) positively regulates the expression of defence-related genes including *PR* and *chitinase* and produces sharp eyespot (*Rhizoctonia cerealis*) resistance in wheat (Shan et al. 2016). The *chitinase* genes code for chitinase enzyme which degrades fungal chitin. Thus *chitinases* are particularly involved in resistance against a range of fungus species (Singh et al. 2007). Another MYB TF *TaPIMP2* positively regulates the PR genes, including PR1a, PR2, PR5 and PR10 against root rot (*Bipolaris sorokiniana*) in wheat (Wei et al. 2017).

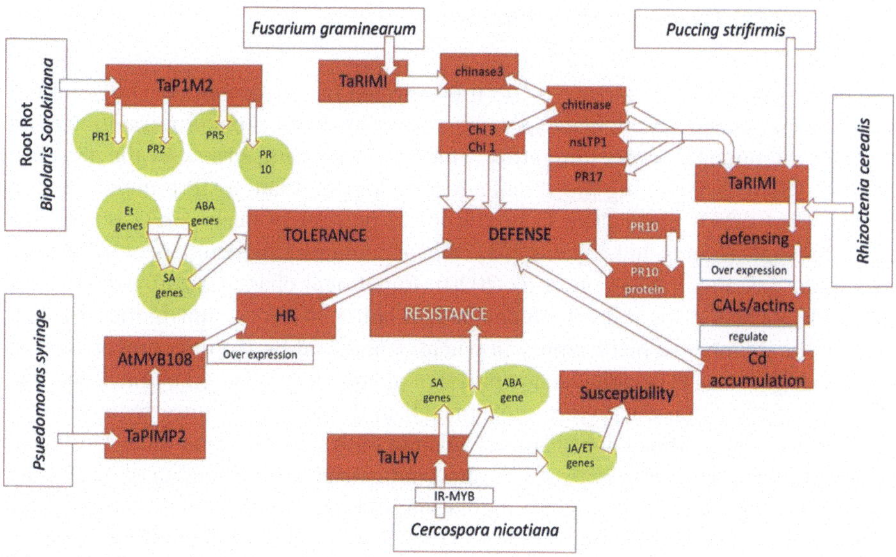

Fig. 7.2 Regulation of disease resistance by MYB transcription factors

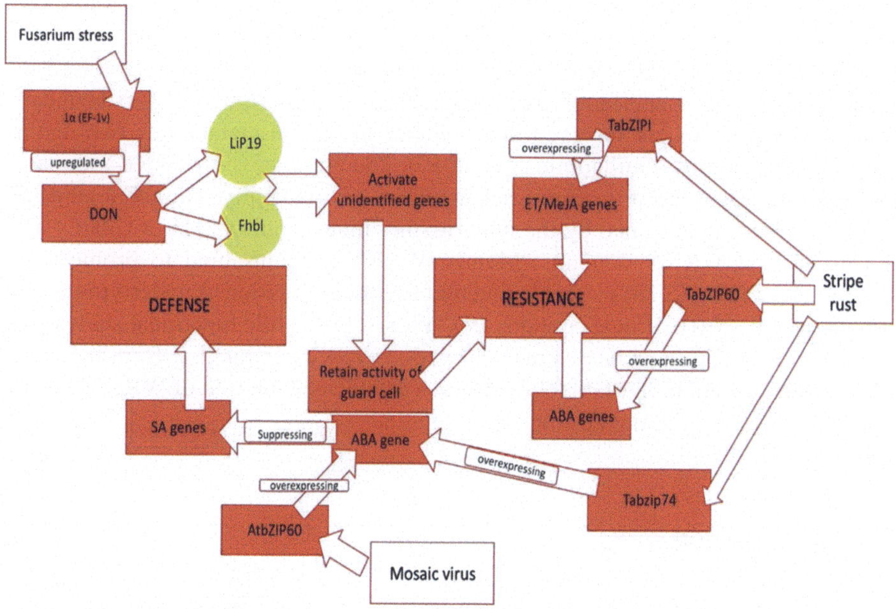

Fig. 7.3 Regulation of disease resistance by bZIP transcription factors

7.3.3 bZIP Transcription Factors

The basic region/leucine zipper (bZIP) transcription factors constitute a large family with numerous members in plant species performing diverse functions ranging from organ development to stress mitigation. This family is characterized by the presence of N-x7-R/K DNA-binding motif and a leucine repeat. Plant bZIP proteins bind to G-box (GACGTG) motif. The G-box motif is present in many stress-responsive genes in plants (Landschulz et al. 1988). Wheat contains 186 members of bZIP family (Table 7.2, Fig. 7.3) (Tian et al. 2020). Functional characterization of several bZIP TFs revealed their involvement in regulation of biotic and biotic stresses. *TabZIP1*, a member of bZIP family in wheat, is induced by *Pst* infection as well as MeJA and ethylene. The enhanced expression of *TabZIP1* is caused by incompatible interaction indication positive role of this TF in diseased resistance regulation by modulating ABA-independent pathway (Zhang et àl. 2008). Some other studies in model plant species (*Arabidopsis*, rice, etc.) indicated the involvement of bZIP gene family in both positive and negative regulations of biotic stress by modulating NPR1 (non-expressor of pathogenesis-related (PR) genes) and SA signalling pathway (Zhang et al. 2018).

7.3.4 WRKY Transcription Factors

The WRKY (single letters that code for four amino acids, i.e. arginine, tryptophan, lysine and tyrosine) superfamily is a large family of transcription factors abundant in plants. These TFs are involved in regulation of several metabolites like lignin, flavanols and tannins (Guillaumie et al. 2010). The presence of the WRKY domain and the zinc-finger motif distinguishes members of this family. The three kinds of WRKY TFs are determined by the type of zinc-finger motif (C2H2 or C2HC) and the number of WRKY domains. Mainly, WRKY domains bind to promoter at W-box – (C/T)TGAC(T/C) – region and regulate the expression of underlying genes (Rushton et al. 2010). Wheat contains 171 WRKY TFs, while functional analysis of some TFs revealed their involvement in various processes ranging from development to stress regulation (Table 7.2). Studies established the role of WRKY TFs in regulation of disease resistance in wheat.

7.4 Conclusion

Diseases are the major yield limiting factors in productivity of wheat. Being a food crop, chemical control of the diseases is not desirable. Thus, a more sustainable solution, i.e. genetic improvement, is sought in this regard. The resistance against diseases is controlled by several genes including transcription factors. Many transcription factors have been evaluated for their role in disease resistance in wheat. The major transcription factor families including NAC, WRKY, bZIP and MYB have been found critical in modulating disease resistance in wheat. Overall, most of these transcription factors are involved in modulation of PR genes or ABA signalling pathway. Overexpression of some TFs resulted in enhanced disease resistance in wheat. Thus, such TFs could be used as functional markers in breeding programs for the improvement of disease resistance.

References

Akio Amorim LL, da Fonseca dos Santos R, Pacifico Bezerra Neto J, Guida-Santos M, Crovella S, Maria Benko-Iseppon A (2017) Transcription factors involved in plant resistance to pathogens. Curr Prot Peptide Sci 18(4):335–351

Andersen EJ, Nepal MP, Purintun JM, Nelson D, Mermigka G, Sarris PF (2020) Wheat disease resistance genes and their diversification through integrated domain fusions. Front Genet 11:898

Ansari KI, Walter S, Brennan JM, Lemmens M, Kessans S, McGahern A, Egan D and Doohan FM, (2007). Retrotransposon and gene activation in wheat in response to mycotoxigenic and non-mycotoxigenic-associated fusarium stress. Theoretical and Applied Genetics 114: 927–937.

Baillo EH, Kimotho RN, Zhang Z, Xu P (2019) Transcription factors associated with abiotic and biotic stress tolerance and their potential for crops improvement. Genes 10(10):771

Branlard G, Dardevet M, Saccomano R, Lagoutte F, Gourdon J (2001) Genetic diversity of wheat storage proteins and bread wheat quality. Euphytica 119(1):59–67

Century K, Reuber TL, Ratcliffe OJ (2008) Regulating the regulators: the future prospects for transcription-factor-based agricultural biotechnology products. Plant Physiol 147(1):20–29

Dubos C, Stracke R, Grotewold E, Weisshaar B, Martin C, Lepiniec L (2010) MYB transcription factors in Arabidopsis. Trends Plant Sci 15(10):573–581

El Baidouri, M., Murat, F., Veyssiere, M., Molinier, M., Flores, R., Burlot, L., ... & Salse, J. (2017). Reconciling the evolutionary origin of bread wheat (Triticum aestivum). New Phytologist, 213(3), 1477–1486.

Feng H, Duan X, Zhang Q, Li X, Wang B, Huang L, Wang X, Kang Z (2014) The target gene of tae-miR164, a novel NAC transcription factor from the NAM subfamily, negatively regulates resistance of wheat to stripe rust. Mol Plant Pathol 15(3):284–296

Guillaumie, S., Mzid, R., Méchin, V., Léon, C., Hichri, I., Destrac-Irvine, A., ... & Lauvergeat, V. (2010). The grapevine transcription factor WRKY2 influences the lignin pathway and xylem development in tobacco. Plant molecular biology, 72(1), 215–234.

Goutam U, Kukreja S, Yadav R, Salaria N, Thakur K, Goyal AK (2015) Recent trends and perspectives of molecular markers against fungal diseases in wheat. Front Microbiol 6:861

Guérin C, Roche J, Allard V, Ravel C, Mouzeyar S, Bouzidi MF (2019) Genome-wide analysis, expansion and expression of the NAC family under drought and heat stresses in bread wheat (T. aestivum L.). PLoS One 14(3):e0213390

Hobert O (2008) Gene regulation by transcription factors and microRNAs. Science 319(5871):1785–1786

Johnson PF, McKnight SL (1989) Eukaryotic transcriptional regulatory proteins. Annu Rev Biochem 58(1):799–839

Lafiandra D, Riccardi G, Shewry PR (2014) Improving cereal grain carbohydrates for diet and health. J Cereal Sci 59(3):312–326

Landschulz WH, Johnson PF, McKnight SL (1988) The leucine zipper: a hypothetical structure common to a new class of DNA binding proteins. Science 240(4860):1759–1764

Lindhout P (2002) The perspectives of polygenic resistance in breeding for durable disease resistance. Euphytica 124(2):217–226

Li H, Wu J, Shang X, Geng M, Gao J, Zhao S, Yu X, Liu D, Kang Z and Wang X (2020). Wrky transcription factors shared by bth-induced resistance and npr1-mediated acquired resistance improve broad-spectrum disease resistance in wheat. Molecular Plant-Microbe Interactions 33: 433–443.

Liu H, Zhou X, Dong N, Liu X, Zhang H and Zhang Z (2011). Expression of a wheat myb gene in transgenic tobacco enhances resistance to ralstonia solanacearum, and to drought and salt stresses. Functional & Integrative Genomics 11: 431–443.

Liu R, Lu J, Zhou M, Zheng S, Liu Z, Zhang C, Du M, Wang M, Li Y, Wu Y (2020) Developing stripe rust resistant wheat (Triticum aestivum L.) lines with gene pyramiding strategy and marker-assisted selection. Genet Resour Crop Evol 67(2):381–391

Miedaner T, Juroszek P (2021) Climate change will influence disease resistance breeding in wheat in Northwestern Europe. Theor Appl Genet:1–15

Pena RJ (2002) Wheat for bread and other foods. In: Curtis BC, Rajaram S, Macpherson HG (eds) Bread wheat—improvement and production. FAO Plant Production and Protection Series, Rome, p 30

Perochon A, Kahla A, Vranić M, Jia J, Malla KB, Craze M, Wallington E, Doohan FM (2019) A wheat NAC interacts with an orphan protein and enhances resistance to Fusarium head blight disease. Plant Biotechnol J 17(10):1892–1904

Rind R, Baloch A, Jatoi W, Asad M, Khokhar A, Nizamani F, Rind M, Nizamani A, Nizamani M (2019) Genetic diversity analysis in Pakistani commercial and landrace genotypes of bread wheat. Asian J Agr Biol 7(2):251–262

Rushton PJ, Somssich IE, Ringler P, Shen QJ (2010) WRKY transcription factors. Trends Plant Sci 15(5):247–258

Saira S, Muhammad K, Rabail A, Muazim A (2019) Multivariate assessment to determine drought tolerant genotypes to combat drought risk in wheat (Triticum aestivum L.). Asian J Agr Biol 7(4):519–530

Shan T, Rong W, Xu H, Du L, Liu X, Zhang Z (2016) The wheat R2R3-MYB transcription factor TaRIM1 participates in resistance response against the pathogen Rhizoctonia cerealis infection through regulating defense genes. Sci Rep 6(1):1–14

Shewry PR, Hey S (2015) Do "ancient" wheat species differ from modern bread wheat in their contents of bioactive components? J Cereal Sci 65:236–243

Singh A, Kirubakaran SI, Sakthivel N (2007) Heterologous expression of new antifungal chitinase from wheat. Protein Expr Purif 56(1):100–109

Singh RP, Singh PK, Rutkoski J, Hodson DP, He X, Jørgensen LN, Hovmøller MS, Huerta-Espino J (2016) Disease impact on wheat yield potential and prospects of genetic control. Annu Rev Phytopathol 54:303–322

Singh S, Koyama H, Bhati KK, Alok A (2021) The biotechnological importance of the plant-specific NAC transcription factor family in crop improvement. J Plant Res:1–21

Stuthman D, Leonard K, Miller-Garvin J (2007) Breeding crops for durable resistance to disease. Adv Agron 95:319–367

Tian F, Yang D-C, Meng Y-Q, Jin J, Gao G (2020) PlantRegMap: charting functional regulatory maps in plants. Nucleic Acids Res 48(D1):D1104–D1113

Wang J, Tao F, An F, Zou Y, Tian W, Chen X, Xu X and Hu X (2017). Wheat transcription factor tawrky70 is positively involved in high-temperature seedling plant resistance to puccinia striiformis f. Sp. Tritici. Molecular plant pathology 18: 649–661.

Wang B, Wei J, Song N, Wang N, Zhao J, Kang Z (2018) A novel wheat NAC transcription factor, TaNAC30, negatively regulates resistance of wheat to stripe rust. J Integr Plant Biol 60(5):432–443

Wang F, Lin R, Li Y, Wang P, Feng J, Chen W and Xu S (2019). Tabzip74 acts as a positive regulator in wheat stripe rust resistance and involves root development by mrna splicing. Frontiers in Plant Science 10: 1551.

Wang H, Zou S, Li Y, Lin F and Tang D (2020). An ankyrin-repeat and wrky-domain-containing immune receptor confers stripe rust resistance in wheat. Nature communications 11: 1–11.

Wei X, Shan T, Hong Y, Xu H, Liu X, Zhang Z (2017) TaPIMP2, a pathogen-induced MYB protein in wheat, contributes to host resistance to common root rot caused by Bipolaris sorokiniana. Sci Rep 7(1):1–15

Wei Q, Chen R, Wei X, Liu Y, Zhao S, Yin X, Xie T (2020) Genome-wide identification of R2R3-MYB family in wheat and functional characteristics of the abiotic stress responsive gene TaMYB344. BMC Genomics 21(1):1–16

Xia N, Zhang G, Liu X-Y, Deng L, Cai G-L, Zhang Y, Wang X-J, Zhao J, Huang L-L and Kang Z-S, (2010). Characterization of a novel wheat nac transcription factor gene involved in defense response against stripe rust pathogen infection and abiotic stresses. Molecular biology reports 37: 3703–3712.

Xia N, Zhang G, Liu XY, Deng L, Cai GL, Zhang Y, Wang XJ, Zhao J, Huang LL, Kang ZS (2010a) Characterization of a novel wheat NAC transcription factor gene involved in defense response against stripe rust pathogen infection and abiotic stresses. Mol Biol Rep 37(8):3703–3712

Xia N, Zhang G, Sun Y-F, Zhu L, Xu L-S, Chen X-M, Liu B, Yu Y-T, Wang X-J, Huang L-L (2010b) TaNAC8, a novel NAC transcription factor gene in wheat, responds to stripe rust pathogen infection and abiotic stresses. Physiol Mol Plant Pathol 74(5–6):394–402

Zhang Y, Zhang G, Xia N, Wang X-J, Huang L-L, Kang Z-S (2008) Cloning and characterization of a bZIP transcription factor gene in wheat and its expression in response to stripe rust pathogen infection and abiotic stresses. Physiol Mol Plant Pathol 73(4–5):88–94

Zhang Z, Chen J, Su Y, Liu H, Chen Y, Luo P, Du X, Wang D, Zhang H (2015) TaLHY, a 1R-MYB transcription factor, plays an important role in disease resistance against stripe rust fungus and ear heading in wheat. PLoS One 10(5):e0127723

Zhang H, Lv S, Wang C, Ji W (2018) The role of transcription factor in wheat defense against pathogen and its prospect in breeding. J Plant Biol Crop Res 1:1005

Zhang Y, Geng H, Cui Z, Wang H and Liu D, (2021). Functional analysis of wheat nac transcription factor, tanac069, in regulating resistance of wheat to leaf rust fungus. Frontiers in plant science 12: 604797.

Zhou, F. 2011. Review of compendium of wheat diseases and pests. J. Agric. Food Inf. 12:210.

Zhu K, Huang S, Peng W, Qian H, Zhou H (2010) Effect of ultrafine grinding on hydration and antioxidant properties of wheat bran dietary fiber. Food Res Int 43(4):943–948

Chapter 8
Potential Transcription Factors for Biotic Stress Tolerance in Sugarcane

Moutoshi Chakraborty, Saurab Kishore Munshi, Tofazzal Islam, and Muhammad J. A. Shiddiky

Abstract Sugarcane is an economically important crop plant under the family of Poaceae. It provides approximately 80% of sugar required for human consumption in the world. It is also used as a raw substance for bioethanol production that is a renewable energy source alternative to hazardous fossil fuels. However, various biotic stresses caused by insects, and fungal, bacterial, and other microbial pathogens may restrict sugar yield from sugarcane to a large extent. Crop plants possess a range of signal transduction and perception networks as a complex defense mechanism in response to these biotic stresses. Especially, transcription factors (TFs), which are triggered by various signal transduction pathways, can potentially improve crop yields by regulating the transcription efficacy of target gene/genes via indirect or direct interaction with *cis*-acting factors. Nevertheless, literature on TFs in stress tolerance in sugarcane is limited though several TFs of various other plants of the Poaceae family have been revealed over the years as important regulators of the responses to various biotic stresses. This chapter provides significant insights into the key TF families like WRKY, NAC, MYB, AP2/ERF, and bZIP that are known to have important functions in gene regulation of plant in response to various biotic stresses, and their potential contribution in the development and improvement of biotic stress-tolerance in sugarcane.

Moutoshi Chakraborty and Saurab Kishore Munshi contributed equally with all other contributors.

M. Chakraborty (✉) · T. Islam
Institute of Biotechnology and Genetic Engineering (IBGE), Bangabandhu Sheikh Mujibur Rahman Agricultural University, Gazipur, Bangladesh

S. K. Munshi
Department of Microbiology, Stamford University Bangladesh, Dhaka, Bangladesh

M. J. A. Shiddiky
School of Environment and Science (ESC) and Queensland Micro- and Nanotechnology Centre (QMNC), Griffith University, Nathan, Australia

© The Author(s), under exclusive license to Springer Nature Switzerland AG 2022
S. H. Wani et al. (eds.), *Transcription Factors for Biotic Stress Tolerance in Plants*, https://doi.org/10.1007/978-3-031-12990-2_8

Keywords Biotic stress · Crop improvement · Gene regulation · *Saccharum* spp. · Transcription factor

8.1 Introduction

Sugarcane (*Saccharum* spp.) is the key provider of the world's sugar, contributing about 80% of sugar production globally (http://www.sucden.com/). It is a C4 crop which belongs to the Poaceae family. It is more photosynthetically capable of producing higher dry mass content, making it an ideal feedstock for green renewable energy. Sugarcane cultivation for commercial purposes began over 2500 years ago in China and India, and it was tamed in Western Europe in the eighteenth century (Grivet et al. 2004). Several *Saccharum* species are found all over the world, such as *Saccharum officinarum, S. barberi, S. edule, S. sinense, S. robustum,* and *S. spontaneum* (Khan et al. 2021). Sugarcane provides the largest crop tonnage, with an approximated ability to reserve sugar approximately 62% dry weight and 25% fresh weight of the stalk, with an average of 40 t/ha dry stalk and trash (Waclawovsky et al. 2010; FAOSTAT 2018). Because of sugarcane's complex genome, polyploid-aneuploid, and heterogeneous nature, the genetic enhancement of contemporary sugarcane varieties is hindered by conventional hybrid breeding (Ali et al. 2019). The increasing global need for energy has heightened interest in producing novel high-yielding sugarcane varieties for usage as the bioenergy feedstock (Khan et al. 2021).

Sugarcane production is continuously threatened by diverse biotic stresses (Goebel and Sallam 2011). These comprise fungal infections by various species like *Colletotrichum falcatum* (red rot disease), *Sporisorium scitamineum* (sugarcane smut), *Ceratocystis paradoxa* (pineapple disease), and *Fusarium* spp., bacterial species like *Xanthomonas albilineans* (sugarcane leaf scald), *Acidovorax avenae* (red stripe), and *Phytoplasma* (sugarcane grassy shoot disease), virus infections by species like yellow leaf virus (ScYLV) and sugarcane mosaic virus (ScMV), and insect attacks by species like *Eldana saccharina* (African sugarcane stalk borer) *Sphenophorus levis* (sugarcane weevil), and *Diatraea saccharalis* (sugarcane stem borer). Biotic stresses are severe environmental risks that affect crop yields significantly (Foyer et al. 2016; Cohen and Leach 2019). Plants have developed complex fast responses to deal with diverse stresses, yet agricultural productivity remains substantially hindered. Biotic stresses, for example, have been shown to cause yield drops of up to 35% (Savary et al. 2012; Savary et al. 2019). Plants have developed multiple stress-response pathways in response to adverse environmental conditions, like cellular signal transduction and perception, triggering expression of particular subsets of defense-related genes, and thus triggering the entire defense response, contributing significantly to the phenotype (Fraire-Velázquez et al. 2011; Baillo et al. 2019).

Transcription factors (TFs) play a crucial role in gene expression regulation in all living entities. They are involved in the development of plant, cell signaling, cell cycle, and several stress responses (Gonzalez 2016). TFs control gene expression by binding to the distal and local *cis*-elements of their target gene, which may be influenced through genomic properties, TF interactions, and DNA structure (Inukai et al. 2017). Around 10% of genes in plants encode TFs that regulate specific signaling-mediated pathways at different stages (Gonzalez 2016). There exist currently several TF databases that offer comprehensive information on different TF families in multiple species of plants (Table 8.1). Main TF families including WRKY, MYB, NAC, AP2/ERF, and bZIP are critical regulators of many genes associated with diverse biotic stresses, making them an ideal candidate for genetic engineering to promote plant tolerance toward various stress stimuli (Table 8.2) (Wang et al. 2016b). Several TFs have been discovered in *Saccharum spp.* (672), *Triticum aestivum* (3437), *Oryza sativa* (2389), and *Hordeum vulgare* (2620) based on plant TF database. Till now, 39 WRKY, 38 MYB, 44 NAC, and 73 AP2/ERF TF gene families have been identified in sugarcane (http://planttfdb.gao-lab.org/).

TF genes have been identified and their responses to different biotic stresses have been studied extensively over the past two decades. Because most of such genes are responsive to stress and regulate a myriad of genes, which are downstream, developing plant stress resistance by modulating TF gene expressions has emerged as a prominent field of research. As a consequence, crops for better stress tolerance could be engineered by genome editing technologies (Hoang et al. 2017b). The overexpression of multiple TF genes has resulted in significant advancement in this field. Several TFs engaged in response to diverse biotic stresses were investigated utilizing contemporary molecular methods like transcriptomics during the last decade in sugarcane (Mustafa et al. 2018). This chapter delineates the regulatory role of TFs under biotic stresses and the expression pattern and role of key TF families in response to multiple biotic stresses in diverse crops with special reference to sugarcane, and finally illustrates the potential features of TFs in mediating biotic stress tolerance in sugarcane.

8.2 Plant TFs' Regulatory Mechanisms Concerning Biotic Stresses

Plant growth and development are affected by various biotic stresses; nevertheless, plants have generated fast response mechanisms to adverse circumstances, which entail interlinked pathways at the molecular scale regulated by signal cascades. Signal transduction, signal perception, and stress-responsive gene expression are fundamental elements of stress responses (Kosová et al. 2015). When a plant cell receives a stress signal, sensors or receptors in the cell membrane identify stress stimuli, which triggers a fast reaction, which converts external signal into intracellular signals. Along with cytoplasmic kinase cascades, signal cascades involving

Table 8.1 Databases of plant transcription factor

Transcription factor databases	Plant species covered	Website
Plant Transcription Factor Database (PlantTFDB)	*Arabidopsis thaliana, Oryza sativa, Saccharum spp., Populus trichocarpa,* etc.	http://planttfdb.gao-lab.org/
Plant Transcription Factor Databases (PlnTFDB)	*Arabidopsis thaliana, Oryza sativa subsp. japonica, Saccharum spp., Populus trichocarpa,* etc.	http://plntfdb.bio.uni-potsdam.de/v3.0/
Grass Transcription Factor Database (GrassTFDB)	*Saccharum* spp., *Oryza sativa, Zea mays. Sorghum bicolor*, and *Brachypodium* spp.	https://grassius.org/grasstfdb.php
Phytozome Database	Many plant species	https://phytozome.jgi.doe.gov/pz/portal.html#
Sorghum Functional Genomic Database	*Sorghum bicolor*	http://structuralbiology.cau.edu.cn/sorghum/index.html
MOROKOSHI Sorghum Transcriptome Database	*Sorghum bicolor*	http://sorghum.riken.jp/morokoshi/Home.html
iTAK-transcription Factor Database	Many plant species	http://itak.feilab.net/cgi-bin/itak/index.cgi
DBD: Transcription factor prediction database	*Arabidopsis thaliana, Zea mays, Oryza sativa, Sorghum bicolor, Vitis vinifera, Populus trichocarpa*, and *Lotus japonicus*	http://www.transcriptionfactor.org
LegumeTFDB	*Medicago truncatula, Lotus japonicas, Glycine max*	http://legumetfdb.psc.riken.jp
PlantTFcat	Many plant species	http://plantgrn.noble.org/PlantTFcat/
The Plant *cis*-Acting Regulatory Element (Plant CARE)	Many plant species	http://bioinformatics.psb.ugent.be/webtools/plantcare/html/
RIKEN *Arabidopsis* Transcription Factor Database (RARTF)	*Arabidopsis thaliana*	http://rarge.gsc.riken.jp/rartf/
PLACE or Plant *cis*-acting Regulatory DNA Elements dDatabase	Vascular plants and *Chlamydomonas reinhardtii*	http://www.dna.affrc.go.jp/PLACE/index.html
ITAK	Many plant species	http://bioinfo.bti.cornell.edu/cgi-bin/itak/index.cgi
The Database of Poplar Transcription Factors (DPTF)	*Populus trichocarpa*	http://dptf.cbi.pku.edu.cn/
AGRIS	*Arabidopsis thaliana*	http://arabidopsis.med.ohio-state.edu
The Database of *Arabidopsis* Transcription Factors (DATF)	*Arabidopsis thaliana*	http://datf.cbi.pku.edu.cn/

(continued)

Table 8.1 (continued)

Transcription factor databases	Plant species covered	Website
The dDatabase of Tobacco Transcription fFactors (TOBFAC)	*Nicotiana tabacum*	http://compsysbio.achs.virginia.edu/tobfac/
The Database of Rice Transcription Factors (DRTF)	*Oryza sativa* L. ssp. *Indica* and *Oryza sativa* L. ssp. *Japonica*	http://drtf.cbi.pku.edu.cn/
Athamap	*Arabidopsis thaliana*	http://www.athamap.de/

Table 8.2 General properties of different plant TF families

Family	*Cis*-acting element	DNA-binding domain	Structural characteristics	References
WRKY	W-box (TTGACT/C)	WRKYGQK	Comprising ~60 amino acid surplus, with a zinc-finger structure in their C-terminus	Eulgem et al. (2000), Chen et al. (2017)
NAC	NACRS (TCNACACGCATGT)	NAC	Comprising 150 amino acid surplus in N-terminal	Jensen et al. (2010), Puranik et al. (2012)
MYB	MYBR (TAACNA/G)	MYB	Comprising several repeats; each repeat has nearly 52 amino acids, creating helix-turn-helix structure	Roy (2016), Mmadi et al. (2017)
AP2/ ERF	GCC box (AGCCGCC) and (TACCGACAT)	AP2/ERF	Comprising 60 amino acids and conserved domain containing putative amphiphilic α-helix and three parallel β-sheets	Sakuma et al. (2002), Song et al. (2013)
bZIP	A-box (TACGTA), C-box (GACGTC), G-box (CACGTG), GLM (GTGAGTCAT), and PB-like (TGAAAA)	bZIP	Comprising ~16 amino acid surplus that consists of nuclear localization signal accompanied by N-x7-R/K motif interacts the DNA	Ali et al. (2016), Agarwal et al. (2019)

intracellular ions or molecules are activated. Calcium ions (Ca2+) and reactive oxygen species (ROS) are engaged with major cascades. Phytohormones like salicylic acid (SA), jasmonic acid (JA), abscisic acid (ABA), and ethylene (ET) are potent second messengers that aid in the coordination of signal transduction systems during stressful situations. Multiple concurrent transduction pathways are activated by these signals, which frequently include protein kinases and phosphatases (Kosová et al. 2015). Plants initiate two main signal cascades after the first stage of signal-perceived notion: the mitogen-activated protein kinase (MAPK) as well as the calcium-dependent protein kinase (CDPK) cascades (Hernandez-Garcia and Finer 2014; Erpen et al. 2018). Ultimately, protein kinases and phosphatases upregulate or downregulate specific TFs, and the transcription factors attach to cis-elements of stress-related genes to either stimulate or inhibit transcription. TFs play a significant role in plants because they form regulatory networks which regulate the transcription of stress-responsive genes through a variety of interactions. Fundamentally, transcription factors (TFs) control the expression of important downstream genes. Recent reports on the key TF families are summarized in the sections below.

8.3 WRKY Transcription Factors

8.3.1 WRKY Genes Family: Classes and Diversity

WRKY is a group of plant transcription factors that has been extensively studied; it governs several metabolic, physiological, and developmental mechanisms in plants (Chen et al. 2017). Although WRKY TF was initially found in *Ipomoea batatas* in the 1990s, WRKY genes were originally believed to be specific TFs to plants (Ishiguro and Nakamura 1994; Baillo et al. 2019). Other eukaryotic species, like fungus, diplomonads, and amoebae, have WRKY proteins in their genome, according to numerous studies. According to distribution pattern studies, ancient genetic modification events were assumed to be the source of such non-plant WRKYs (Rinerson et al. 2015). Several WRKY TFs have been discovered experimentally in a plethora of plant species, like *Arabidopsis* (Chen et al. 2017; Erpen et al. 2018), barley (Erpen et al. 2018), rice (Erpen et al. 2018), wheat (Erpen et al. 2018; Kumar et al. 2018), cotton (Erpen et al. 2018), soybean (Yang et al. 2017), poplar (Erpen et al. 2018), and so on (Table 8.3).

Plant-related WRKY TFs are designated by 60 amino acids in a DNA-binding sequence that is highly conserved (known as WRKY domain) according to the most certified and widely acknowledged classification strategy depending on Arabidopsis genomic characterization (Goyal et al. 2020). Moreover, WRKY domains contain an extremely conserved WRKYGQK motif at N terminus which serves as a protein to protein interaction interface, as well as zinc finger sequence at the C terminus that has an affinity for DNA binding, either C-X_7-C-X_{23}-H-X-C or C-X_{4-5}-C-X_{22-23}-H-X-H (Eulgem et al. 2000). WRKY proteins have been categorized into three classes (Group I, II, and III) depending on the existence of WRKY domains as well as the zinc finger sequence (Brand et al. 2013). A zinc finger motif (C-X4-5-C-X22-23-H-X-H) is identified in two WRKY domains located alongside unusual in Group I. There is only one WRKY domain in Group II and III WRKYs. The finger motif of WRKYs in I and II groups is same, but WRKYs in group III have C-X_7-C-X_{23}-H--X-C motif at C-terminus. According to the zinc-finger motif sequence, Group II

WRKYs are classified into five subgroups (IIa, IIb, IIc, IId, and IIe), whereas Group III WRKYs are categorized into two groups, i.e., IIIb (C-X7-C-Xn-H-X-C, n 24) and IIIa (C-X7-C-X23-H-X-C) (Eulgem et al. 2000). However, in *S. spontaneum*, a new WRKY Group IV has been proposed, as shown by genes with an incomplete region (merely the WRKYGQK motif has been found), indicating that they might have ceased their role as WRKYs (Li et al. 2020a, b). The potential of all WRKY groups to bind precisely to W-box in promoter sequence of target genes and control gene expression has been related to their biological activities (Ciolkowski et al. 2008).

8.3.1.1 Role and Expression Pattern of the WRKY Genes in Response to Biotic Stresses

Plants experience total or partial regulation of various signal transduction networks, including plant hormones under biotic stress, which leads to the activation of various associated transcriptional genes, culminating in a positive reaction toward the adverse environmental condition (Fraire-Velázquez et al. 2011). WRKYs of plant have been implicated in microbe-associated molecular pattern-triggered immunity (PAMP-triggered immunity), the system-acquired resistance (SAR), or effector-triggered immunity (ETI) (Chen et al. 2017). *CsWRKY50*, for instance, has an important role in *Cucumis sativus* infection stress resistance against *Pseudopernospora cubensis* (Luan et al. 2019). Stress resistance response toward downy mildew of *Vitis vinifera* is positively regulated by a JA pathway-related gene *VvWRKY1* (Marchive et al. 2013). *CaWRKY27*, another JA-regulated gene, provides tolerance in tobacco against *Ralstonia solanacearum* infection (Dang et al. 2014). *GhWRKY44*, a key TF in the cotton-pathogen interaction, enhances tobacco tolerance to fungal and bacterial infections. Overexpression of *GhWRKY44* in plants through disease resistance signaling transduction pathways SA and JA showed a reduced amount of ROS accumulation in research (Li et al. 2015). When plants were subjected to pathogen (*Puccinia striiformis* sp. *tritici*) infection, *TaWRKY70* transcription factor increased tolerance significantly via SA and ET driven signal transduction cascades (Wang et al. 2018c). *WsWRKY1* (Singh et al. 2017), *GmWRKY31* (Dong et al. 2019), and *AcWRKYs* (Jing and Liu 2018) have all been shown to have a multifaceted role in pathogen responses in various crop species. ScWRKY3 of sugarcane was persistently expressed in smut-tolerance cultivar but repressed in the smut-susceptible cultivar during the initial phases (0–72 h) of smut pathogen (*Sporisorium scitamineum*) infection. Nevertheless, it has been proposed that in *N. benthamiana* this gene functions as a negative regulator when infected with *Ralstonia solanacearum* or *Fusarium solani* var. *coeruleum* (Wang et al. 2018d). Recently, the majority of *WRKY33* alleles were shown to be highly upregulated toward *Xanthomonas albilineans* infections in sugarcane (Ntambo et al. 2019).

Table 8.3 Function of different TF gene families in response to biotic stresses in plant species

Plants	Stress (disease)	Gene expression		Reference
		Upregulated	Downregulated	
Arabidopsis *(Arabidopsis thaliana)*	*Botrytis cinerea* (gray mold)	*AtERF1*$^{+}$, *AtERF14*$^{+}$	–	Baillo et al. (2019)
	Botryosphaeria dothidea (ripe rot)	*MdERF11*$^{+}$	–	Wang et al. (2020)
	Erysiphe cruciferarum (powdery mildew)	*AtbZIP10*$^{+}$	–	Erpen et al. (2018)
	Fusariuum oxysporum (fusarium wilt)	*AtERF2*$^{+}$	*AtERF4*-	Baillo et al. (2019)
	Pseudomonas syringae (bacterial leaf spot)	*AtWRKY38*-, *AtERF014*$^{+}$, *AtWRKY41*$^{+}$, *AtWRKY62*-, *AtNAC72*$^{+}$, *AtNAC55*†, *AtNAC19*$^{+}$, *AtMYB30*$^{+}$, *AtMYB96*$^{+}$, *AtMTB44*$^{+}$, *CBNAC/NTL9*-, *AtNAC042/JUB1*-, *CabZIP*$^{+}$	*AtWRKY22*$^{+}$, *AtWRKY29*$^{+}$	Segarra et al. (2009), Chen et al. (2017), Baillo et al. (2019), Erpen et al. (2018), Yuan et al. (2019b)
	Heterodera schachtii (cyst nematode)	*AtMYB12*$^{+}$	*AtWRKY6*$^{+}$, *AtWRKY23*$^{+}$, *AtWRKY11*$^{+}$, *AtWRKY17*$^{+}$, *AtWRKY33*$^{+}$	Hamamouch et al. (2020)
	Meloidogyne incognita (root-knot nematodes)	–	*AtMYB12*$^{+}$	Hamamouch et al. (2020)
	Tobacco mosaic virus (TMV)	*AtWRKY8*$^{+}$, *ATAF2*$^{+}$, *AtWRKY61*$^{+}$	–	Chen et al. (2017), Erpen et al. (2018)
	Pieris brassicae (cabbage moth)	*AtMYB75*$^{+}$	–	Shen et al. (2018)
	Myzus persicae (green-peach aphids)	*AtMYB102*-	–	Zhu et al. (2018b)

(continued)

Table 8.3 (continued)

Plants	Stress (disease)	Gene expression		Reference
		Upregulated	Downregulated	
Rice *(Oryza sativa)*	*Magnaporthe oryzae, Pyricularia oryzae* (rice blast)	*OsWRKY22*⁺, *OsWRKY58*⁺, *OsWRKY7*⁺, *OsWRKY62*⁺, *OsWRKY45*⁺, *OsWRKY76*⁺, *OsWRKY64*⁺, *OsNAC6*⁺, *OsNAC66*⁺, *OsNAC122*⁺, *OsNAC19*⁺, *OsNAC131*⁺	–	Tolosa and Zhang (2020), Erpen et al. (2018)
	Xanthomonas oryzae (bacterial blight)	*OsWRKY45*⁺, *OsWRKY6*⁺, *OsWRKY13*⁺, *OsWRKY67*⁺, *OsWRKY71*⁺, *OsNAC66*⁺, *OsNAC58*⁺, *OsEREBP1*⁺	–	Chen et al. (2017), Baillo et al. (2019), Erpen et al. (2018), Yuan et al. (2019b)
	Rice stripe mosaic virus (RSMV)	*OsMYB4*⁺	–	Erpen et al. (2018)
	Rice dwarf virus (RDV)	*OsNAC*⁺	–	Yuan et al. (2019b)
	Rhizoctonia solani (sheath blight)	*OsWRKY80*⁺, *OsWRKY4*⁺	–	Erpen et al. (2018)
	Nilaparvata lugens (brown plant hopper)	–	*OsWRKY45*⁺	Huang et al. (2016a)
	Chilo suppressalis (striped stem borer)	–	*OsWRKY53*⁺	Hu et al. (2016a)
	Diuraphis noxia (Russian wheat aphid)	*OsERF3*⁺	*TaWRKY53*⁺	Van Eck et al. (2014); Lu et al. (2011)
Wheat (*Triticum aestivum*)	*Puccinia triticina* (leaf rust)	*TaWRKY1B*⁺	–	Kumar et al. (2018)

(continued)

Table 8.3 (continued)

Plants	Stress (disease)	Gene expression		Reference
		Upregulated	Downregulated	
	Puccinia striiformis (yellow rust/ stripe rust)	*TaWRKY62*+, *TaWRKY70*+, *TaNAC4*+, *TaNAC1*-, *TaNAC8*+, *TaNAC30*-, *TaNAC21/22*-, *TabZIP74*+	*TaWRKY49*+	Xia et al. (2010), Erpen et al. (2018); Wang et al. (2017), Wang et al. (2018c)
	Bipolaris sorokiniana, Rhizoctonia cerealis (root rot)	*TaPIEP1*+, *TaRIM1*+	*TaERF3*-	Shan et al. (2016), Dong et al. (2010), Baillo et al. (2019)
	Sitobion avenae (English grain aphid)	–	*TaMYB2*+, *TaMYB44*+, *TaMYB19*+	Shen et al. (2018)
	Erysiphe cruciferarum (powdery mildew)	*TaNAC21/22*-, *TaNAC6*+	*TaNAC30*+	Yuan et al. (2019b)
Maize (*Zea mays*)	*Colletotrichum graminicola* (anthracnose)	*ZmNAC41*+, *ZmNAC100*+	–	Voitsik et al. (2013)
	Colletotrichum sublineolum (anthracnose)	*y1MYB*+	–	Ibraheem et al. (2015)
Barley *(Hordium vulgare)*	*Bipolaris sorokiniana* (spot blotch)	*HvMYB6*+	–	Baillo et al. (2019)
	Ralstonia solanacearum (bacterial wilt)	*HvRAF*+	–	Jung et al. (2007)
	Blumeria gramini (powdery mildew)	*HvWRKY19*+, *HvWRKY10*+, *HvWRKY28*+	*HvNAC6*-	Erpen et al. (2018)
Sugarcane *(Saccharum officinarum)*	*Sporisorium scitamineum* (sugarcane smut)	–	*ScWRKY3*-	Wang et al. (2018d)
	Colletotrichum falcatum (red rot)	*SobZIP4*+	*SoNACH*+, *SobZIP15*+	Muthiah et al. (2013)
	Xanthomonas albilineans (leaf scald)	*WRKY33*+	–	Ntambo et al. (2019)
Cotton *(Gossypium hirsutum)*	*Rhizoctonia solani* (sheath blight)	*GhWRKY39-1*+	–	Erpen et al. (2018)

(continued)

Table 8.3 (continued)

Plants	Stress (disease)	Gene expression		Reference
		Upregulated	Downregulated	
Tomato (*Solanum lycopersicum*)	*Rhizopus nigricans* (rhizopus soft rot)	*SlERF1*+	–	Baillo et al. (2019)
	Botrytis cinerea (gray mold)	–	*SlSRN1*-	Yuan et al. (2019b)
	Plectosphearella cucumerina (tomato wilt)	*SlERF1*+	–	Baillo et al. (2019)
	Ralstonia solanacearum (bacterial wilt)	*SlERF3*+, *SlERF5*+, *SlNAC35*+	–	Erpen et al. (2018)
	Xanthomonas campestris (bacterial spot)	*SlERF1*+, *SlNAC35*+	–	Erpen et al. (2018)
	Tomato yellow leaf curl virus (TYLCV)	*SlNAC20*+, *SlNAC47*+, *SlNAC24*+, *SlNAC61*+	–	Huang et al. (2017), Yuan et al. (2019a)
	Meloidogyne javanica (root knot nematode)	*SlWRKY3*+, *SlWRKY45*-,	*SlWRKY70*+	Chinnapandi et al. (2019), Chinnapandi et al. (2017)
Potato (*Solanum tubersum*)	*Phytophthora infestans* (late blight)	*StNAC4*+, *StNAC18*+, *StNAC5*+, *StNAC81*+, *StNAC48*+, *StERF3*-	–	Tolosa and Zhang (2020), Baillo et al. (2019)
Cucumber (*Cucumis sativus*)	*Pseudopernospora cubensis* (downy mildew)	*CsWRKY50*+	–	Luan et al. (2019)
Grapevine (*Vitis vinifera*)	*Botrytis cinerea* (gray mold)	*VvERF20*+	–	Baillo et al. (2019)
	Erysiphe necator (powdery mildew)	*VdMYB1*+	–	Yu et al. (2019)
	Pseudomonas syringae (bacterial leaf spot)	*VvERF20*+	–	Erpen et al. (2018), Chen et al. (2017)
	Plasmopara viticola (downy mildew)	*VvWRKY1*+	–	Marchive et al. (2013)
Cassava (*Manihot esculenta*)	*Xanthomonas axonopodis* (bacterial blight)	*MebZIP3*+, *MebZIP5*+	–	Erpen et al. (2018), Li et al. (2017)

(continued)

Table 8.3 (continued)

Plants	Stress (disease)	Gene expression		Reference
		Upregulated	Downregulated	
Lettuce (*Lactuca sativa*)	*Pseudomonas cichorii* (bacterial midrib rot)	*LsNAC069*+	–	Meisrimler et al. (2019)
Soybean (*Glycine max*)	*Phakospora pachyrhizi* (soybean rust)	*GmbZIP1*+, *GmbZIP62*+, *GmbZIP2*+, *GmbZIP105*+	–	Baillo et al. (2019), Alves et al. (2015)
	Phytophthora sojae (root rot)	*GmERF113*+, *GmERF5*+	–	Baillo et al. (2019)
	Heterodera glycines (soybean cyst nemadtode)	*GmWRKY53*+, *GmWRKY136*+, *GmWRKY86*+	–	Yang et al. (2017)
	Ralstonia solanacearum (bacterial wilt)	*GmERF3*↑	–	Erpen et al. (2018)
Stiff brome (*Brachpodium distachyon*)	*Fusarium graminearum* (fusarium head blight)	*BdWRKY34*+, *BdWRKY8*+, *BdWRKY50*+, *BdWRKY69*+, *BdWRKY70*+	+	Erpen et al. (2018)
Poplar (*Populus trichocarpa*)	*Melampsora medusae* (popular leaf rust)	*PtrWRKY35*+, *PtrWRKY18*+, *PtrWRKY89*+	–	Erpen et al. (2018)
Pepper (*Capsicum annuum*)	*Xanthomonas axonopodis* (bacterial spot)	*CaWRKY58*-	–	Erpen et al. (2018)
	Bacillus thuringiensis (pepper root rot)	*CaPF1*+	–	Chen et al. (2017)
	Ralstonia solanacearum (bacterial wilt)	*CaWRKY6*+, *CaWRKY27*+, *CaPHL8*+	–	Noman et al. (2019), Erpen et al. (2018)
Tobacco (*Nicotiana benthamiana*)	*Fusarium solani* (root rot)	–	*ScWRKY3*−	Wang et al. (2018d)
	Colletotrichum orbicular (anthracnose)	–	*NbWRKY8*+	Erpen et al. (2018)
	Ralstonia solanacearum (bacterial wilt)	*GhWRKY44*+	*ScWRKY3*−	Wang et al. (2018d), Li et al. (2015)
	Rhizoctonia solani (Target spot)	*GhWRKY44*+	–	Li et al. (2015)

(continued)

Table 8.3 (continued)

Plants	Stress (disease)	Gene expression		Reference
		Upregulated	Downregulated	
Tobacco *(Nicotiana tabacum)*	*Ralstonia solanacearum* (bacterial wilt)	*CaWRKY27*⁺, *NtWRKY50*⁺, *CaERF5*⁺	–	Dang et al. (2014), Chen et al. (2017), Lai et al. (2014)
	Xanthomonas axonopodis (bacterial blight)	*MebZIP3*⁺, *MebZIP5*⁺	–	Li et al. (2017)
	Tobacco mosaic virus (TMV)	*NtERF5*⁺, *WRKY8*⁺	–	Erpen et al. (2018), Chen et al. (2017)
Chrysanthemum *(Chrysanthemum* spp.)	*Aphidodea* (aphid)	*CmMYB19*⁺, *CmMYB15*⁺	–	An et al. (2019)

"+" symbol denotes positive function of TFs; "–" symbol denotes negative function of TFs under stress conditions

8.3.2 NAC Transcription Factors

8.3.2.1 NAC Gene Family: Classes and Diversity

NAC is the most important and broadest stress-responsive plant's TF family (Jensen et al. 2010). The first NAC proteins were Ataf1/2 from the *Arabidopsis* and NAM from *Petunia hybrida* (Aida et al. 1997; Sablowski and Meyerowitz 1998). In sugarcane as well as other major crops, several NAC genes were discovered (Table 8.3). NAC TFs are identified by a varied transcriptional regulatory sequence (TR) in the C terminal, as well as an N-terminal loaded with 150–160 amino acids and a DNA-binding NAC domain (Ooka et al. 2003; Olsen et al. 2005). The extremely conserved NAC domain is then subdivided into five distinct subdomains (A–E). Moreover, homo/heterodimer formation, DNA binding, and nuclear localization have been correlated with activities of the NAC domain, whereas the TR region has been linked with transcription regulation as a repressor or activator (Olsen et al. 2005). NAC TFs are classified into two groups depending on their structure: atypical and typical NAC TFs. Typical NAC TFs have a diverging C-region and a NAC domain at the N terminus (Olsen et al. 2005), however atypical NAC TFs include additional motifs/domains in C-terminal areas or C-terminus is absent (Puranik et al. 2012). NTLs are atypical NAC TFs identified by the presence of a transmembrane (TM) motif in C-region (Ernst et al. 2004). Specifically, the TM motif is thought to have a role in plasma membrane anchoring, and it might be released by proteolysis to carry out its function (Kim et al. 2007; Liang et al. 2015).

8.3.2.2 Role and Expression Pattern of the NAC TFs in Response to Biotic Stresses

A plethora of NAC TFs have been shown to have dual activities in plant defense immunity against different pathogens via ETI and hypersensitive responses in a number of studies (Yuan et al. 2019a, b). TaNAC8, wheat TF, plays an important role in protecting plants from stripe rust pathogen invasion (Xia et al. 2010). To ensure maize resistance against *Colletotrichum graminicola*, the JA and SA pathways activated the *ZmNAC41* and *ZmNAC100* genes, respectively (Voitsik et al. 2013). TaNAC2 (Zhang et al. 2018a, b) and TaNAC30 (Wang et al. 2018b), on the other hand, negatively modulated the defense system against biotic stresses. Virus-induced gene silencing (VIGS) study in tomatoes revealed that *SlNAC61* has a positive role toward infection stress mediated by TYLCV (tomato yellow leaf curl virus) (Huang et al. 2017). ONAC131 and ONAC122 TFs play critical roles in disease resistance responses in rice by regulating the expression of signaling and defense-associated genes including *OsLOX*, *OsWRKY45*, *OsNH1*, and *OsPR1a* (Sun et al. 2013).

Two identical NAC TFs (JA2 and JA2L) influenced stomatal closure and reopening in tomatoes in distinct ways during pathogen invasion. Specifically, through regulating the expression of an ABA biosynthetic gene, JA2 enhanced stomatal closure, while JA2L increased JA/COR (JA/coronatine)-mediated stomatal reopening by controlling the transcription of JA metabolism genes (Du et al. 2014). Interestingly, overexpression of NAC4 increased hypersensitive cell death in *Arabidopsis* in reaction to bacterial infections (Lee et al. 2017). There was no difference in sensitivity to *Bremia Lactucae* among the *LsNAC069*-silenced lettuce cultivars; however, there was an increase in resistance to *Pseudomonas cichorii* (Meisrimler et al. 2019). The study of TFs in sugarcane is still in its early stages. Thus, NAC TFs play an essential role in safeguarding plants against various biotic stresses via multiple signal-mediated cascades.

8.3.3 MYB Transcription Factors

8.3.3.1 MYB Gene Family: Classes and Diversity

Eukaryotes have a large and functionally varied protein class known as the MYB family. Protein-protein interactions, DNA binding, and protein regulatory function management are all important functions of this protein TFs family (Roy 2016). Multiple MYB proteins were shown to regulate different cellular mechanisms in diverse crop species, including cell morphogenesis, cell cycle, and stress responses (Ambawat et al. 2013). In *Zea mays*, The MYB gene *colored1* (*c1*), which codes for a MYB-protein domain involved in anthocyanin synthesis in maize seeds' aleurone layer, was found (Paz-Ares et al. 1987). MYB family proteins are divided into four categories depending on the repetition number in their sequences (ranging from

1–4): 1R-MYB (one repeat), 3R-MYB (three repeats), R2R3-MYB (two repeats), and 4R-MYB (four repeats) (Hajiebrahimi et al. 2017). Each repetition is formed of three-helices, each comprising 50–53 amino acids, having 2nd and 3rd helices producing the helix-turn-helix fold (HTH). The HTH fold is made up of 3 tryptophans which are evenly spaced and form a hydrophobic core (Ogata et al. 1996). According to various N- and C-terminal domains, subfamily R2R3-MYB have also been divided into 30–38 groups (Mmadi et al. 2017). MYB TFs have been extensively researched in a wide range of plant species, making them important regulators of biotic stress responses (Table 8.3).

8.3.3.2 Role and Expression Pattern of the MYB TFs in Response to Biotic Stresses

MYB transcription factors have been discovered to have a role in the defense against biotic stresses. In *Arabidopsis*, AtMYB96 serves as a crucial molecular linkage between ABA and SA crosstalk, enhancing pathogen resistance (Seo and Park 2010). Beneficial microbes also induce plant defensive responses, and MYB72 (a root-specific signaling pathway) acts as a convergence node in Arabidopsis (Segarra et al. 2009). AtMYB102 was shown to increase vulnerability to GPA (green-peach aphid) infestation in Arabidopsis (Zhu et al. 2018b). In transgenic wheat, overexpression of TaRIM1 enhanced tolerance against *Rhizoctonia cerealis* infestation (Shan et al. 2016). In maize, 3-deoxyanthocyanidin phytoalexins were produced by a MYB TF *y1* (*yellow seed 1*) of the sorghum against invasion of *Colletotrichum sublineolum* (Ibraheem et al. 2015). Furthermore, activation of MYB TFs in response to the insect attack has been noticed in the chrysanthemum, like the overexpression of *CmMYB15* by lignin formation, which can suppress the growth of aphids (An et al. 2019). Similarly, MdMYB30, an MYB TF that regulates wax biosynthesis in apples, could improve disease resistance (Zhang et al. 2019b). CaPHL8, a new MYB TF, has been shown to promote pepper plant resistance toward *Ralstonia solanacerum* infestation (Noman et al. 2019). In response to *Erysiphe necator* fungal stress in grapevine, VdMYB1, a part of the R2R3-MYB TF, was revealed to be a positive stimulator of the defensive reaction by promoting the expression of *stilbene synthase gene 2* (*VdSTS2*) –(Yu et al. 2019). Overall, MYB TFs have a crucial role in increasing plant tolerance to biotic stresses.

8.3.4 AP2/ERF Transcription Factors

8.3.4.1 AP2/ERF Gene Family: Classes and Diversity

Every domain of the AP2/ERF (ethylene response element-binding factors/APETALA2) family genes has a DNA-binding motif that is highly conserved, comprising 60–70 amino acids in each (Song et al. 2013). Furthermore, the *cis*-element binding is regulated by the presence of aspartate and alanine at positions 9 and 14, respectively (Sakuma et al. 2002). However, AP2/ERF may be categorized into five primary sub-groups depending on the domain number (single or double) found in genes: RAV (related to ABI3/VP1), AP2, DREB (dehydration response element-binding protein), ERF (ethylene responsive factors), and others (Sakuma et al. 2002). To date, AP2/ERF TFs have been identified in various plant species (Table 8.3).

8.3.4.2 Role and Expression Pattern of the AP2/ERF TFs in Response to Biotic Stresses

The modification of disease tolerance in plants has been associated with AP2/ERF transcription factors. Overexpression of *TaPIE1* (Zhu et al. 2014), *OsEREBP1* (Jisha et al. 2015), *Soly106* (Huang et al. 2016b), *OsERF83* (Tezuka et al. 2019), and *GmERF113* (Zhao et al. 2017) genes concerning JA, SA, or ET-related signaling cascades were thought to be beneficial in the fight against pathogens. Under biotic (*R. solanacearum*) stresses, *HvRAF*, a new AP2/ERF TF in barley, has modulatory roles (Jung et al. 2007). In response to *Bipolaris sorokiniana* invasion, TaPIEP1 overexpression was significantly induced in wheat. *TaPIEP1* overexpression in transgenic plants resulted in considerably higher tolerance towards fungal stress (Dong et al. 2010). In response to TYLCV, numerous AP2/ERF TFs, including WRKY and NAC, were differentially expressed in tolerant and susceptible tomato cultivars (Chen et al. 2013). *CaERF5* from peppers proved helpful in preventing transgenic tobacco plants from *R. solanacearum* infestation (Lai et al. 2014). Regarding *Botryosphaeria dothidea*, ectopic expression of *MdERF11* in apple resulted in considerably higher endurance in *Arabidopsis* more recently (Wang et al. 2020). Using several stress-mediated signal transduction mechanisms, AP2/ERF TFs collectively serve a crucial function in biotic stress tolerance.

8.3.5 bZIP Transcription Factors

8.3.5.1 bZIP Gene Family: Classes and Diversity

The basic leucine zipper (bZIP) TFs belong to one of the most varied TF families. The highly conserved bZIP dimerization domain is composed of a basic region and a leucine zipper region that is poorly conserved. The basic domain of bZIP contains a nuclear localization signal and nearly 16 DNA-binding amino acid residues; the leucine zipper domain is required for bZIP's capacity to dimerize (Ali et al. 2016; Agarwal et al. 2019). A-box (TACGTA), G-box (CACGTG), C-box (GACGTC), GLM (GTGAGTCAT), and PB-like (TGAAAA) regions at the *cis*-element of different stress response genes have been found to bind particularly to bZIP proteins in previous studies (Ali et al. 2016). Several bZIP members have been discovered in several plants including sugarcane (Muthiah et al. 2013), soybean (Baillo et al. 2019), *Arabidopsis* (Erpen et al. 2018), wheat (Erpen et al. 2018; Wang et al. 2018c), etc. (Table 8.3). The identification of bZIP TFs in a wider range of species, including watermelon, cassava, and peanuts, has recently been made possible by genome-wide studies (Hu et al. 2016b; Mehmet et al. 2018; Wang et al. 2019b). Multiple biological processes, especially the sensitivity to biotic stresses, are associated with bZIP TFs (Alves et al. 2013; Sornaraj et al. 2016).

8.3.5.2 Role and Expression Pattern of bZIP TFs in Response to Biotic Stresses

bZIP TFs have been associated with plant responses toward pathogen infection in a few studies. Noman et al. (2019) reviewed the known roles of bZIP TFs as negative or/and positive modulators of disease tolerance and characterized plant bZIP TF responsiveness against various pathogens. Alves and co-workers (2013) summarized the response of plants' bZIP TFs against phytopathogens, as well as bZIPs' molecular interaction partners and the signal transduction mechanism following pathogen infestation. They observed that bZIP proteins contribute to the defensive mechanism against Asian soybean rust disease (ASR) by modulating ASR-related genes expression, and identified 4 bZIP genes (*GmbZIP62, GmbZIP2, GmbZIPE1*, and *GmbZIP105*) in soybean. Salicylic acid, hydrogen peroxide, and *Xanthomonas axonopodis* Pv. *manihotis* were found to induce MebZIP5 and MebZIP3 expression in cassava (Li et al. 2017). Furthermore, transgenic tobacco that overexpressed *MebZIP5* and *MebZIP3* showed enhanced resistance to cassava bacterial blight. In transgenic plants, however, silencing *MebZIP5* and *MebZIP3* decreased the transcript levels of defense-related genes, resulting in a disease-susceptible trait. Several bZIP TFs have diverse functions and are linked to different biotic stress conditions. For instance, transgenic *Arabidopsis* plants that overexpressed *CabZIP* of pepper demonstrated improved *Pseudomonas syringae* pv. tomato DC3000 tolerance (Lee et al. 2006). Furthermore, there is a scarcity of research on bZIP TFs involved in plant responses to insects and other biotic stresses. Because of their devastating effect on several major crops, including sugarcane, future research will be required to understand the roles of bZIP TFs in relation to nematodes and related pests.

8.4 Genome Editing Tools in Modulating TFs for Biotic Stress Tolerance

Plant genome editing is now feasible with the utilization of sequence-specific nucleases (SSNs), like ZFNs (zinc finger nucleases), TALENs((transcription activator-like effector nucleases), and CRISPR-CAS9v(clustered regularly interspaced short palindromic repeats (CRISPR)/CRISPR-associated (Cas) 9) (Baltes and Voytas

2015; Weeks et al. 2016). Identifying programmable nucleases which generate double-strand cuts has significantly altered molecular biology for targeted genome editing; ZFNs was pioneered this achievement, with the TALEN expanding genome-modifying capability (Chandrasegaran and Carroll 2016). CRISPR-Cas9 has been acknowledged by researchers around the world for its obvious advantages over ZFN and TALEN (Mao et al. 2013). CRISPR/Cas9 have been shown to be most powerful SSN to date, and it has been utilized to alter the genome of major crops including rice (Miao et al. 2013; Endo et al. 2015; Wei et al. 2021), maize (Zhu et al. 2016; Zhang et al. 2020), wheat (Upadhyay et al. 2013; Shan et al. 2014; Li et al. 2021), sorghum (Jiang et al. 2013; Char et al. 2020), tomato (Ito et al. 2015; Martínez et al. 2020), soybean (Jacobs et al. 2015), potato (Kieu et al. 2021), etc. CRISPR-Cas9 is a cheap, simple, quick, and efficient method for advanced gene screening, knockout of gene, live-cell tagging of chromosomal loci, endogenous gene expression, and ssRNA edition in cells (Khan et al. 2019). CRISPR-Cas9 technology has the potential to discover different biotic stress tolerance genes, as well as provide molecular understanding and genome editing to help crops develop stress tolerance (Khatodia et al. 2016). The utilization of CRISPR-Cas9 to study gene function has resulted in disease models. It is believed that CRISPR-Cas9 would improve our understanding of infection progression and its management by regulating TF genes which react to various biotic stresses (Borrelli et al. 2018; Jaganathan et al. 2018). The loss of functionality of *VvWRKY52* TF gene developed resistance to *Botrytis cinerea* in grape using CRISPR-Cas9 (Wang et al. 2018a). This technique was also utilized to improve resistance to *Phytophthora palmivora* and *P. tropicalis* in papaya and cacao (Fister et al. 2018; Gumtow et al. 2018). In rice, CRISPR-Cas9 effectively targeted the TF gene *OsERF922* toward blast fungus resistance (Wang et al. 2016). CRISPR-Cas9 have been also utilized to disrupt the genomes of several viruses, including TYLCV (tomato yellow leaf curl virus) and TYLCSV (tomato yellow leaf curl Sardinia virus) (Zaidi et al. 2016). sgRNA and FnCas9 have been used to develop RNA viral genome modification technologies for TMV (tobacco mosaic virus) and CMV (cucumber mosaic virus). As a result, sgRNA/FnCas9 expression in *Arabidopsis* and tobacco provided molecular tolerance against the RNA viruses (Zhang et al. 2018a, b). These findings show that CRISPR-Cas9 has enormous potential for improving biotic stress resistance in sugarcane by regulating several TF genes.

8.5 Sugarcane Response to Biotic Stresses: Transcriptomics Research

A vast number of genes involved in governing major biological pathways have been identified as a consequence of multiple transcriptomic studies conducted in recent decades (Augustine et al. 2015; Mustafa et al. 2018). Genes discovered using transcriptomic methods might be utilized as DNA markers or to produce transgenic crops (Li et al. 2016; Mustafa et al. 2018). Transcriptome analysis uses different in silico approaches, such as probe hybridization arrays, ESTs (expressed sequenced tags), or identified genes from related crops, to provide the necessary information about genes. The Brazilian EST repository of sugarcane is among the biggest databases, with over 238,000 ESTs obtained through 26 distinct cDNA libraries generated using the tissue of a range of Brazilian sugarcane cultivars (Ma et al. 2004; Cardoso-Silva et al. 2014). The ESTs were grouped into 43,141 putative distinct transcripts with 16,338 singletons and 26,803 contigs, all collectively known as sugarcane-assembled sequences (Vettore et al. 2003). There are 282,683 ESTs and 499 cDNA sequences in the sugarcane gene index (version 3.0), including 121,342 unigenes. Nevertheless, there are an estimated 10,000 coding genes of sugarcane that have yet to be discovered (Xu et al. 2018). The transcriptomes of 59 F1 individuals (*S. robustum* and *S. officinarum*) were sequenced recently, revealing 8998 and 11,157 single-nucleotide polymorphisms (SNPs), as well as 105 and 83 linkage groups, respectively (Zhang et al. 2019a). However, predicting gene function and utilizing the transcriptome dataset are challenging due to the absence of a complete and accurate sugarcane genome as a reference (Xu et al. 2018). Because of the relatively high similarity (95%) in the genomic sequences of sorghum and sugarcane genomes, the genome of *Sorghum bicolor* as reference is often utilized in transcriptome research of sugarcane (Grivet et al. 1994; Wang et al. 2010). 47% of transcriptome data unigenes of sugarcane have been linked to Sorghum bicolor proteins in the top BLASTx hits, whereas just 2% exhibit significant resemblance to the sugarcane hybrid line R570, demonstrating high genetic variability across sugarcane cultivars (Xu et al. 2018). In eukaryotic transcriptome studies, high-throughput RNA-Seq has been frequently employed (Mutz et al. 2013). On the other hand, short reads generated by second-generation sequencing strategies need relatively extensive computational assemblies therefore unable to cover full-length transcripts, lowering the accuracy of gene model prediction (Wang et al. 2016a). As a result, single-molecule long-read sequencing techniques, such as Pacific Biosciences' long-read isoform sequencing (Iso-Seq), have emerged as a viable alternative for sequencing even more comprehensive transcriptomes and efficiently verifying and forecasting gene models (Wang et al. 2016a). The Iso-Seq method has also been used to study the sugarcane long-read transcriptome (Hoang et al. 2017a; Thirugnanasambandam et al. 2019).

8.6 Fungal Infections

Plants have evolved complex defense mechanisms toward biotic stresses such as diseases and pests. In the context of sugarcane and fungus interactions, Muthiah and co-workers (2013) studied the potential role of transcription factors (TFs) in the regulation of defensive systems against *Colletotrichum falcatum*, the causative agent of sugarcane red rot. Five distinct groups of transcription factors (WRKY, MYB, NAC, and bZIP) have been tested for differential expression in two parallel studies. The differential modulation of 24 transcription factors following phytopathogen exposure and the differential modulation of 15 transcription factors following SAR (systemic acquired resistance) activator stimulation have been observed among the 41 transcription factors studied. Overall, the findings imply that early TF induction may entail actively coordinating or promoting pathogen tolerance. The ESTs produced after infecting two sugarcane-resistant and susceptible cultivars with *C. falcatum* inoculum were studied by Sathyabhama et al. (2016). The improved forward subtraction has been used to estimate the differential expression. Through cloning and sequencing, 136 EST sequences have been assembled in 10 clusters in the final phase of subtraction. These clusters have been discovered to have a role in plant reactive oxygen species signaling, secretion, and defense mechanisms, as well as programmed cell death owing to allergic reactions. Prasanth et al. (2017) produced a huge number of transcript readings (24,732) that predicted about 13,320 genes related to *C. falcatum*. The virulence genes were classified as transition-specific transporters, potential effectors, secondary metabolites, peptidases, and proteases, indicating that *C. falcatum*'s transcript encodes a range of membrane transporters and secondary metabolites. Furthermore, a relative transcriptome analysis of potential secretory effector proteins of *C. falcatum* in sugarcane infection revealed that these anticipated secretory proteins may contribute in the host system's stabilization of fungal secretory proteins during pathogenesis (Prasanth et al. 2019).

Smut, which is caused by *Sporisorium scitamineum*, is among the most devastating fungal infections of sugarcane. Using differential expression data from SSH (suppression subtractive hybridization) databases and quantitative real-time PCR (qRT-PCR), Huang et al. (2018) discovered some key mechanisms as response elements toward *S. scitamineum* infection in sugarcane, including threonine/ serine kinases, mitogen-activated protein genes, Ca2+ sensors, and some NBS-LRR (nucleotide-binding site leucine-rich repeat) genes, and specifically genes related to plant hormone signaling cascades. Few differentially expressed genes (DEGs) associated with the biosynthesis of cell walls, phenylpropanoid cascade, plant hormones signaling pathways, and the disease tolerance genes were also discovered by McNeil et al. (2018).

Another major sugarcane disease is brown rust that is induced by *Puccinia melanocephala*. In sugarcane, potential resistance-linked genes of 11 of the 217 unigenes in the subtractive database were activated in response to this pathogen (Avellaneda et al. 2018). *Fusarium verticillioides* have also been linked to pokkah boeng infection. Lin and co-workers (2016) observed that 1779 transcripts out of 13,999 genes were expressed differentially in *F. verticillioides* cultured with varied nitrogen sources. All of these transcripts were engaged in the transport, assimilation, and metabolism of nitrogen. Multiple TFs were associated to the usage of nitrogen in different biological activities, whereas several genes were found to be linked with pathogenicity. In susceptible and resistant cultivars inoculated with *F. verticillioides*, Wang and co-workers (2019a) revealed that main DEGs important for tolerance were strongly connected to nitrogenous metabolism, cutin, phenylpropanoid, suberine, and wax formation, as well as plant-pathogen interactions.

8.7 Bacterial Diseases

In response to the red stripe causal pathogen *Acidovorax avenae* subsp. *avenae* infection, Santa Brigida et al. (2016) revealed 467 DEPs and several metabolic pathways in sugarcane. Genes in the pattern recognition receptors (PRRs), jasmonate (JA) and ethylene (ET) biosynthesis, NBS-LRR genes, SAR triggered genes, oxidative burst genes, fortification genes of cell membrane, and pathogenesis-related genes (PR) were all upregulated, according to differential study.

Another prominent bacterial sugarcane disease is ratoon stunting disease, which is induced by *Leifsonia xyli* subsp. *xyli* (Lxx). Sugarcane infected with Lxx caused changes in gibberellic acid (GA3), auxin (IAA), and abscisic acid (ABA) synthesis, according to Zhang et al. (2016). In comparison to the control, Lxx-infected sugarcane plants had reduced plant height, single stalk weight, stalk diameter, and water potential, but amino acid content and membrane permeability were higher. Additionally, in response to Lxx infection, the expression of phenylalanine ammonia-lyase (PAL), zinc finger protein (ZFP), and NBS-LRR genes were upregulated. Following that, Zhu et al. (2019) investigated the function of the membrane protein gene Lxx18460 (anti-sigma K) that was transferred into *N. tabacum*, hypothesizing that the Lxx 18,460 has a detrimental impact on the growth and development of tobacco by lowering photosynthesis, damaging defense enzyme function, and influencing endogenous hormone levels. Cia et al. (2018) found that sugarcane infected by Lxx affects 150 proteins and 267 DEGs associated with the development of plant, signal transduction, biosynthesis of hormones, and defense mechanisms.

8.8 Viral Diseases

Depending on RNA-seq data, a few studies have depicted sugarcane-virus interactions. Sugarcane mosaic disease is caused by two major viruses in China: sugarcane steak mosaic virus (SCSMV) and sorghum mosaic virus (Luo et al. 2016). Dong et al. (2017) found that 50 DEGs were downregulated and 3791 DEGs were upregulated, and the three major KEGG processes, proteasome, ubiquitin proteolytic, and translational processes in the endoplasmic reticulum of SCSMV-infected sugarcane cultivars. In addition, from the RNA-seq data, Ling et al. (2018) have discovered 481 DEGs as well as 51 homologous regions of the potyvirus host interactor (PHI) genes, suggesting that defense-associated genes, reactive oxygen species (ROS), endoplasmic reticulum, phytohormone signaling, an ethylene-inducible TF gene, and a calmodulin-related protein gene were linked to the regulations of sugarcane response toward SrMV infection. These findings might aid in understanding the molecular pathways behind the interaction between sugarcane and viruses.

8.9 TFs' Ductility and Flexibility in Carrying Out Their Roles

Transcription factors are highly versatile proteins in nature, and this characteristic is crucial for their regulatory role. In eukaryotic species, extended areas of disordered/ductile residues are predicted to exist in 83–94% of TFs. The overall TFs number, the amount of spliced variants, and the entire aberrant residue contents are all found to have a positive and significant association with organismic complexity. The families of TFs involved in the cell size, cell differentiation, cell cycle, and proliferation of cells are more versatile having more disordered residues. These findings show that increasing TFs is essential for organismic complexity development (Yruela et al. 2017). By functioning as protein chaperones or by conserving other cellular components and structures, TF ductility assists plants in dealing with a range of biotic stress reactions. To effectively adapt to environmental changes, TFs have complicated and flexible networks. TF dysregulation is crucial in plants because it provides them with a quick strategy for developing interconnected, complex, and flexible molecular pathways (Yruela 2015).

8.10 Future Perspectives and Concluding Remarks

Since the global population is forecasted to surpass 9 billion by 2050, advanced technologies for the enhancement of stress endurance in crop plants are essential for satisfying the world's projected food and energy demands (World Population Prospects 2013). The TFs can be modified as important stress mediators to improve crops' tolerance to different biotic stresses. TFs have crucial roles in transcriptional regulation, either inhibiting or activating genes in response to several stresses. TFs regulate genes at transcriptional level, accounting for about 7% of the coding ability of the vascular plant genome (Rushton et al. 2008). Thousands of transcription factors (TFs) have been discovered in plants. Important TF families (MYB, WRKY, NAC, AP2/ERF, bZIP) have been used to deal with biotic stresses in a variety of crops via various signal transduction mechanisms over the last two decades. To maintain food security, however, a more comprehensive field study is required to uncover the mechanisms of the TF genes for developing stress-resistant, high-yielding crops. TF responses toward stress may be highly complex, as evidenced by the literature. A single gene can be regulated by numerous TFs attaching to its *cis*-elements, and a single TF can respond to a range of stresses; thus overexpression of a single TF gene can activate or repress a huge number of downstream genes (Inukai et al. 2017). Because of its complicated polyploid genome, AP2/ERF and MYB TFs have yet to be found in *Saccharum* species on a genome-wide scale to date, except from other gene families (Li et al. 2020a, b).

Individual molecular characterization of millions of TF genes might be a huge challenge. As a result, rather than studying a single TF and stress, to understand the cross-talk across many TFs, future studies might employ combinatorial methodologies to investigate several TFs and diverse stresses. The availability of whole-genome sequences for a growing number of species, as well as advancements in sequencing technology, has facilitated the discovery and study of TFs. Furthermore, publicly accessible genomic databases may enable in silico analysis of genome-wide annotation data in order to identify transcription factors. Next-generation sequencing (NGS) and chromatin immunoprecipitation with highly parallel sequencing (CHIP-Seq) have substantially assisted genome identification strategies. Moreover, plant epigenetics, a conserved regulatory cascade in gene expression, encompasses histone modification, DNA methylation, non-coding RNA, chromatin remodeling, and other processes; and it is a sophisticated technique for thoroughly understanding the biological systems involved in sugarcane environmental responses.

CRISPR/Cas9 is a revolutionary toolkit for modifying the genome of any plants for promoting stress resistance and the translational and transcriptional regulation of the genes (Haque et al. 2018; Islam 2019; Bao et al. 2019; Oz et al. 2021). This fast-evolving technology is becoming a user-friendly tool for also editing TF regulators, which could be utilized in sugarcane for producing stress-tolerant variety (Molla et al. 2020). The functional redundancy of TF genes will need to be addressed in future research. Furthermore, while initial studies of gene overexpression of TF in response to particular stress have been exceptionally beneficial, research relying on crop yield is required. So, future studies should need to determine whether stress-related TF gene overexpression in the transgenic plants increases growth and stress resistance and whether it has a detrimental impact on-field production. Many research has been conducted to verify and characterize the function of TFs in various stress responses; However, the molecular pathways of several transcription factors remain unclear. Therefore, more study is needed for precise understanding of the molecular roles of these transcription factors.

References

Agarwal P, Baranwal VK, Khurana P (2019) Genome-wide analysis of bZIP transcription factors in wheat and functional characterization of a TabZIP under abiotic stress. Sci Rep 9:1–18

Aida M, Ishida T, Fukaki H, Fujisawa H, Tasaka M (1997) Genes involved in organ separation in *Arabidopsis*: An analysis of the cup-shaped cotyledon mutant. Plant Cell 9:841–857

Ali Z, Sarwat SS, Karim I, Jaskani MJ, Khan AA (2016) Functions of plant's bZIP transcription factors. Pak J Agric Sci 53:303–314

Ali A, Khan M, Sharif R, Mujtaba M, Gao S-J (2019) Sugarcane omics: An update on the current status of research and crop improvement. Plan Theory 8:344

Alves MS, Dadalto SP, Gonçalves AB, De Souza GB, Barros VA, Fietto LG (2013) Plant bZIP transcription factors responsive to pathogens: a review. Int J Mol Sci 14:7815–7828

Alves MS, Soares ZG, Vidigal PMP, Barros EG, Poddanosqui AMP, Aoyagi LN, Abdelnoor RV, Marcelino-Gumaraes FC, Fietto LG (2015) Differential expression of four soybean bZIP genes during *Phakopsora pachyrhizi* infection. Funct Integr Genomic 15:685–696

Ambawat S, Sharma P, Yadav NR, Yadav RC (2013) MYB transcription factor genes as regulators for plant responses: An overview. Physiol Mol Biol Plants 19:307–321

An C, Sheng L, Du X, Wang Y, Zhang Y, Song A, Jiang J, Guan Z, Fang W, Chen F (2019) Overexpression of *CmMYB15* provides chrysanthemum resistance to aphids by regulating the biosynthesis of lignin. Hortic Res 6:1–10

Augustine SM, Narayan JA, Syamaladevi DP, Appunu C, Chakravarthi M, Ravichandran V, Tuteja N, Subramonian N (2015) Overexpression of *EaDREB2* and pyramiding of *EaDREB2* with the pea DNA helicase gene (*PDH45*) enhance drought and salinity tolerance in sugarcane (*Saccharum* spp. hybrid). Plant Cell Rep 34:247–263

Avellaneda MC, Parco AP, Hoy JW, Baisakh N (2018) Putative resistance-associated genes induced in sugarcane in response to the brown rust fungus, *Puccinia melanocephala* and their use in genetic diversity analysis of Louisiana sugarcane clones. Plant Gene 14:20–28

Baillo EH, Kimotho RN, Zhang Z, Xu P (2019) Transcription factors associated with abiotic and biotic stress tolerance and their potential for crops improvement. Genes 10(10):771

Baltes NJ, Voytas DF (2015) Enabling plant synthetic biology through genome engineering. Trends Biotechnol 33(2):120–131

Bao A, Burritt DJ, Chen H, Zhou X, Cao D, Tran LSP (2019) The CRISPR/Cas9 system and its applications in crop genome editing. Crit Rev Biotechnol 39:321–336

Borrelli VMG, Brambilla V, Rogowsky P, Marocco A, Lanubile A (2018) The enhancement of plant disease resistance using CRISPR/Cas9 technology. Front Plant Sci 9:1245

Brand LH, Fischer NM, Harter K, Kohlbacher O, Wanke D (2013) Elucidating the evolutionary conserved DNA-binding specificities of WRKY transcription factors by molecular dynamics and in vitro binding assays. Nucleic Acids Res 41:9764–9778

Cardoso-Silva CB, Costa EA, Mancini MC, Balsalobre TWA, Canesin LEC, Pinto LR, Carneiro MS, Garcia AAF, de Souza AP, Vicentini R (2014) De novo assembly and transcriptome analysis of contrasting sugarcane varieties. PLoS One 9:e88462

Chandrasegaran S, Carroll D (2016) Origins of programmable nucleases for genome engineering. J Mol Biol 428:963

Char SN, Wei J, Mu Q, Li X, Zhang ZJ, Jianming Y, Yang B (2020) An agrobacterium-delivered CRISPR/Cas9 system for targeted mutagenesis in sorghum. Plant Biotechnol J 18(2):319–321

Chen T, Lv Y, Zhao T, Li N, Yang Y, Yu W, He X, Liu T, Zhang B (2013) Comparative transcriptome profiling of a resistant vs. susceptible tomato (*Solanum lycopersicum*) cultivar in response to infection by tomato yellow leaf curl virus. PLoS One 8:1–12

Chen F, Hu Y, Vannozzi A, Wu K, Cai H, Qin Y, Mullis A, Lin Z, Zhang L (2017) The WRKY transcription factor family in model plants and crops. Crit Rev Plant Sci 36:311–335

Chinnapandi B, Bucki P, Braun Miyara S (2017) SlWRKY45, nematode-responsive tomato WRKY gene, enhances susceptibility to the root knot nematode; *M javanica* infection. Plant Signaling Behav 12(12):e1356530

Chinnapandi B, Bucki P, Fitoussi N, Kolomiets M, Borrego E, Braun Miyara S (2019) Tomato SlWRKY3 acts as a positive regulator for resistance against the root-knot nematode *Meloidogyne javanica* by activating lipids and hormone mediated defense-signaling pathways. Plant Signal Behav 14(6):1601951

Cia MC, de Carvalho G, Azevedo RA, Monteiro-Vitorello CB, Souza GM, Nishiyama-Junior MY, Lembke CG, Antunes de Faria RS, Marques JPR, Melotto M (2018) Novel insights into the early stages of ratoon stunting disease of sugarcane inferred from transcript and protein analysis. Phytopathology 108:1455–1466

Ciolkowski I, Wanke D, Birkenbihl RP, Somssich IE (2008) Studies on DNA-binding selectivity of WRKY transcription factors lend structural clues into WRKY-domain function. Plant Mol Biol 68:81–92

Cohen SP, Leach JE (2019) Abiotic and biotic stresses induce a core transcriptome response in rice. Sci Rep 9:6273

Dang F, Wang Y, She J, Lei Y, Liu Z, Eulgem T, Lai Y, Lin J, Yu L, Lei D (2014) Overexpression of *CaWRKY27*, a subgroup IIe WRKY transcription factor of *Capsicum annuum*, positively regulates tobacco resistance to *Ralstonia solanacearum* infection. Physiologia Plant 150:397–411

Dong N, Liu X, Lu Y, Du L, Xu H, Liu H, Xin Z, Zhang Z (2010) Overexpression of *TaPIEP1*, a pathogen-induced ERF gene of wheat, confers host-enhanced resistance to fungal pathogen *Bipolaris sorokiniana*. Funct Integr Genomic 10:215–226

Dong M, Cheng G, Peng L, Xu Q, Yang Y, Xu J (2017) Transcriptome analysis of sugarcane response to the infection by Sugarcane steak mosaic virus (SCSMV). Trop Plant Biol 10:45–55

Dong H, Tan J, Li M, Yu Y, Jia S, Zhang C, Wu Y, Liu Y (2019) Transcriptome analysis of soybean WRKY TFs in response to *Peronospora manshurica* infection. Genomics 111:1412–1422

Du M, Zhai Q, Deng L, Li S, Li H, Yan L, Huang Z, Wang B, Jiang H, Huang T (2014) Closely related NAC transcription factors of tomato differentially regulate stomatal closure and reopening during pathogen attack. Plant Cell 26:3167–3184

Endo M, Mikami M, Toki S (2015) Multigene knockout utilizing off-target mutations of the CRISPR/Cas9 system in rice. Plant Cell Physiol 56(1):41–47

Ernst HA, Olsen AN, Skriver K, Larsen S, Leggio LL (2004) Structure of the conserved domain of ANAC, a member of the NAC family of transcription factors. EMBO Rep 5:297–303

Erpen L, Devi HS, Grosser JW, Dutt M (2018) Potential use of the DREB/ERF, MYB, NAC and WRKY transcription factors to improve abiotic and biotic stress in transgenic plants. Plant Cell Tissue Organ Culture (PCTOC) 132:1–25

Eulgem T, Rushton PJ, Robatzek S, Somssich IE (2000) The WRKY superfamily of plant transcription factors. Trends Plant Sci 5:199–206

FAOSTAT (2018) Food and agriculture data. Available at: http://faostat.fao.org

Fister AS, Landherr L, Maximova SN, Guiltinan MJ (2018) Transient expression of CRISPR/Cas9 machinery targeting TcNPR3 enhances defense response in *Theobroma cacao*. Front Plant Sci 9:268

Foyer CH, Rasool B, Davey JW, Hancock RD (2016) Cross-tolerance to biotic and abiotic stresses in plants: a focus on resistance to aphid infestation. J Exp Bot 67:2025–2037

Fraire-Velázquez S, Rodríguez-Guerra R, Sánchez-Calderón L (2011) Abiotic and biotic stress response crosstalk in plants. In: Abiotic stress response in plants—physiological, biochemical and genetic perspectives. IntechOpen, London, pp 3–26

Goebel F-R, Sallam N (2011) New pest threats for sugarcane in the new bioeconomy and how to manage them. Curr Opinion Environ Sustain 3(1–2):81–89

Gonzalez DH (2016) Introduction to transcription factor structure and function. In: Plant transcription factors. Elsevier, Amsterdam, pp 3–11

Goyal P, Manzoor MM, Vishwakarma RA, Sharma D, Dhar MK, Gupta S (2020) A comprehensive transcriptome-wide identification and screening of WRKY gene family engaged in abiotic stress in *Glycyrrhiza glabra*. Sci Rep 10:1–18

Grivet L, D'Hont A, Dufour P, Hamon P, Roques D, Glaszmann J-C (1994) Comparative genome mapping of sugar cane with other species within the Andropogoneae tribe. Heredity 73:500

Grivet L, Daniels C, Glaszmann JC, D'Hont A (2004) A review of recent molecular genetics evidence for sugarcane evolution and domestication. Ethnobotany Res Appl 2:9–17

Gumtow R, Wu D, Uchida J, Tian M (2018) A *Phytophthora palmivora* extracellular cystatin-like protease inhibitor targets papain to contribute to virulence on papaya. Mol Plant Microbe 31:363–373

Hajiebrahimi A, Owji H, Hemmati S (2017) Genome-wide identification, functional prediction, and evolutionary analysis of the R2R3-MYB superfamily in *Brassica napus*. Genome 60:797–814

Hamamouch N, Winkel BS, Li C, Davis EL (2020) Modulation of *Arabidopsis* flavonol biosynthesis genes by cyst and root-knot nematodes. Plan Theory 9:253

Haque E, Taniguchi H, Hassan MM, Bhowmik P, Karim MR, Śmiech M, Zhao K, Rahman M, Islam T (2018) Application of CRISPR/Cas9 genome editing technology for the improvement of crops cultivated in tropical climates: recent progress, prospects, and challenges. Front Plant Sci 9:617

Hernandez-Garcia CM, Finer JJ (2014) Identification and validation of promoters and *cis*-acting regulatory elements. Plant Sci 217:109–119

Hoang NV, Furtado A, Mason PJ, Marquardt A, Kasirajan L, Thirugnanasambandam PP, Botha FC, Henry RJ (2017a) A survey of the complex transcriptome from the highly polyploid sugarcane genome using full-length isoform sequencing and de novo assembly from short read sequencing. BMC Genomics 18:395

Hoang XLT, Nhi DNH, Thu NBA, Thao NP, Tran L-SP (2017b) Transcription factors and their roles in signal transduction in plants under abiotic stresses. Curr Genomics 18:483–497

Hu L, Ye M, Li R, Lou Y (2016a) *OsWRKY53*, a versatile switch in regulating herbivore-induced defense responses in rice. Plant Signal Behav 11(4):e1169357

Hu W, Yang H, Yan Y, Wei Y, Tie W, Ding Z, Zuo J (2016b) Genome-wide characterization and analysis of bZIP transcription factor gene family related to abiotic stress in cassava. Nature 7:22783

Huang Y, Li M-Y, Wu P, Xu Z-S, Que F, Wang F, Xiong A-S (2016a) Members of WRKY group III transcription factors are important in TYLCV defense signaling pathway in tomato (*Solanum lycopersicum*). BMC Genomics 17:788

Huang Y, Zhang B-L, Sun S, Xing G-M, Wang F, Li M-Y, Tian Y-S, Xiong A-S (2016b) AP2/ERF transcription factors involved in response to tomato yellow leaf curly virus in tomato. Plant Genom 9:1–15

Huang Y, Li T, Xu Z-S, Wang F, Xiong A-S (2017) Six NAC transcription factors involved in response to TYLCV infection in resistant and susceptible tomato cultivars. Plant Physiol Biochem 120:61–74

Huang N, Ling H, Su Y, Liu F, Xu L, Su W, Wu Q, Guo J, Gao S, Que Y (2018) Transcriptional analysis identifies major pathways as response components to *Sporisorium scitamineum* stress in sugarcane. Gene 678:207–218

Ibraheem F, Gaoor I, Tan Q, Shyu C-R, Chopra S (2015) A sorghum MYB transcription factor induces 3-deoxyanthocyanidins and enhances resistance against leaf blights in maize. Molecules 20:2388–2404

Inukai S, Kock KH, Bulyk ML (2017) Transcription factor–DNA binding: beyond binding site motifs. Curr Opin Genet Dev 43:110–119

Ishiguro S, Nakamura K (1994) Characterization of a cDNA encoding a novel DNA-binding protein, SPF1, that recognizes SP8 sequences in the 50 upstream regions of genes coding for sporamin and β-amylase from sweet potato. Mol Gen Genet 244:563–571

Islam T (2019) CRISPR-Cas technology in modifying food crops. CAB Rev 14:1–16

Ito Y, Nishizawa-Yokoi A, Endo M, Mikami M, Toki S (2015) CRISPR/Cas9-mediated mutagenesis of the RIN locus that regulates tomato fruit ripening. Biochem Biophys Res Commun 467(1):76–82

Jacobs TB, LaFayette PR, Schmitz RJ, Parrott WA (2015) Targeted genome modifications in soybean with CRISPR/Cas9. BMC Biotechnol 15:16

Jaganathan D, Ramasamy K, Sellamuthu G, Jayabalan S, Venkataraman G (2018) CRISPR for crop improvement: An update review. Front Plant Sci 9:985

Jensen MK, Kjaersgaard T, Nielsen MM, Galberg P, Petersen K, O'shea C, Skriver K (2010) The Arabidopsis thaliana NAC transcription factor family: structure–function relationships and determinants of ANAC019 stress signalling. Biochem J 426:183–196

Jiang W, Zhou H, Bi H, Fromm M, Yang B, Weeks DP (2013) Demonstration of CRISPR/Cas9/sgRNA-mediated targeted gene modification in Arabidopsis, tobacco, sorghum and rice. Nucleic Acids Res 41(20):e188

Jing Z, Liu Z (2018) Genome-wide identification of WRKY transcription factors in kiwifruit (*Actinidia* spp.) and analysis of WRKY expression in responses to biotic and abiotic stresses. Genes Genom 40:429–446

Jisha V, Dampanaboina L, Vadassery J, Mithöfer A, Kappara S, Ramanan R (2015) Overexpression of an AP2/ERF type transcription factor *OsEREBP1* confers biotic and abiotic stress tolerance in rice. PLoS One 10:1–24

Jung J, Won SY, Suh SC, Kim H, Wing R, Jeong Y, Hwang I, Kim M (2007) The barley ERF-type transcription factor *HvRAF* confers enhanced pathogen resistance and salt tolerance in Arabidopsis. Planta 225:575–588

Khan N, Bano A, Rahman MA, Rathinasabapathi B, Babar MA (2019) UPLC-HRMS-based untargeted metabolic profiling reveals changes in chickpea (*Cicer arietinum*) metabolome following long-term drought stress. Plant Cell Environ 42:115

Khan MS, Mustafa G, Joyia FA, Mirza SA (2021) Sugarcane as future bioenergy crop: potential genetic and genomic approaches, sugarcane-biotechnology for biofuels, Khan MS, IntechOpen. Available from: https://www.intechopen.com/chapters/76581

Khatodia S, Bhatotia K, Passricha N, Khurana SMP, Tuteja N (2016) The CRISPR/Cas genome-editing tool: application in improvement of crops. Front Plant Sci 7:506

Kieu NP, Lenman M, Wang ES, Petersen BL, Andreasson E (2021) Mutations introduced in susceptibility genes through CRISPR/Cas9 genome editing confers increased late blight resistance in potatoes. Sci Rep 11:4487

Kim S-Y, Kim S-G, Kim Y-S, Seo PJ, Bae M, Yoon H-K, Park C-M (2007) Exploring membrane-associated NAC transcription factors in *Arabidopsis*: implications for membrane biology in genome regulation. Nucleic Acids Res 35:203–213

Kosová K, Vítámvás P, Urban MO, Klíma M, Roy A, Tom Prášil I (2015) Biological networks underlying abiotic stress tolerance in temperate crops-a proteomic perspective. Int J Mol Sci 16:20913–20942

Kumar A, Jaiswal JP, Sharma N, Gupta S, Kumar A (2018) Understanding the molecular basis of differential grain protein accumulation in wheat (*Triticum aestivum* L) through expression profiling of transcription factors related to seed nutrients storage 3. Biotech 8:112

Lai Y, Dang F, Lin J, Yu L, Lin J, Lei Y, Chen C, Liu Z, Qiu A, Mou S (2014) Overexpression of a pepper CaERF5 gene in tobacco plants enhances resistance to *Ralstonia solanacearum* infection. Funct Plant Biol 41:758–767

Lee SC, Choi HW, Hwang IS, Choi DS, Hwang BK (2006) Functional roles of the pepper pathogen-induced bZIP transcription factor, CAbZIP1, in enhanced resistance to pathogen infection and environmental stresses. Planta 224:1209–1225

Lee MH, Jeon HS, Kim HG, Park OK (2017) An Arabidopsis NAC transcription factor NAC4 promotes pathogen-induced cell death under negative regulation by microRNA164. New Phytol 214:343–360

Li J, Wang J, Wang N, Guo X, Gao Z (2015) *GhWRKY44*, a WRKY transcription factor of cotton, mediates defense responses to pathogen infection in transgenic *Nicotiana benthamiana*. Plant Cell Tissue Organ Cult 121:127–140

Li M, Liang Z, Zeng Y, Jing Y, Wu K, Liang J, He S, Wang G, Mo Z, Tan F (2016) De novo analysis of transcriptome reveals genes associated with leaf abscission in sugarcane (*Saccharum officinarum* L.). BMC Genomics 17(195)

Li X, Fan S, Hu W, Liu G, Wei Y, He C, Shi H (2017) Two Cassava Basic Leucine Zipper (bZIP) transcription factors (MebZIP3 and MebZIP5) confer disease resistance against cassava bacterial blight. Front Plant Sci 8:1–11

Li Z, Hua X, Zhong W, Yuan Y, Wang Y, Wang Z, Ming R, Zhang J (2020a) Genome-wide identification and expression profile analysis of WRKY family genes in the autopolyploid *Saccharum spontaneum*. Plant Cell Physiol 61:616–630

Li J, Jiao G, Sun Y, Chen J, Zhong Y, Yan L, Jiang D, Ma Y, Xia L (2020b) Modification of starch composition, structure and properties through editing of TaSBEIIa in both winter and spring wheat varieties by CRISPR/Cas9. Plant Biotechnol J 19(5):937–951

Liang M, Li H, Zhou F, Li H, Liu J, Hao Y, Wang Y, Zhao H, Han S (2015) Subcellular distribution of NTL transcription factors in *Arabidopsis thaliana*. Traffic 16:1062–1074

Lin Z, Wang J, Bao Y, Guo Q, Powell CA, Xu S, Chen B, Zhang M (2016) Deciphering the transcriptomic response of *Fusarium verticillioides* in relation to nitrogen availability and the development of sugarcane pokkah boeng disease. Sci Rep 6:29692

Ling H, Huang N, Wu Q, Su Y, Peng Q, Ahmed W, Gao S, Su W, Que Y, Xu L (2018) Transcriptional insights into the sugarcane-sorghum mosaic virus interaction. Trop Plant Biol 11:163–176

Lu J, Ju H, Zhou G, Zhu C, Erb M, Wang X, Wang P, Lou Y (2011) An EAR motif- containing ERF transcription factor affects herbivore-induced signaling, defense and resistance in rice. Plant J 68:583–596

Luan Q, Chen C, Liu M, Li Q, Wang L, Ren Z (2019) *CsWRKY50* mediates defense responses to Pseudoperonospora cubensis infection in *Cucumis sativus*. Plant Sci 279:59–69

Luo Q, Ahmad K, Fu HY, Wang JD, Chen RK, Gao SJ (2016) Genetic diversity and population structure of Sorghum mosaic virus infecting *Saccharum* spp. hybrids. Ann Appl Biol 169:398–407

Ma H-M, Schulze S, Lee S, Yang M, Mirkov E, Irvine J, Moore P, Paterson A (2004) An EST survey of the sugarcane transcriptome. Theor Appl Genet 108:851–863

Mao Y, Zhang H, Xu N, Zhang B, Gou F, Zhu J-K (2013) Application of the CRISPR-Cas system for efficient genome engineering in plants. Mol Plant 6:2008

Marchive C, Léon C, Kappel C, Coutos-Thévenot P, Corio-Costet M-F, Delrot S, Lauvergeat V (2013) Over-expression of *VvWRKY1* in grapevines induces expression of jasmonic acid pathway-related genes and confers higher tolerance to the downy mildew. PLoS One 8:1–8

Martínez SMI, Bracuto V, Koseoglou E et al (2020) CRISPR/Cas9-targeted mutagenesis of the tomato susceptibility gene *PMR4* for resistance against powdery mildew. BMC Plant Biol 20:284

McNeil MD, Bhuiyan SA, Berkman PJ, Croft BJ, Aitken KS (2018) Analysis of the resistance mechanisms in sugarcane during *Sporisorium scitamineum* infection using RNA-seq and microscopy. PLoS One 13:e0197840

Mehmet N, Fadime U, Yasin C, Yasemin K, Altunoglu C, Cengiz M (2018) Comparative identification, characterization, and expression analysis of bZIP gene family members in watermelon and melon genomes. Plant Growth Regul 87:227–243

Meisrimler CN, Pelgrom AJ, Oud B, Out S, Van den Ackerveken G (2019) Multiple downy mildew effectors target the stress-related NAC transcription factor LsNAC069 in lettuce. Plant J 99:1098–1115

Miao J, Guo D, Zhang J, Huang Q, Qin G, Zhang X et al (2013) Targeted mutagenesis in rice using CRISPR-Cas system. Cell Res 23(10):1233–1236

Mmadi MA, Dossa K, Wang L, Zhou R, Wang Y, Cisse N, Sy MO, Zhang X (2017) Functional characterization of the versatile MYB gene family uncovered their important roles in plant development and responses to drought and waterlogging in sesame. Genes 8:362

Molla KA, Karmakar S, Islam MT (2020) Wide horizons of CRISPR-Cas-derived technologies for basic biology, agriculture, and medicine. In: Islam MT, Bhowmik PK, Molla KA (eds) CRISPR-Cas methods. Springer Protocols Handbooks/Humana, New York

Mustafa G, Joyia FA, Anwar S, Parvaiz A, Khan MS (2018) Biotechnological interventions for the improvement of sugarcane crop and sugar production. In: Sugarcane-technology and research. IntechOpen, London, pp 113–138

Muthiah M, Ramadass A, Amalraj RS, Palaniyandi M, Rasappa V (2013) Expression profiling of transcription factors (TFs) in sugarcane X *Colletotrichum falcatum* interaction. J Plant Biochem Biotechnol 22:286–294

Mutz KO, Heilkenbrinker A, Lonne M, Walter JG, Stahl F (2013) Transcriptome analysis using next-generation sequencing. Curr Opin Biotechnol 24:22–30

Noman A, Hussain A, Adnan M, Khan MI, Ashraf MF, Zainab M, Khan KA, Ghramh HA, He S (2019) A novel MYB transcription factor CaPHL8 provide clues about evolution of pepper immunity against soil borne pathogen. Microb Pathogen 137:103758

Ntambo MS, Meng J-Y, Rott PC, Henry RJ, Zhang H-L, Gao S-J (2019) Comparative transcriptome profiling of resistant and susceptible sugarcane cultivars in response to infection by *Xanthomonas albilineans*. Int J Mol Sci 20:6138

Ogata K, Kanei-Ishii C, Sasaki M, Hatanaka H, Nagadoi A, Enari M, Nakamura H, Nishimura Y, Ishii S, Sarai A (1996) The cavity in the hydrophobic core of Myb DNA-binding domain is reserved for DNA recognition and trans-activation. Nat Struct Biol 3:178

Olsen AN, Ernst HA, Leggio LL, Skriver K (2005) NAC transcription factors: structurally distinct, functionally diverse. Trends Plant Sci 10:79–87

Ooka H, Satoh K, Doi K, Nagata T, Otomo Y, Murakami K, Matsubara K, Osato N, Kawai J, Carninci P (2003) Comprehensive analysis of NAC family genes in *Oryza sativa* and *Arabidopsis thaliana*. DNA Res 10:239–247

Oz MT, Altpeter A, Karan R, Merotto A, Altpeter F (2021) CRISPR/Cas9-mediated multi-allelic gene targeting in sugarcane confers herbicide tolerance. Front Genome Ed 3:673566

Paz-Ares J, Ghosal D, Wienand U, Peterson P, Saedler H (1987) The regulatory c1 locus of *Zea mays* encodes a protein with homology to myb proto-oncogene products and with structural similarities to transcriptional activators. EMBO J 6:3553–3558

Prasanth CN, Viswanathan R, Krishna N, Malathi P, Sundar AR, Tiwari T (2017) Unraveling the genetic complexities in gene set of sugarcane red rot pathogen *Colletotrichum falcatum* through transcriptomic approach. Sugar Tech 19:604–615

Prasanth CN, Viswanathan R, Malathi P, Sundar AR (2019) Comparative transcriptome analysis of candidate secretory effector proteins from *Colletotrichum falcatum* infecting sugarcane. Agri Gene 13:100089

Puranik S, Sahu PP, Srivastava PS, Prasad M (2012) NAC proteins: regulation and role in stress tolerance. Trends Plant Sci 17:369–381

Rinerson CI, Rabara RC, Tripathi P, Shen QJ, Rushton PJ (2015) The evolution of WRKY transcription factors. BMC Plant Biol 15:66

Roy S (2016) Function of MYB domain transcription factors in abiotic stress and epigenetic control of stress response in plant genome. Plant Signal Behav 11:e1117723

Rushton PJ, Bokowiec MT, Laudeman TW, Brannock JF, Chen X, Timko MP (2008) TOBFAC: the database of tobacco transcription factors. BMC Bioinform 9:53

Sablowski RW, Meyerowitz EM (1998) A homolog of no apical meristem is an immediate target of the floral homeotic genes APETALA3/PISTILLATA. Cell 92:93–103

Sakuma Y, Liu Q, Dubouzet JG, Abe H, Shinozaki K, Yamaguchi-Shinozaki K (2002) DNA-binding specificity of the ERF/AP2 domain of *Arabidopsis* DREBs, transcription factors involved in dehydration-and cold-inducible gene expression. Biochem Biophys Res Commun 290:998–1009

Santa Brigida AB, Rojas CA, Grativol C, de Armas EM, Entenza JO, Thiebaut F, Lima MF, Farrinelli L, Hemerly AS, Lifschitz S (2016) Sugarcane transcriptome analysis in response to infection caused by *Acidovorax avenae* subsp. *avenae*. PLoS One 11:e0166473

Sathyabhama M, Viswanathan R, Malathi P, Sundar AR (2016) Identification of differentially expressed genes in sugarcane during pathogenesis of *Colletotrichum falcatum* by suppression subtractive hybridization (SSH). Sugar Tech 18:176–183

Savary S, Ficke A, Aubertot JN, Hollier C (2012) Crop losses due to diseases and their implications for global food production losses and food security. Food Secur 4:519–537

Savary S, Willocquet L, Pethybridge SJ, Esker P, McRoberts N, Nelson A (2019) The global burden of pathogens and pests on major food crops. Nat Ecol Evol 3:430

Segarra G, Van der Ent S, Trillas I, Pieterse C (2009) MYB72, a node of convergence in induced systemic resistance triggered by a fungal and a bacterial beneficial microbe. Plant Biol 11:90–96

Seo PJ, Park CM (2010) MYB96-mediated abscisic acid signals induce pathogen resistance response by promoting salicylic acid biosynthesis in *Arabidopsis*. New Phytol 186:471–483

Shan Q, Wang Y, Li J, Gao C (2014) Genome editing in rice and wheat using the CRISPR/Cas system. Nat Protoc 9(10):2395–2410

Shan T, Rong W, Xu H, Du L, Liu X, Zhang Z (2016) The wheat R2R3-MYB transcription factor TaRIM1 participates in resistance response against the pathogen *Rhizoctonia cerealis* infection through regulating defense genes. Sci Rep 6:1–14

Shen X-J, Wang Y-Y, Zhang Y-X, Guo W, Jiao Y-Q, Zhou X-A (2018) Overexpression of the wild soybean R2R3-MYB transcription factor GsMYB15 enhances resistance to salt stress and Helicoverpa armigera in transgenic *Arabidopsis*. Int J Mol Sci 19(12):3958

Singh AK, Kumar SR, Dwivedi V, Rai A, Pal S, Shasany AK, Nagegowda DA (2017) A WRKY transcription factor from *Withania somnifera* regulates triterpenoid with anolide accumulation and biotic stress tolerance through modulation of phytosterol and defense pathways. New Phytol 215:1115–1131

Song X, Li Y, Hou X (2013) Genome-wide analysis of the AP2/ERF transcription factor superfamily in *Chinese cabbage* (*Brassica rapa* ssp. *pekinensis*). BMC Genomics 14(573)

Sornaraj P, Luang S, Lopato S, Hrmova M (2016) Basic leucine zipper (bZIP) transcription factors involved in abiotic stresses: a molecular model of a wheat bZIP factor and implications of its structure in function. Biochim Biophys Acta 1860:46–56

Sun L, Zhang H, Li D, Huang L, Hong Y, Ding XS, Nelson RS, Zhou X, Song F (2013) Functions of rice NAC transcriptional factors, ONAC122 and ONAC131, in defense responses against *Magnaporthe grisea*. Plant Mol Biol 81:41–56

Tezuka D, Kawamata A, Kato H, Saburi W, Mori H, Imai R (2019) The rice ethylene response factor OsERF83 positively regulates disease resistance to *Magnaporthe oryzae*. Plant Physiol Biochem 135:263–271

Thirugnanasambandam PP, Mason PJ, Hoang NV, Furtado A, Botha FC, Henry RJ (2019) Analysis of the diversity and tissue specificity of sucrose synthase genes in the long read transcriptome of sugarcane. BMC Plant Biol 19:160

Tolosa LN, Zhang Z (2020) The role of major transcription factors in Solanaceous food crops under different stress conditions: current and future perspectives. Plan Theory 9(1):56

Upadhyay SK, Kumar J, Alok A, Tuli R (2013) RNA-guided genome editing for target gene mutations in wheat. G3 3(12):2233–2238

Van Eck L, Davidson RM, Wu S, Zhao BY, Botha A-M, Leach JE, Lapitan NL (2014) The transcriptional network of WRKY53 in cereals links oxidative responses to biotic and abiotic stress inputs. Funct Integr Genomics 14:351–362

Vettore AL, da Silva FR, Kemper EL, Souza GM, da Silva AM, Ferro MIT, Henrique-Silva F, Giglioti ÉA, Lemos MV, Coutinho LL (2003) Analysis and functional annotation of an expressed sequence tag collection for tropical crop sugarcane. Genome Res 13:2725–2735

Voitsik A-M, Muench S, Deising HB, Voll LM (2013) Two recently duplicated maize NAC transcription factor paralogs are induced in response to *Colletotrichum graminicola* infection. BMC Plant Biol 13:85

Waclawovsky AJ, Sato PM, Lembke CG, Moore PH, Souza GM (2010) Sugarcane for bioenergy production: an assessment of yield and regulation of sucrose content. Plant Biotech J 8:263–276

Wang J, Roe B, Macmil S, Yu Q, Murray JE, Tang H, Chen C, Najar F, Wiley G, Bowers J (2010) Microcollinearity between autopolyploid sugarcane and diploid sorghum genomes. BMC Genomics 11:261

Wang F, Wang C, Liu P, Lei C, Hao W, Gao Y, Liu Y-G, Zhao K (2016) Enhanced rice blast resistance by CRISPR/Cas9-targeted mutagenesis of the ERF transcription factor gene *OsERF922*. PLoS One 11:e0154027

Wang B, Tseng E, Regulski M, Clark TA, Hon T, Jiao Y, Lu Z, Olson A, Stein JC, Ware D (2016a) Unveiling the complexity of the maize transcriptome by single-molecule long-read sequencing. Nat Commun 7:11708

Wang H, Wang H, Shao H, Tang X (2016b) Recent advances in utilizing transcription factors to improve plant abiotic stress tolerance by transgenic technology. Front Plant Sci 7:67

Wang J, Tao F, An F, Zou Y, Tian W, Chen X, Xu X, Hu X (2017) Wheat transcription factor TaWRKY70 is positively involved in high-temperature seedling plant resistance to *Puccinia striiformis* f. sp. *tritici*. Mol Plant Pathol 18:649–661

Wang X, Tu M, Wang D, Liu J, Li Y, Li Z, Wang Y, Wang X (2018a) CRISPR/Cas9-mediated efficient targeted mutagenesis in grape in the first generation. Plant Biotechnol J 16:844

Wang B, Wei J, Song N, Wang N, Zhao J, Kang Z (2018b) A novel wheat NAC transcription factor, TaNAC30, negatively regulates resistance of wheat to stripe rust. J Integr Plant Biol 60:432–443

Wang Y, Zhang Y, Zhou R, Dossa K, Yu J, Li D, Liu A, Mmadi MA, Zhang X, You J (2018c) Identification and characterization of the bZIP transcription factor family and its expression in response to abiotic stresses in sesame. PLoS One 13(7):e0200850

Wang L, Liu F, Zhang X, Wang W, Sun T, Chen Y, Dai M, Yu S, Xu L, Su Y et al (2018d) Expression characteristics and functional analysis of the *ScWRKY3* gene from sugarcane. Int J Mol Sci 19:4059

Wang Z, Li Y, Li C, Song X, Lei J, Gao Y, Liang Q (2019a) Comparative transcriptome profiling of resistant and susceptible sugarcane genotypes in response to the airborne pathogen *Fusarium verticillioides*. Mol Biol Rep 46:3777–3789

Wang Z, Yan L, Wan L, Huai D, Kang Y, Shi L, Jiang Y, Lei Y (2019b) Genome-wide systematic characterization of bZIP transcription factors and their expression profiles during seed development and in response to salt stress in peanut. BMC Genomics 20:1–14

Wang JH, Gu KD, Han PL, Yu JQ, Wang CK, Zhang QY, You CX, Hu DG, Hao YJ (2020) Apple ethylene response factor MdERF11 confers resistance to fungal pathogen *Botryosphaeria dothidea*. Plant Sci 291:110351

Weeks DP, Spalding MH, Yang B (2016) Use of designer nucleases for targeted gene and genome editing in plants. Plant Biotechnol J 14(2):483–495

Wei Z, Abdelrahman M, Gao Y, Ji Z, Mishra R, Sun H, Sui Y, Wu C, Wang C, Zhao K (2021) Engineering broad-spectrum resistance to bacterial blight by CRISPR-Cas9-mediated precise homology directed repair in rice. Mol Plant 14(8):1215–1218

World population prospects: the 2012 revision; United Nations: New York (2013)

Xia N, Zhang G, Sun Y-F, Zhu L, Xu L-S, Chen X-M, Liu B, Yu Y-T, Wang X-J, Huang L-L (2010) *TaNAC8*, a novel NAC transcription factor gene in wheat, responds to stripe rust pathogen infection and abiotic stresses. Physiol Mol Plant Pathol 74:394–402

Xu S, Wang J, Shang H, Huang Y, Yao W, Chen B, Zhang M (2018) Transcriptomic characterization and potential marker development of contrasting sugarcane cultivars. Sci Rep 8:1683

Yang Y, Zhou Y, Chi Y, Fan B, Chen Z (2017) Characterization of soybean WRKY gene family and identification of soybean WRKY genes that promote resistance to soybean cyst nematode. Sci Rep 7:1–13

Yruela I (2015) Plant development regulation: overview and perspectives. J Plant Physiol 182:62–78

Yruela I, Oldfield CJ, Niklas KJ, Dunker AK (2017) Evidence for a strong correlation between transcription factor protein disorder and organismic complexity. Genome Bio Evo 9:1248–1265

Yu Y, Guo D, Li G, Yang Y, Zhang G, Li S, Liang Z (2019) The grapevine R2R3-type MYB transcription factor VdMYB1 positively regulates defense responses by activating the *stilbene synthase gene 2* (*VdSTS2*). BMC Plant Biol 19:478

Yuan X, Wang H, Cai J, Bi Y, Li D, Song F (2019a) Rice NAC transcription factor ONAC066 functions as a positive regulator of drought and oxidative stress response. BMC Plant Biol 19(1):1–19

Yuan X, Wang H, Cai J, Li D, Song F (2019b) NAC transcription factors in plant immunity. Phytopathol Res 1(1):1–3

Zaidi SS-E-A, Tashkandi M, Mansoor S, Mahfouz MM (2016) Engineering plant immunity: using CRISPR/Cas9 to generate virus resistance. Front Plant Sci 7:1673

Zhang X, Chen M, Liang Y, Xing Y, Yang L, Chen M, Comstock JC, Li Y, Yang L (2016) Morphological and physiological responses of sugarcane to *Leifsonia xyli* subsp. *xyli* infection. Plant Dis 100:2499–2506

Zhang X-M, Zhang Q, Pei C-L, Li X, Huang X-L, Chang C-Y, Wang X-J, Huang L-L, Kang Z-S (2018a) TaNAC2 is a negative regulator in the wheat-stripe rust fungus interaction at the early stage. Physiol Mol Plant P 102:144–153

Zhang T, Zheng Q, Yi X, An H, Zhao Y, Ma S, Zhou G (2018b) Establishing RNA virus resistance in plants by harnessing CRISPR immune system. Plant Biotechnol J 16:1415

Zhang J, Zhang Q, Li L, Tang H, Zhang Q, Chen Y, Arrow J, Zhang X, Wang A, Miao C (2019a) Recent polyploidization events in three Saccharum founding species. Plant Biotechnol J 17:264–274

Zhang Y-L, Zhang C-L, Wang G-L, Wang Y-X, Qi C-H, Zhao Q, You C-X, Li Y-Y, Hao Y-J (2019b) The R2R3 MYB transcription factor MdMYB30 modulates plant resistance against pathogens by regulating cuticular wax biosynthesis. BMC Plant Biol 19:362

Zhang J, Zhang X, Chen R, Yang L, Fan K, Liu Y, Wang G, Ren Z, Liu Y (2020) Generation of transgene-free Semidwarf maize plants by gene editing of *Gibberellin-Oxidase20-3* using CRISPR/Cas9. Front Plant Sci 11:1048

Zhao Y, Chang X, Qi D, Dong L, Wang G, Fan S, Jiang L, Cheng Q, Chen X, Han D (2017) A novel soybean ERF transcription factor, GmERF113, increases resistance to *Phytophthora sojae* infection in soybean. Front Plant Sci 8:299

Zhu X, Qi L, Liu X, Cai S, Xu H, Huang R, Li J, Wei X, Zhang Z (2014) The wheat ethylene response factor transcription factor pathogen-induced ERF1 mediates host responses to both the necrotrophic pathogen *Rhizoctonia cerealis* and freezing stresses. Plant Physiol 164:1499–1514

Zhu J, Song N, Sun S, Yang W, Zhao H, Song W et al (2016) Efficiency and inheritance of targeted mutagenesis in maize using CRISPR-Cas9. J Genet Genomics 43(1):25–36

Zhu L, Guo J, Ma Z, Wang J, Zhou C (2018b) Arabidopsis transcription factor MYB102 increases plant susceptibility to aphids by substantial activation of ethylene biosynthesis. Biomol Ther 8:39

Zhu K, Shao M, Zhou D, Xing Y-X, Yang L-T, Li Y-R (2019) Functional analysis of *Leifsonia xyli* subsp. *xyli* membrane protein gene Lxx18460 (anti-sigma K). BMC Microbiol 19(2)

Chapter 9
The Role of Transcription Factors in Response to Biotic Stresses in Maize

Jyoti Prakash Sahoo, Deepali Dash, Ankit Moharana, Manaswini Mahapatra, Amit Kumar Sahoo, and Kailash Chandra Samal

Abstract Maize is a significant food grain grown globally for consumption by people, animal feed and biofuels. Unfortunately, regular incidences of stress factors like abiotic and biotic stress have been noticed as a result of water scarcity and change in the climatic condition. This has been a persistent danger in boosting global maize production and yield. Plants generally use transcription factors to respond to the effect of biotic and abiotic stresses. The transcription factors are the groups of genes that encode for particular proteins. Target genes of transcriptional regulators are part of a regulon that controls the suppression or activation of the involved genes in response to both the stresses in maize. This is why a thorough investigation of each TF family of maize implicated in various biotic stress responses of maize is crucial and critical. The complete genome sequence of maize is now available. It allows the scientific community to make significant progress in understanding transcription factors and the processes regulating the expression of genes associated with maize. This chapter covers the essential elements of transcription factors associated with maize and their responses to biotic and abiotic stress.

Keywords Maize · Transcription factors · TFs · Biotic stresses

J. P. Sahoo (✉) · K. C. Samal
Department of Agricultural Biotechnology, Odisha University of Agriculture and Technology, Bhubaneswar, India

D. Dash
Department of Plant Physiology, Odisha University of Agriculture and Technology, Bhubaneswar, India

A. Moharana
Department of Seed Science and Technology, Odisha University of Agriculture and Technology, Bhubaneswar, India

M. Mahapatra
Department of Nematology, Odisha University of Agriculture and Technology, Bhubaneswar, India

A. K. Sahoo
Department of Bioscience and Bioinformatics, Khallikot University, Berhampur, India

© The Author(s), under exclusive license to Springer Nature Switzerland AG 2022

S. H. Wani et al. (eds.), *Transcription Factors for Biotic Stress Tolerance in Plants*, https://doi.org/10.1007/978-3-031-12990-2_9

9.1 Introduction

Maize is an important and essential cereal crop in the world. The world maize yield was recorded at 1148 million thousand tonnes during year 2019 (Fig. 9.1). The world output of maize grew by an average yearly rate of 3.41% from 265,000 tonnes in year 1970 to 1148,000 tonnes in year 2019 (Knoema, 2021). In both developing and established countries, maize has grown considerably. Maize is an essential food stuff in many regions of the globe (Wang et al. 2013). It can be eaten raw, used for feeding stuff and produced various processed maize products. Maize has been used as a model plant to investigate several biological processes, including paramutation, conversion mechanisms, heterosis breeding and diversity analysis (Bennetzen and Hake 2009; Perlack 2005). Biotic and abiotic stresses are constantly present between 400 and 580 degrees north across the maize-growing areas in the world (Gong et al. 2014). Salt stress, drought stress, nutrient shortage and temperature extremities are leading environmental variables affecting maize productivity. Drought, flood and extreme temperature influence corn production considerably (Ahuja et al. 2010). Extreme temperature, toxicity to heavy metal and osmotic stress are some of the abiotic stresses maize plants have to cope (Suzuki et al. 2014). The initial reactions induced by stressors lead to metabolic reprogramming in plant's

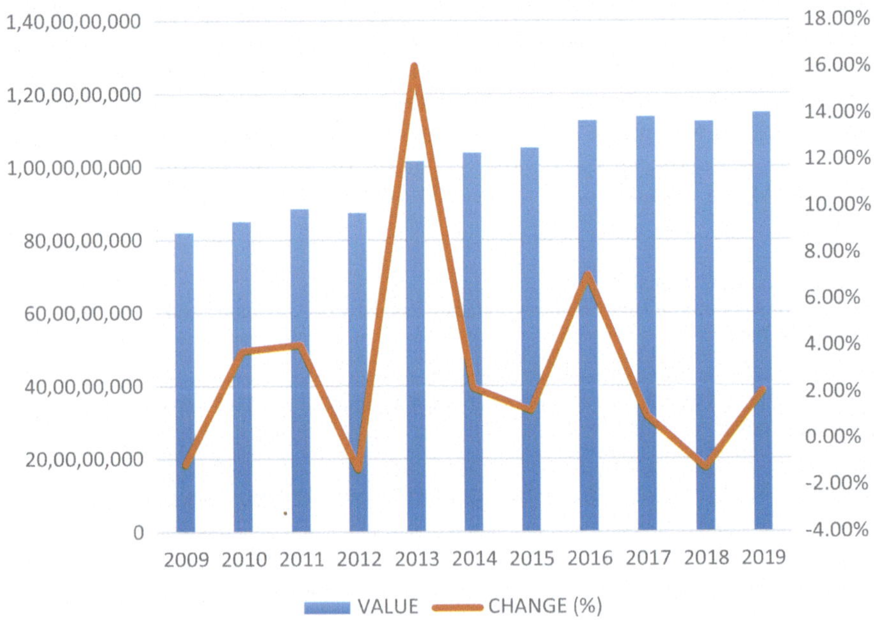

Fig. 9.1 World maize production quantity. (Source: Knoema 2021)

defence system by considerable changes in ion fluid, phytohormones and reactive oxygen species (ROS) (Bartoli et al. 2016). Antioxidative systems typically remove ROS rapidly, but the rise in ROS in cells can hinder this process (Ashtamker et al. 2007).

Another routes generally involved in the response to abiotic and biotic stresses in plants are the MAPK, i.e., mitogen-activated protein kinase, cascades (Wurzinger et al. 2011). MAPK cascades are activated and govern the stress response pathways when stress stimuli are recognised and experienced. In plant biotic and abiotic stress reactions, hormonal signalling is essential. The plant hormone ABA, i.e., abscisic acid, is the most prominent hormone involved in stress signalling (Kimotho et al. 2019). An increase of ABA in plant system under the influence of abiotic stress commands the control network against abiotic stress. The biotic stress response is mediated through the antagonism of another stress hormone such as jasmonic acid, salicylic acid and ethylene (Xiong et al. 2002). Gene expression regulation is essential for all biological activities. Regulation of transcription is one of the most common means of controlling gene expression for eukaryotes. One common approach is to tie a specific type of protein into three to eight base-pair long DNA sequence motifs with transcription factors (TFs). These factors are found in the promoter region of the genes. These are usually organised into the regulatory modules, which account for the overall regulatory response of the gene. In a hierarchical gene-controlled network, TFs are organised where the effects of one regulation protein on the expression of another TF are positive or negative (Davidson 2001). This reveals many regulatory reasons that leads to an extensive architecture that divides genetic regulation networks when integrated into regulation modules (Yu and Gerstein 2006). This chapter summarises the structures and functions of maize transcription factors in response to biotic and abiotic stresses during maize cultivation.

9.2 Structures of Plant Transcription Factors

The three-dimensional structure is also expected to be preserved for transcription factors (TFs). It is preserved across the kingdom, and meaningful forecasts may be made for crucial DNA residues like MYB TF domains. The structure of the animal c-MYB protein R2R3-MYB region was solved, which shares approx. 50% of the R2R3-MYB protein (Ogata et al. 1994). A helix-turn-helix fold and a third α-helix fold are applied to each MYB repeat to make base-pair interactions. There are substantial structural variations despite the identical sequence between the plant MYB domains and the animal MYB domains. The most substantial evidence of these variations is that in animal MYB areas have one Cys residue, while two proximal, highly preserved Cys residues may form an intramolecular disulphide (S-S) bond under non-reduction circumstances in most R2R3-MYB plant areas (Heine et al. 2004). The *Antirrhinum majus* protein structure of RADIALIS (RAD) has recently been resolved to 1.9 Å (Stevenson et al. 2006). RAD's structure is significantly different from conventional MYB repeats and a considerably larger third α-helix according to RAD as part and consistent with a distinctive set of individual MYB repeating proteins (Stevenson et al. 2006). Several TFs present having motifs similar to MYB (3R or R2R3) and those of MYB (Stevenson et al. 2006). The GARP division shows a hallmark of 60 amino acids in typical MYB repeats, called the B-motif. NMR has established the B-motif structure of the *Arabidopsis* ARR10 transcription factor, which is a His-to-Asp signal transduction pathway response regulator (Hosoda et al. 2002). Furthermore, the B-motif includes a signal of nuclear localisation, which makes it a domain of multifunctionality. MYBs also illustrate a category of the regulatory protein sequence that has probably arisen before the plant-animal division (Lipsick 1996).

The WRKY, NAM and TCP families represent over 10% of all plant TFs identified so far. From the ~46,000 PDB structures available in October 2007, only ~14 matched the plant diagnostic protein, which showed that structural study of plant diagnostics had to grow. The structure of the single-stranded plant defence DNA-binding component TF PBF-2, p24, solved at 2.3 AR, provides a further example. This protein comes from an all-new family of the WHIRLY family and has a quaternary structure (Hosoda et al. 2002). The p24 subunits are unique for ssDNA-binding proteins in the non-crystallographic C_4-symmetry arrangement and can precisely describe the selective binding activity of PBF-2. The predictive positioning in binding promoter areas structurally also promotes the ability of PBF-2 to regulate gene expression (Desveaux et al. 2004). *Arabidopsis* crystal structures have been identified for WRKY4 and WRKY1 proteins (Wang et al. 2007). The new structure of the WRKY4 comprises with consecutively four β-sheet with a zinc-bounding bag at one end part of the β-sheet (Wright et al. 2005). The *Arabidopsis* ERF1 proteins are part of the family of ethylene-responsive element-binding proteins (EREBPs) (Ohme-Takagi and Shinshi 1995). In *Arabidopsis* ANAC which is the DNA-binding NAC domain is another important plant-specific TF that has a typical helix-turn-helix motif.

The NAC TF family is involved in a wide variety of plant reactions, including developing the apical shoot meristem, lateral shooting, floral bodies, monitoring and defence of plant hormones (Ernst et al. 2004). The protein acts as a functional dimer. NMR has been used to solve the structure of several different TF plants. Various plant-specific TFs include auxin-regulated factors, and ABA-regulated transcription shares a B3 DNA-compliance domain. This TF has a substantial structural resemblance to EcoRI and is expected to be comparable at the same place between the DNA-binding residuals (Yamasaki et al. 2004). Ethylene insensitive 3 (EIN3) and EIN3/EIL proteins are the most critical TFs for this signal transduction. The TFs bind to downstream gene promoters and work with a range of stress reactions. Using chemical shift studies, DNA binding has been indicated for a region containing the site altered on the EIN3 allele (Yamasaki et al. 2004). Among plant TFs, another family is the SBP TF family, identified by separating the floral meristem gene of *Antirrhinum majus* SQUAMOSA (Klein et al. 1996).

9.3 Functions of Plant Transcription Factors

A product contains genes that directly enable cells to withstand environmental stress, e.g. osmotic regulative protein, abundant late embryogenesis (LEAs), proline-synthesising enzymes and other osmo-regulators. The plant genes regulated by transcription are directly controlled by transcription factor-binding sites (TFBS) containing TF networks (Ciarmiello et al. 2011; Chaves and Oliveira 2004). TFs are typically two domain containing proteins, namely, the activation domain (DA) and the DNA-binding domain (DB). A transcription factor (TF) attaches itself to the elements which are cis-acting of a gene responsible for stress induction in the promoter region with the help of binding domain (Yamasaki et al. 2013). The activation domain approaches the target gene, which causes this gene to be repressed or activated. Approximately 10% of plant genes primarily encode TFs. TFs allow or inhibit RNA polymerase activity, which regulates the action of gene (Franco-Zorrilla et al. 2014). It may be necessary to regulate many genes, including the TF genes, because TF can affect numerous genes involved in various stress tolerances by itself (Riechmann et al. 2000). A complete survey on all TFs in conjunction with the biotic and abiotic stress control pathways in maize would be very lucrative, for example, by accessing the Sub1 site encoding an ET response factor TF and activating about 900 genes with an effect on stress. Rice-sensitive flood genotypes were successfully converted to those vulnerable to flood (Xu et al. 2006). In addition, *Arabidopsis* and many other plants disclosed several different responses to biotic and abiotic stress independently. They indicated the possible controls of the susceptibility or tolerance of biomass stress and abiotic stress on the transcriptome level by an advanced gene regulation network (GRN) (Umezawa et al. 2006; Honório et al. 2009). ABA-independent regulon is included in the regulon CA (CUC, NAM, ATAF) and HFD (Honório et al. 2009). The different TFs typically operate independently of stress, although among these TFs, a considerable degree of cross-linkage can occur.

Many studies show that ABA can converge in many unexpected places. These convergence locations show transcriptional repression and enhancers involving both the DRE/C repeater and the ABA-responsive element (ABRE) concurrently or directly. The functions and potential uses of abiotic stress-effective TFs for future molecular breeding and improvement of different plant species have been proven recently. Many steps to understand transcription, gene expression and signal translation of plant responses to abiotic stress have been taken (Zhu 2016). The rise, for instance, has increased the overexpression and drought tolerance of a transgenic rice gene (SNAC1) (Hu et al. 2006). In transgenic *Arabidopsis*, excessive expression of glycine soybean NACTF in seedlings and mature plants led to alkaline stress despite a reduction in ABA's sensitivity due to transgenic plants (Cao et al. 2017). Similarly, the *Pyrus betulifolia* NAC transcription factor gene functional research showed that the gene regulates cold stress and drought stress regulation (Du et al. 2017; Yu et al. 2016). In finger millet, bZIP transcription factor gene EcbZIP17 exhibited better germination, higher content of biomass and improved durability rates in transgenic plants (Ramakrishna et al. 2018). In transgenic tobacco plants, seed yields have been enhanced in proportion to control plants. Also, it was shown as a positive regulation for salinity stress tolerance that GmbZIP110 works with transgenic *Arabidopsis* and soybean plantings, which upregulate soybean bZIP TF, as GmbTZIP110. The GmbZIP110 functional transgenic *Arabidopsis* research shows that, this gene may connect the ACGT motif with multiple downstream target genes ((Xu et al. 2006; Cao et al. 2017). The activation of SWPA2 promoter, ABF3, and a bZIP TF, in transgenic alfalfa, for sweet potato oxidative stress led to increased growth due to drought stress (Wang et al. 2016).

The overexpression in hot pepper of the CaBZ1 significantly increased the endurance of dehydration stress without any ad hoc impact on plant growth or production, and at the same time, overexpression of OsMYB55 led to enhanced plant growth, dehydration and unfavourable high-temperature impacts (Moon et al. 2015; Casaretto et al. 2016). The cloned *Cichorium intybus* CiMYB3 as a response to various biotic and abiotic stresses was also proved (Jiao et al. 2017). The cold stress tolerance of transgenic plants such as banana (*Musa paradisiaca*) has risen substantially over the expression of an MYB TF gene called MpMYBS3 (Dou et al. 2020). The *Medicago truncatula* MYB TF gene otherwise was utilised to increase salt and drought tolerance by improving transgenic *Arabidopsis* primary root growth (Dong et al. 2017). GaMYB62L overexpression also increases dryness tolerance in transgenic *Arabidopsis* (Butt et al. 2017). In the transgenic *Arabidopsis*, the exogenously expression of the AtDREB1A gene has resulted in increased oxidant and photosynthetic stress in plants with enhanced antioxidant activity (Butt et al. 2017). The overexpression of gene, SbDREB2A in *Salicornia brachiata*, resulted in increased seed germination in the more osmotic stresses in transgenic tobacco (Gupta et al. 2014). OsWRKY 71 from rice in the WYKYTF gene family has revealed that the regulation of several downstream genes such as WSI76 and OsTGFR offers good regulatory cold stress resistance (Kim et al. 2016; Ullah et al. 2017; Liu et al. 2014).

9.4 Plant Transcription Factors and Their Organisation into Family

Based on the availability of preserved DNA fields, TFs are categorised into distinct families; according to different authors, it is difficult to compare from one research to the other. *Arabidopsis* TFs adapt the family organisation (Davuluri et al. 2003).

9.4.1 Heat Shock Factor (HSF) Family

A broad range of organisms reacts to high temperature by synthesising heat shock proteins (HSPs). The genes are regulated by the old TFs, the HSFs, which are maintained between plants and animals. Although just one HSF is present in the yeast, many animals are usually present, and about 20 of them are found in plants. These HSF family members comprise N-terminal DNA in the first region known as the HSF domain (PFAM PF00447). The cDNA or genomic sequences for at least 22 maize HSFs were discovered during PlantaGDB database (Fu et al. 2006). They were called after rice homologues, 16 of whom maintained the genetic organisation of each species (intron sites). The remaining family members are incompletely genomically sequenced. Plant HSFs comprise HR-A and HR-B spin-coil domains that interact with heat shock protein two cytoplasmic regulators; HSBP2. Variation in non-winded flanking areas allows the protein to evolve in this interaction.

9.4.2 MYB Family

MYB factors are a heterogeneously ubiquitous collection of eukaryotic proteins. As a result, MYB proteins generally categorise the number of repeat proteins corresponding to an MYB domain. Most MYB vertebrate proteins are built up of three inefficiently occurring recurrences (R1, R2 and R3) (Rabinowicz et al. 1999). However, the large majority of MYB proteins in plants are R2R3-MYB, which are characterised by their presence of two MYB repeats, R2 and R3. The R2R3-MYB is large and has around 130 members (Stracke et al. 2001). It was suggested that R1 was lost from the old 3R-MYB precursor from the R2R3-MYB genes (Dias et al. 2003). However, 450–200 years ago, possibly when the earth was overwhelmed by vegetation, the R2R3 family was amplified (Rabinowicz et al. 1999). Furthermore, some subsets of R2R3-MYB genes (Rabinowicz et al. 1999) still seem to extend the herb, which is linked to the variety of the plant metabolism pathway featuring one repetition of MYB, most likely of the protein R2R3-MYB, certain vegetable factor MYB, such *Arabidopsis* CAPRICE (CPC) and TRIPTYCHON (TRY) proteins (Molina and Grotewold 2005).

9.4.3 MADS Family

A MADS domain is a DNA-binding/dimerising zone maintained by many TF kingdoms (MCM1, AGAMOUS, DEPHICIENS and SRF [serum response factor]). MADS-box genes are a large multigene family of vascular plants (e.g. at least 64 loci in rice). Angiosperms involve many MADS family genes to determine the organ's floral meristem and identity in various developmental stages (e.g. AGAMOUS and DEFICIENS). The roles of MADS boxes are nevertheless not restricted to developing reproductive plant structures (Riechmann and Meyerowitz 1997).

9.4.4 bHLH Family

The fundamental protein helix (bHLH) family collects functionally diverse TFs discovered in plants and animals (at least 144 loci in rice). Early in eukaryotes, these proteins appear to work before animals divide in plant- or animal-specified processes. The bHLH proteins are regulated for animals in a wide range of critical developmental stages. In plants, on the other hand, BHLH proteins are not adequately studied. Features also include those characterised by anthocyanin biosynthesis, globulin expression, fruit dehiscence and carpeting and epidermis. These TFs feature a highly retained bHLH that makes inactive monomers trans-activating dimers during the correct design stages easier for the WRKY family to bind. WRKY TF members of a wide range of superior plants were found and associated with pathogen responses (Li et al. 2021). Roughly they have been found in rice, including disease resistance, salicylic and jasmonic acid responses, seed-mediated growth and germination with gibberellin, development of processes including senescence and abiotic stressors and abscisic acid reactions (Ross et al. 2007).

9.4.5 AP2-EREBP and GLK (G2-Like) Family

AP2 (APETALA2)/EREBP (ethylene-responsive protein element-binding protein) includes numerous TFs of developmental and physiological importance (Aharoni et al. 2004). One hundred sixty-four AP2/EREBP loci are thought to be present in rice and responsible for making it the most prominent family of the TF genes. Two distinct subfamilies are split into AP2/EREBP genes, AP2 and EREBP, with a single AP2/ERF (ethylene-responsive binding element factor). MicroRNA miR172 focuses on the expression of the AP2 gene, with gymnosperm AP2 homologues the target site of miR172, which implies over 300 million years since gymnosperm and flowering plantation lines have been conserved for microRNA regulatory mechanisms of this TF (Shigyo et al. 2006). In the beginning, plant insulation discovered the Golden2 (G2) (or bundle sheath defective 1) gene (Hall et al. 1998). G2, which is behind the development of bundle sheath cells in maize, is crucial for the differentiation of chloroplasts. G2-like genes defines a plant-like G2 (GLK) family from other maize and rice. The reason for these TFs in HLH DNA is known as the GLK/C terminal cabinet (Rossini et al. 2001). GLK factors in angiosperms control the development of at least three kinds of chloroplasts. Retention of moss chloroplast-mediated GLK is one of the earliest preservative regulating mechanisms in the plant sector (Yasumura et al. 2005). Recent research that has overexpressed the *Arabidopsis* GLK1 has demonstrated that it regulates a spectrum of genes associated with pathogenic response and detoxification, making them useful in farm plants for resistant illnesses.

9.5 Maize TFs for Plant-Parasitic Nematodes

The plant-parasitic nematodes (PPNs), which cause severe harm and decrease agricultural outputs, are compulsory biotrophic parasites. In different crops, several economically critical genera of plant-parasitic nematodes serve as parasites. The root-knot nematode, root lesion and cyst nematodes are the most prevalent harmful genera in crops in the family Heteroderidae. In commercially essential plants, it is necessary to develop different management techniques for PPNs. The response of plants to PPNs includes numerous proteins that play an essential role in plant-nematode interaction, including several transcription factors. Interestingly, there have been variations in response to nematode infection from monotonous and dicotyledonous hosts regarding alterations in gene expression which encode these proteins (Jammes et al. 2005; Bao et al. 2014). The most commonly analysed transcription factors (TFs) are essential in controlling gene expression and are viewed as downstream genes that respond to biotic and abiotic stress situations (Xing et al. 2017).

An essential set of transcription factors functioning as positive or negative regulators for the two compounds of plant immunity are the WRKY family: pathogen-associated molecular pattern-caused immunity (PTI) and effector-induced immunity (ETI) (Negi and Khurana 2021). The participation of WRKYs in host resistance against RKN infection has also been observed (Rushton et al. 2010). WRKY53 is a critical factor in WRKY and regulates plant growth, among other variables (Zentgraf and Doll 2019). This factor was also found in the degradation of salicylic acid by signals of jasmonic acid and ethylene in *Arabidopsis* (Dewitte et al. 2007). In addition, WRKY53 is induced by chitin oligosaccharides into the rice and promotes PR protein and peroxidase expression (Chujo et al. 2007). However, it has not been examined its involvement in the response of plants to *Meloidogyne* infection. The elongation factor 1 (EF1) has been developed from the TFs to play a significant role in numerous plant processes. EF1, including a G-protein (EF1a) and an exchange factor of guanine-nucleotide (EF1b), involves several plant processes in regulating, proliferating and differentiating cells (Gao et al. 2019). EF1a is a multifunctional protein that catalyses the aminoacyl tRNA binding to the ribosome site of the acceptor and engages in several other cell activities, such as signals or nuclear protein exports (Suhandono et al. 2014). It is also an essential protein linked with the cytoskeleton and is a binding microtubule with an active actin microfilament (Gungabissoon et al. 2001). There have been reports of EF1a interacting in *Nicotiana benthamiana* infections with the viral RNA-dependent RNA polymerase and 30-terminal TMV genomic RNA of TVB (Gaguancela et al. 2016). Suppression of the gene producing EF1a reduces cell death and changes this host response to the soybean mosaic virus (Li et al. 2016). However, EF1a and EF1b's functions have not yet been defined in plant-RKN interactions. The front line is the cellular wall, whose dry weight consists of cellulose, hemicellulose and pectins in 90% (Malinovsky et al. 2014).

The penultimate stage in pectin breakdown involves polygalacturonates (PGs), although their significance is also discussed in plant growth (Danalache et al. 2018). In addition, PGs were observed to eliminate programmed cell deaths in maize (He 2019). Some results also show the significance of *Glycine max* during *Heterodera* infection in plants. The involvement of pathogenic PGs and plant polygalacturonase-inhibiting proteins (PGIP) was nevertheless mainly concentrated in plant-pathogen interaction research (Haeger et al. 2020). However, the involvement of glycine-rich proteins (GRPs) in the infection process has been reasonably thoroughly investigated. GRPs were hypothesised, among other things, to identify environmental stimuli and engage in signal transduction (Czolpinska and Rurek 2018). Although GRPs have also been discovered as part of the plant defence and repair system, their molecular action method is still unclear (Mousavi and Hotta 2005). In addition, the wide range and structural variety of sub-cellular sites of GRPs show that they take part in numerous separate physiological processes (Sachetto-Martins et al. 2000). GRPs were characterised as extracellular ligands of kinase proteins, RNA-binding proteins and many other activities linked to cell wall function and plant defence response (Mangeon et al. 2010).

RNA metabolic control via glycine-rich RNA protein binding has also been described as crucial for the immune system in plants (Wasee Ullah et al. 2016). The glycine-rich RNA-binding proteins have been recognised to regulate gene expression and RNA processing after transcription, which is part of plants' developmental control (Sanan-Mishra et al. 2002). The plant response to RKN activates various cellular processes, including changes in genetic expression that encode transcription and elongation factors and cell wall organisation-related proteins. The identification of chosen transcription and elongation factors encoding the WRKY53, EF1a and EF1b genes and encoding two cell wall-related proteins (RNA-rich glycine protein, GRP, and polygalacturonase, PG) is essential. Changes in gene expression relative levels that encode these proteins were evaluated using the quantitative PCR reverse transcription reaction. Glycine-rich RNA-binding protein and EF1b are strongly involved in the maize reaction and RKN resistance. During *Arabidopsis thaliana* infection with RKN, a common transcription factor DPE2F-like 1 (DEL1) reduced salicylic acid (SA) accumulated in RKN-induced gall. In root galls, significant salicylic acid build-up occurred in the DEL1-deficient *Arabidopsis* mutant (del1-1) more resistant to RKN infection (Przybylska and Spychalski 2021).

9.6 Engineering of Maize Transcription Factors

The recent discovery of TFs as a tool to modify and generate quantitative characteristics such as drought and salinity has prompted the invention of new technologies based on TFs that benefit genes and boost agricultural products. In these attempts, the creation of TF was a key objective, an approach that has prospects for future regulation of metabolic pathways (Sakuma et al. 2002). They also produce drought-tolerant plants, dispersing the repressive feature using point mutation engineering.

Stress-related TFs are unwanted to change the boom and development from time to time, which leads to decelerated growth or toxicity (Hussain et al. 2011). ZmDREB2A overexpression in transgenic plant life at different places resulted in a significant improvement in drought tolerance under a stress promoter (Chai et al. 2020). A systemic discovery of TF families leads to accurate genes that may improve abiotic and biotic stress tolerances in insignificant crops in the *Arabidopsis* model plant (Riechmann et al. 2000). The ornithologist for maize (ZmNF-YB2) (AtNF-YB1) led to drought-resistant plants in transgenic maize after being overexpressed (Nelson et al. 2007). In order to limit the negative consequences of overexpression of specific TFs, the validation of identified genes in model plant life and necessary plant life must be conducted with the help of a stress-induced promoter (Lan Thi Hoang et al. 2017). TF and other gene families of maize were formerly impeded by using the GenBank membership and EST numbers, which so frequently cease, for example, assigning the alleged copies of the same TF to quite a few features. A famous TF nomenclature has been developed to remedy this problem, which shows similar attempts in other species.

The maize TFs are named by a family descriptor (e.g. bHLH for helix loop helix, HD for the homeodomain) and a range-opening for a species identifier (Zm for maize) with a '1' range (e.g. ZmbHLH1). On the current GRASSIUS server, an in-depth set of TFs from specific weeds together with maize is provided (Grassius 2021). This community support gives suggestions for identifying newly identified TFs for maize which urge the community to make 'TFome' available in various plants as long as feasible. The transcription elements from maize, sugar cane, sorghum and rice are presented on GrassTFDB (Grassius 2021). As sequence statistics become available, other grasses are protected. GRASSIUS will operate as a conduit to enter the Grasses Transcription Factor ORFome Collection (Fig. 9.2), where 2042 (97%) TFome maize and 62 (only 3%) rice TFome numbers were already contributed (Fig. 2; Grassius 2021). Table 9.1 presents several variables in maize transcription (TFs).

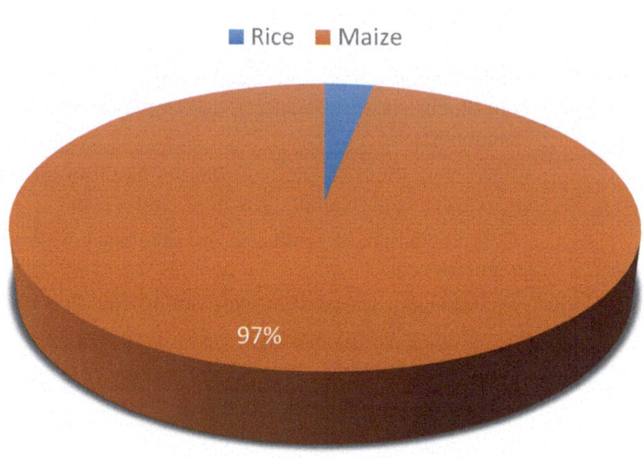

Fig. 9.2 Rice and maize TFome comparison submitted to GRASSIUS database

Table 9.1 Some maize transcription factors (TFs) and their function

Maize TF family	Function
ABI3-VP1	Involved in developing, maturing and germinating seeds
Alvin-like	Growth increases in average and saline circumstances and does not significantly influence shooting growth
AP2-EREBP	Determination or regulation of the identity of the leaf's epidermal cell to form part of plant systems for responding to different forms of biotic and environmental stress
ARF	An essential function in the expression of primary response genes by auxin-regulated genes
ARR-B	Perception and signalling of cytokine and contributed to the transduction of ethylene signal
bHLH	In several biological processes as regulatory components
bZIP	Pathogen defence, signalling of light and stress, seed ripening and growth of flower
BZR	Essential for regular plant growth
C2C2-CO-like	An important function in photoperiod regulating
C2H2	Floral leaf initiation gametogenesis and seed development lateral shoot initiation
CSD	It helps cells acclimatise cooler circumstances of growth
DBP	Allows cells to adjust the factor of transcription in response to stimuli quickly and reversibly
E2F-DP	Produced in early S phase cells with the most excellent in transcript levels
EIL	Limit the reaction from a plant to the ethylene hormone
FAR1-like	Phytochrome light-induced nuclear build-up and light reactions
FLO/LFY	Involvement to regulate homeotic genes, which may be separated from their roles in the floral fate
G2-like	Bundle sheath cell chloroplasts differentiation
GBP	Acting as a repressor of leaf cell fate
GRF (OsGRF1)	Regulatory role in stem elongation
Homeobox (Athb-12)	Some development and the plant response to water stress with gene expression mediated by ABA
Heat shock factors (HSFs)	Heat shock response and transcriptional activators
Mitochondrial termination factor family (mTERF)	Gene expression does not need to operate as transcription factors but blocks antisense transcription or plays a role in ribosome biogenesis
MYB	Secondary plant metabolism and the morphology of the cell
NAC	Auxins to stimulate the growth of lateral root
OVATE	Cell elongation, biosynthesis and secondary wall construction regulation
Required cell differentiation 1 (RDC1)	Included in a sexual differentiation process that has been developed
SHORT INTERNODE/ STYLISH	Auxin biosynthesis

(continued)

Table 9.1 (continued)

Maize TF family	Function
Cycloidea (cyc) and teosinte branched 1 (tb1)	Floral primordia, meristem growth
Trihelix	Function in fruit and seed development
WHIRLY	Plant disease resistance responses
WRKY	Pathogen defence, senescence and trichome development
ZF-HD	Regulatory role in floral development
ZIM	Jasmonic acids (JA) are essential plant protection and development hormones, and the ZIM domains are JA signalling repressors for the ZIM domain proteins

9.7 Current Scenario and Post-Genomics Approaches for Maize TFs

Various abiotic and biotic stresses include some specific features that consist of a quantitative pattern. To get awareness of the effective plant responses to a variety of biotic and abiotic stresses at their molecular level, it is thus desirable to develop their expertise in transcriptions regulations. The use of treatment currently evaluates the genetic mechanisms of more than a few abiotic, feature characterisation, genome selection, short RNA and excessive total performance SNP genotyping tools, sequence technologies and various platforms, consisting of drought, salinity and cold in the rush processing activity of the reproduction method in maize (Nepolean et al. 2018). Current science and modification in the field of genetic analyses are techniques of genome modification. RNA interference is a quick, cheap approach for assessing the gene function in precise gene knockdown analyses (Rabara et al. 2014). The harmful elements of this research are that gene inhibition has ceased to be complete and can also lead to unanticipated off-target effects that misinterpret the results (Gaj et al. 2013). 'CRISPR' is the most environmentally friendly technique for modifying plant genomes among the focused approaches presently available (Cong et al. 2013). A method for adjusting in vivo gene expression in a plant, called CRISPR (CRISPR-ATFs), is becoming increasingly recognisable (Lowder et al. 2018). The novel plant-based glaucoma and legislative methods have been established which are known as the CRISPR-Act2.0 and the mTALE-Act (Lowder et al. 2018), and some more structures are nucleases and zinc-finger activators equivalent (Boch et al. 2009; Kim and Morr 1996). Any other method that is understood in deliberate genomics in plants is to target adjacent lesions in genomes (TILLING). For example, the TILLING eco-tilling approach is used to identify changes in natural populations to effectively identify TFs in rice related to drought resistance (Yu et al. 2012). In maize surveys, it is vital to determine higher genotypes and target genes for abiotic stress resistance.

In contemporary times, the usefulness of machine learning in the search for TF's GRN is another technique. TFBS and related TFTGs using the machine learning approach contributed to the search of GRNs, in particular FATTs (Cui et al. 2014). In order to be aware of the mechanisms implicated in methods for gene law anywhere in biotic and abiotic plant stress restoration, it is essential to understand the interplay between TF, TFBS and TFTG (Fujita et al. 2006). Several programmes of the software supply the computer algorithms. Increased use of them has shown that while specific techniques have been developed for one species, the same strategies can review the information set for each species (Cui et al. 2014). For example, yeast-tested *Escherichia coli* algorithms (Faith et al. 2007), t-testing algorithms analysed on *Escherichia coli* and networks of learning modules have been utilised to identify oxidative stress management TFs in *Arabidopsis* (Faith et al. 2007). Transcriptional interactions to regulate root physiology and improvement procedures were later used for *Arabidopsis* (González-Morales et al. 2016).

Gene regulatory networks provide insight into connections between TFs and their target genes (Koryachko et al. 2015). A computer approach for learning about the TF-gene interactions in microbial TF-GRNs utilised in the assessment is a statistical verification of the TF-GRNs, essential for improving blooming in *Arabidopsis*, i.e. network element analysis (Ni et al. 2016). While severe TFs have been expressed in several organs, an evaluation of several different levels revealed that many TFs are regulatory. Furthermore, 76.6% of the genes in all maize tissues are found. 54.46% were recorded in all four tissues, of the 2.587 TF identified in the GRASSIUS maize (Meng et al. 2013), while 86.63% were expressed in at least one of the four tissues. To understand how TFs influence gene expression in response to unique abiotic and biotic stresses, it is essential to understand GRN methods (Penfold and Wild 2011). The recent introduction of a publicly available TFORF collection including 2034 clones, equates to 2017 unique co-regulatory TFs (CoREGs) (Burdo et al. 2014). The data for the synthesis, the sequences and URL requests are publicly available for TFome maize data via GRASSIUS. Finally, adapting the available crop foundations such as Gramene (Tello-Ruiz et al. 2018) and GRASSIUS to maize in the technology sector will also help store and implement new databases, allowing scientists to access information like the Wheat Information System (Shikha et al. 2017).

9.8 Conclusion

The world's population is anticipated to grow, and essential crop productivity in combination with rapid climate change has to be stepped up urgently. Recognising molecular methods and mining demanding genes that regulate the plant responses to several abiotic and biotic stimuli are a prerequisite for high-performance crop types that are stress-resistant. In order to sustain food supply worldwide, reasonably desirable plants like maize have to be developed in severe climates. However, recent advancements have dramatically broadened the potential for various stress tolerances in the reproduction of maize, genomes and practical gene analyses combined with high-performance technological gene sequencing. These changes are expected to be made to subtropical and tropical maize in developing nations, which are the essential crops in food security. TFs as a group are significant components of the genes for protein encoding. However, it was formerly challenging to make their functions apparent; particularly those that control procedures no longer produce rapid phenotypic changes. The availability to the related grass of whole-genome sequence and associated assets gives a unique opportunity to become involved in comparative genomic rules or evaluate the improvement of the TF function in a short evolution time.

Conflict of Interest The authors declare no conflict of interest.

References

Aharoni A, Dixit S, Jetter R, Thoenes E, Van Arkel G, Pereira A (2004) The SHINE clade of AP2 domain transcription factors activates wax biosynthesis, alters cuticle properties, and confers drought tolerance when overexpressed in Arabidopsis. Plant Cell 16(9):2463–2480
Ahuja I, de Vos RC, Bones AM, Hall RD (2010) Plant molecular stress responses face climate change. Trends Plant Sci 15(12):664–674
Ashtamker C, Kiss V, Sagi M, Davydov O, Fluhr R (2007) Diverse subcellular locations of cryptogein-induced reactive oxygen species production in tobacco Bright Yellow-2 cells. Plant Physiol 143(4):1817–1826
Bao Y, Vuong T, Meinhardt C, Tiffin P, Denny R, Chen S, Young ND (2014) Potential of association mapping and genomic selection to explore PI 88788 derived soybean cyst nematode resistance. Plant Genome 7(3) plantgenome2013-11
Bartoli A, Cavicchioli D, Kremmydas D, Rozakis S, Olper A (2016) The impact of different energy policy options on feedstock price and land demand for maize silage: the case of biogas in Lombardy. Energy Policy 96:351–363
Boch J, Scholze H, Schornack S, Landgraf A, Hahn S, Kay S, Bonas U (2009) Breaking the code of DNA binding specificity of TAL-type III effectors. Science 326(5959):1509–1512
Burdo B, Gray J, Goetting-Minesky MP, Wittler B, Hunt M, Li T, Grotewold E (2014) The Maize TF ome–development of a transcription factor open reading frame collection for functional genomics. Plant J 80(2):356–366
Butt HI, Yang Z, Gong Q, Chen E, Wang X, Zhao G, Li F (2017) GaMYB85, an R2R3 MYB gene, in transgenic Arabidopsis plays an important role in drought tolerance. BMC Plant Biol 17(1):1–17
Cao, D., Wang, X., Luo, X., Liu, G., & Zheng, H. (2017). Effects of polystyrene microplastics on the fitness of earthworms in an agricultural soil. In IOP conference series: earth and environmental science (Vol. 61, 1, p. 012148). IOP Publishing
Casaretto JA, El-Kereamy A, Zeng B, Stiegelmeyer SM, Chen X, Bi YM, Rothstein SJ (2016) Expression of OsMYB55 in maize activates stress-responsive genes and enhances heat and drought tolerance. BMC Genomics 17(1):1–15
Chai M, Cheng H, Yan M, Priyadarshani SVGN, Zhang M, He Q, Qin Y (2020) Identification and expression analysis of the DREB transcription factor family in pineapple (Ananas comosus (L.) Merr.). PeerJ 8:e9006
Chaves MM, Oliveira MM (2004) Mechanisms underlying plant resilience to water deficits: prospects for water-saving agriculture. J Exp Bot 55(407):2365–2384

Chujo T, Takai R, Akimoto-Tomiyama C, Ando S, Minami E, Nagamura Y, Yamane H (2007) Involvement of the elicitor-induced gene OsWRKY53 in the expression of defense-related genes in rice. Biochim Biophys Acta (BBA)-Gene Structure and Expression 1769(7–8):497–505
Ciarmiello LF, Woodrow P, Fuggi A, Pontecorvo G, Carillo P (2011) Plant genes for abiotic stress. Abiotic Stress Plants–Mech Adapt:283–308
Cong L, Ran FA, Cox D, Lin S, Barretto R, Habib N, Zhang F (2013) Multiplex genome engineering using CRISPR/Cas systems. Science 339(6121):819–823
Cui D, Guo YQ, Lee HS, Wu WM, Liang B, Wang AJ, Cheng HY (2014) Enhanced decolorization of azo dye in a small pilot-scale anaerobic baffled reactor coupled with biocatalyzed electrolysis system (ABR–BES): a design suitable for scaling-up. Bioresour Technol 163:254–261
Czolpinska M, Rurek M (2018) Plant glycine-rich proteins in stress response: an emerging, still prospective story. Front Plant Sci 9:302
Danalache F, Mata P, Alves VD, Moldão-Martins M (2018) Enzyme-assisted extraction of fruit juices. In: Fruit juices. Academic Press, pp 183–200
Davidson EH (2001) Genomic regulatory systems: in development and evolution. Elsevier
Davuluri RV, Sun H, Palaniswamy SK, Matthews N, Molina C, Kurtz M, Grotewold E (2003) AGRIS: Arabidopsis gene regulatory information server, an information resource of Arabidopsis cis-regulatory elements and transcription factors. BMC Bioinform 4(1):1–11
Desveaux D, Subramaniam R, Després C, Mess JN, Lévesque C, Fobert PR, Brisson N (2004) A "Whirly" transcription factor is required for salicylic acid-dependent disease resistance in Arabidopsis. Dev Cell 6(2):229–240
Dewitte W, Scofield S, Alcasabas AA, Maughan SC, Menges M, Braun N, Murray JA (2007) Arabidopsis CYCD3 D-type cyclins link cell proliferation and endocycles and are rate-limiting for cytokinin responses. Proc Natl Acad Sci 104(36):14537–14542
Dias L, Pereira-da-Silva S, Tavares M, Malfeito-Ferreira M, Loureiro V (2003) Factors affecting the production of 4-ethylphenol by the yeast Dekkera bruxellensis in enological conditions. Food Microbiol 20(4):377–384
Dong J, Ni W, Yu R, Deng XW, Chen H, Wei N (2017) Light-dependent degradation of PIF3 by SCFEBF1/2 promotes a photomorphogenic response in Arabidopsis. Curr Biol 27(16):2420–2430
Dou T, Shao X, Hu C, Liu S, Sheng O, Bi F, Yi G (2020) Host-induced gene silencing of Foc TR4 ERG6/11 genes exhibits superior resistance to Fusarium wilt of banana. Plant Biotechnol J 18(1):11
Du, S. S., Jin, C., Lee, J. D., Jordan, M. I., Poczos, B., & Singh, A. (2017). Gradient descent can take exponential time to escape saddle points. arXiv preprint arXiv:1705.10412
Ernst HA, Nina Olsen A, Skriver K, Larsen S, Lo Leggio L (2004) Structure of the conserved domain of ANAC, a member of the NAC family of transcription factors. EMBO Rep 5(3):297–303
Faith JJ, Hayete B, Thaden JT, Mogno I, Wierzbowski J, Cottarel G, Gardner TS (2007) Large-scale mapping and validation of Escherichia coli transcriptional regulation from a compendium of expression profiles. PLoS Biol 5(1):e8
Franco-Zorrilla JM, López-Vidriero I, Carrasco JL, Godoy M, Vera P, Solano R (2014) DNA-binding specificities of plant transcription factors and their potential to define target genes. Proc Natl Acad Sci 111(6):2367–2372
Fu S, Rogowsky P, Nover L, Scanlon MJ (2006) The maize heat shock factor-binding protein paralogs EMP2 and HSBP2 interact non-redundantly with specific heat shock factors. Planta 224(1):42–52
Fujita M, Fujita Y, Noutoshi Y, Takahashi F, Narusaka Y, Yamaguchi-Shinozaki K, Shinozaki K (2006) Crosstalk between abiotic and biotic stress responses: a current view from the points of convergence in the stress signaling networks. Curr Opin Plant Biol 9(4):436–442
Gaguancela OA, Zúñiga LP, Arias AV, Halterman D, Flores FJ, Johansen IE et al (2016) The IRE1/bZIP60 pathway and bax inhibitor 1 suppress systemic accumulation of potyviruses and potexviruses in Arabidopsis and Nicotiana benthamiana plants. Mol Plant-Microbe Interact 29(10):750–766

Gaj T, Gersbach CA, Barbas CF III (2013) ZFN, TALEN, and CRISPR/Cas-based methods for genome engineering. Trends Biotechnol 31(7):397–405

Gao Z, Deng W, Zhu F (2019) Reference gene selection for quantitative gene expression analysis in black soldier fly (Hermetia illucens). PLoS One 14(8):e0221420

Gong F, Yang L, Tai F, Hu X, Wang W (2014) "Omics" of maize stress response for sustainable food production: opportunities and challenges. Omics: J Integr Biol 18(12):714–732

González-Morales SI, Chávez-Montes RA, Hayano-Kanashiro C, Alejo-Jacuinde G, Rico-Cambron TY, de Folter S, Herrera-Estrella L (2016) Regulatory network analysis reveals novel regulators of seed desiccation tolerance in Arabidopsis thaliana. Proc Natl Acad Sci 113(35):E5232–E5241

Grassius (2021) Welcome to the transcription factor database. Available at: https://www.grassius.org/grasstfdb.php. Accessed on: 11th Sept 2021

Gungabissoon RA, Khan S, Hussey PJ, Maciver SK (2001) Interaction of elongation factor 1α from Zea mays (ZmEF-1α) with F-actin and interplay with the maize actin severing protein, ZmADF3. Cell Motil Cytoskeleton 49(2):104–111

Gupta K, Jha B, Agarwal PK (2014) A dehydration-responsive element binding (DREB) transcription factor from the succulent halophyte Salicornia brachiata enhances abiotic stress tolerance in transgenic tobacco. Mar Biotechnol 16(6):657–673

Haeger A, Alexander S, Vullings M, Kaiser FM, Veelken C, Flucke U, Friedl P (2020) Collective cancer invasion forms an integrin-dependent radioresistant niche. J Exp Med 217(1)

Hake, S. (2009). Handbook of maize (pp. 693-713). J. L. Bennetzen (Ed.). New York: Springer

Hall LN, Rossini L, Cribb L, Langdale JA (1998) GOLDEN 2: a novel transcriptional regulator of cellular differentiation in the maize leaf. Plant Cell 10(6):925–936

He JH (2019) Lagrange crisis and generalized variational principle for 3D unsteady flow. Int J Numer Meth Heat Fluid Flow

Heine VM, Maslam S, Zareno J, Joëls M, Lucassen PJ (2004) Suppressed proliferation and apoptotic changes in the rat dentate gyrus after acute and chronic stress are reversible. Eur J Neurosci 19(1):131–144

Honório NA, Nogueira RMR, Codeço CT, Carvalho MS, Cruz OG, de Avelar Figueiredo Mafra Magalhães M et al (2009) Spatial evaluation and modeling of dengue seroprevalence and vector density in Rio de Janeiro, Brazil. PLoS Negl Trop Dis 3(11):e545

Hosoda K, Imamura A, Katoh E, Hatta T, Tachiki M, Yamada H, Yamazaki T (2002) Molecular structure of the GARP family of plant Myb-related DNA binding motifs of the Arabidopsis response regulators. Plant Cell 14(9):2015–2029

Hu Z, Fan C, Oh DS, Marron JS, He X, Qaqish BF, Perou CM (2006) The molecular portraits of breast tumors are conserved across microarray platforms. BMC Genomics 7(1):1–12

Hussain A, Villarreal-Barajas JE, Dunscombe P, Brown DW (2011) Aperture modulated, translating bed total body irradiation. Med Phys 38(2):932–941

Jammes F, Lecomte P, de Almeida-Engler J, Bitton F, Martin-Magniette ML, Renou JP et al (2005) Genome-wide expression profiling of the host response to root-knot nematode infection in Arabidopsis a. Plant J 44(3):447–458

Jiao Y, Peluso P, Shi J, Liang T, Stitzer MC, Wang B, Ware D (2017) Improved maize reference genome with single-molecule technologies. Nature 546(7659):524–527

Kim YD, Morr CV (1996) Microencapsulation properties of gum arabic and several food proteins: spray-dried orange oil emulsion particles. J Agric Food Chem 44(5):1314–1320

Kim CY, Vo KTX, Nguyen CD, Jeong DH, Lee SK, Kumar M et al (2016) Functional analysis of a cold-responsive rice WRKY gene, OsWRKY71. Plant Biotechnol Rep 10(1):13–23

Kimotho RN, Baillo EH, Zhang Z (2019) Transcription factors involved in abiotic stress responses in maize (Zea mays L.) and their roles in enhanced productivity in the post genomics era. PeerJ 7:e7211

Klein J, Saedler H, Huijser P (1996) A new family of DNA binding proteins includes putative transcriptional regulators of the Antirrhinum majus floral meristem identity gene SQUAMOSA. Mol Gen Genet MGG 250(1):7–16

Knoema (2021) World – Maize production quantity. Available at: https://knoema.com/atlas/World/topics/Agriculture/Crops-Production-Quantity-tonnes/Maize-production. Accessed on: 30th Aug 2021

Koryachko A, Matthiadis A, Ducoste JJ, Tuck J, Long TA, Williams C (2015) Computational approaches to identify regulators of plant stress response using high-throughput gene expression data. Curr Plant Biol 3:20–29

Lan Thi Hoang X, Du Nhi NH, Binh Anh Thu N, Phuong Thao N, Phan Tran LS (2017) Transcription factors and their roles in signal transduction in plants under abiotic stresses. Curr Genomics 18(6):483–497

Li N, Yin JL, Li C, Wang DG, Yang YQ, Karthikeyan A, Zhi HJ (2016) NB-LRR gene family required for Rsc4-mediated resistance to Soybean mosaic virus. Crop Pasture Sci 67(5):541–552

Li L, Liu Q, Liu T, Cui X, Ning W (2021) Expression of putative luteolin biosynthesis genes and WRKY transcription factors in Taraxacum antungense kitag. Plant Cell Tissue Organ Culture (PCTOC) 145(3):649–665

Lipsick JS (1996) One billion years of Myb. Oncogene 13(2):223–235

Liu X, Chen X, Zhong B, Wang A, Wang X, Chu F, Dong C (2014) Transcription factor achaete-scute homologue 2 initiates follicular T-helper-cell development. Nature 507(7493):513–518

Lowder LG, Zhou J, Zhang Y, Malzahn A, Zhong Z, Hsieh TF, Qi Y (2018) Robust transcriptional activation in plants using multiplexed CRISPR-Act2. 0 and mTALE-Act systems. Mol Plant 11(2):245–256

Malinovsky FG, Fangel JU, Willats WG (2014) The role of the cell wall in plant immunity. Front Plant Sci 5:178

Mangeon A, Junqueira RM, Sachetto-Martins G (2010) Functional diversity of the plant glycine-rich proteins superfamily. Plant Signal Behav 5(2):99–104

Meng Q, Hou P, Wu L, Chen X, Cui Z, Zhang F (2013) Understanding production potentials and yield gaps in intensive maize production in China. Field Crop Res 143:91–97

Molina C, Grotewold E (2005) Genome wide analysis of Arabidopsis core promoters. BMC Genomics 6(1):1–12

Moon SJ, Han SY, Kim DY, Yoon IS, Shin D, Byun MO et al (2015) Ectopic expression of a hot pepper bZIP-like transcription factor in potato enhances drought tolerance without decreasing tuber yield. Plant Mol Biol 89(4):421–431

Mousavi A, Hotta Y (2005) Glycine-rich proteins. Appl Biochem Biotechnol 120(3):169–174

Negi N, Khurana P (2021) A salicylic acid inducible mulberry WRKY transcription factor, Mi WRKY53 is involved in plant defence response. Plant Cell Rep:1–21

Nelson ME, Rejeski WJ, Blair SN, Duncan PW, Judge JO, King AC et al (2007) Physical activity and public health in older adults: recommendation from the American College of Sports Medicine and the American Heart Association. Circulation 116(9):1094

Nepolean T, Kaul J, Mukri G, Mittal S (2018) Genomics-enabled next-generation breeding approaches for developing system-specific drought tolerant hybrids in maize. Front Plant Sci 9:361

Ni Y, Aghamirzaie D, Elmarakeby H, Collakova E, Li S, Grene R, Heath LS (2016) A machine learning approach to predict gene regulatory networks in seed development in Arabidopsis. Front Plant Sci 7:1936

Ogata K, Morikawa S, Nakamura H, Sekikawa A, Inoue T, Kanai H, Nishimura Y (1994) Solution structure of a specific DNA complex of the Myb DNA-binding domain with cooperative recognition helices. Cell 79(4):639–648

Ohme-Takagi M, Shinshi H (1995) Ethylene-inducible DNA binding proteins that interact with an ethylene-responsive element. Plant Cell 7(2):173–182

Penfold CA, Wild DL (2011) How to infer gene networks from expression profiles, revisited. Interf Focus 1(6):857–870

Perlack, R. D. (2005). Biomass as feedstock for a bioenergy and bioproducts industry: the technical feasibility of a billion-ton annual supply. Oak Ridge National Laboratory

Przybylska A, Spychalski M (2021) Changes in the expression level of genes encoding transcription factors and cell wall-related proteins during Meloidogyne arenaria infection of maize (Zea mays). Mol Biol Rep

Rabara RC, Tripathi P, Rushton PJ (2014) The potential of transcription factor-based genetic engineering in improving crop tolerance to drought. Omics: J Integr Biol 18(10):601–614

Rabinowicz PD, Braun EL, Wolfe AD, Bowen B, Grotewold E (1999) Maize R2R3 Myb genes: sequence analysis reveals amplification in the higher plants. Genetics 153(1):427–444

Ramakrishna C, Singh S, Raghavendrarao S, Padaria JC, Mohanty S, Sharma TR, Solanke AU (2018) The membrane tethered transcription factor EcbZIP17 from finger millet promotes plant growth and enhances tolerance to abiotic stresses. Sci Rep 8(1):1–14

Riechmann JL, Meyerowitz EM (1997) MADS domain proteins in plant development. Biol Chem 378(10):1079–1102

Riechmann JL, Heard J, Martin G, Reuber L, Jiang CZ, Keddie J, Yu GL (2000) Arabidopsis transcription factors: genome-wide comparative analysis among eukaryotes. Science 290(5499):2105–2110

Ross CA, Liu Y, Shen QJ (2007) The WRKY gene family in rice (Oryza sativa). J Integr Plant Biol 49(6):827–842

Rossini L, Cribb L, Martin DJ, Langdale JA (2001) The maize golden2 gene defines a novel class of transcriptional regulators in plants. Plant Cell 13(5):1231–1244

Rushton PJ, Somssich IE, Ringler P, Shen QJ (2010) WRKY transcription factors. Trends Plant Sci 15(5):247–258

Sachetto-Martins G, Franco LO, de Oliveira DE (2000) Plant glycine-rich proteins: a family or just proteins with a common motif? Biochim Biophys Acta (BBA)-Gene Structure and Expression 1492(1):1–14

Sakuma Y, Liu Q, Dubouzet JG, Abe H, Shinozaki K, Yamaguchi-Shinozaki K (2002) DNA-binding specificity of the ERF/AP2 domain of Arabidopsis DREBs, transcription factors involved in dehydration-and cold-inducible gene expression. Biochem Biophys Res Commun 290(3):998–1009

Sanan-Mishra N, Tuteja N, Sopory SK (2002) Salinity-and ABA-induced up-regulation and light-mediated modulation of mRNA encoding glycine-rich RNA-binding protein from Sorghum bicolor. Biochem Biophys Res Commun 296(5):1063–1068

Shigyo M, Hasebe M, Ito M (2006) Molecular evolution of the AP2 subfamily. Gene 366(2):256–265

Shikha, M., Kanika, A., Rao, A. R., Mallikarjuna, M. G., Gupta, H. S., & Nepolean, T. (2017). Genomic selection for drought tolerance using genome-wide SNPs in maize. Front Plant

Stevenson CE, Burton N, Costa MM, Nath U, Dixon RA, Coen ES, Lawson DM (2006) Crystal structure of the MYB domain of the RAD transcription factor from Antirrhinum majus. Proteins: Struct Func Bioinform 65(4):1041–1045

Stracke R, Werber M, Weisshaar B (2001) The R2R3-MYB gene family in Arabidopsis thaliana. Curr Opin Plant Biol 4(5):447–456

Suhandono S, Apriyanto A, Ihsani N (2014) Isolation and characterization of three cassava elongation factor 1 alpha (MeEF1A) promoters. PLoS One 9(1):e84692

Suzuki N, Rivero RM, Shulaev V, Blumwald E, Mittler R (2014) Abiotic and biotic stress combinations. New Phytol 203(1):32–43

Tello-Ruiz MK, Naithani S, Stein JC, Gupta P, Campbell M, Olson A, Ware D (2018) Gramene 2018: unifying comparative genomics and pathway resources for plant research. Nucleic Acids Res 46(D1):D1181–D1189

Ullah A, Sun H, Yang X, Zhang X (2017) Drought coping strategies in cotton: increased crop per drop. Plant Biotechnol J 15(3):271–284

Umezawa T, Fujita M, Fujita Y, Yamaguchi-Shinozaki K, Shinozaki K (2006) Engineering drought tolerance in plants: discovering and tailoring genes to unlock the future. Curr Opin Biotechnol 17(2):113–122

Wang Y, Liu C, Li K, Sun F, Hu H, Li X, Liu M (2007) Arabidopsis EIN2 modulates stress response through abscisic acid response pathway. Plant Mol Biol 64(6):633–644

Wang J, Wang Z, Mao H, Zhao H, Huang D (2013) Increasing Se concentration in maize grain with soil-or foliar-applied selenite on the Loess Plateau in China. Field Crop Res 150:83–90

Wang Q, Wang Z, Awasthi MK, Jiang Y, Li R, Ren X, Zhang Z (2016) Evaluation of medical stone amendment for the reduction of nitrogen loss and bioavailability of heavy metals during pig manure composting. Bioresour Technol 220:297–304

Wasee Ullah R, Bin Zahur A, Latif A, Iqbal Dasti J, Irshad H, Afzal M, Qureshi ZUA (2016) Detection of peste des petits ruminants viral RNA in fecal samples of goats after an outbreak in Punjab province of Pakistan: a longitudinal study. BioMed Res Int

Wright SI, Bi IV, Schroeder SG, Yamasaki M, Doebley JF, McMullen MD, Gaut BS (2005) The effects of artificial selection on the maize genome. Science 308(5726):1310–1314

Wurzinger B, Mair A, Pfister B, Teige M (2011) Cross-talk of calcium-dependent protein kinase and MAP kinase signaling. Plant Signal Behav 6(1):8–12

Xing X, Li X, Zhang M, Wang Y, Liu B, Xi Q, Yang T (2017) Transcriptome analysis of resistant and susceptible tobacco (Nicotiana tabacum) in response to root-knot nematode Meloidogyne incognita infection. Biochem Biophys Res Commun 482(4):1114–1121

Xiong L, Schumaker KS, Zhu JK (2002) Cell signaling during cold, drought, and salt stress. The Plant Cell 14(suppl_1):S165–S183

Xu K, Xu X, Fukao T, Canlas P, Maghirang-Rodriguez R, Heuer S, Mackill DJ (2006) Sub1A is an ethylene-response-factor-like gene that confers submergence tolerance to rice. Nature 442(7103):705–708

Yamasaki K, Kigawa T, Inoue M, Tateno M, Yamasaki T, Yabuki T, Yokoyama S (2004) Solution structure of the B3 DNA binding domain of the Arabidopsis cold-responsive transcription factor RAV1. Plant Cell 16(12):3448–3459

Yamasaki K, Chuang VTG, Maruyama T, Otagiri M (2013) Albumin–drug interaction and its clinical implication. Biochim Biophys Acta (BBA)-General Subjects 1830(12):5435–5443

Yasumura Y, Moylan EC, Langdale JA (2005) A conserved transcription factor mediates nuclear control of organelle biogenesis in anciently diverged land plants. Plant Cell 17(7):1894–1907

Yu H, Gerstein M (2006) Genomic analysis of the hierarchical structure of regulatory networks. Proc Natl Acad Sci 103(40):14724–14731

Yu S, Liao F, Wang F, Wen W, Li J, Mei H, Luo L (2012) Identification of rice transcription factors associated with drought tolerance using the ecotilling method. PLoS One 7(2):e30765

Yu X, Liu Y, Wang S, Tao Y, Wang Z, Shu Y, Ma H (2016) CarNAC4, a NAC-type chickpea transcription factor conferring enhanced drought and salt stress tolerances in Arabidopsis. Plant Cell Rep 35(3):613–627

Zentgraf U, Doll J (2019) Arabidopsis WRKY53, a node of multi-layer regulation in the network of senescence. Plan Theory 8(12):578

Zhu JK (2016) Abiotic stress signaling and responses in plants. Cell 167(2):313–324

Chapter 10
The Role of Transcription Factors in Response to Biotic Stresses in Pearl Millet

Jeky Chanwala, Deepak Kumar Jha, I. Sriram Sandeep, and Nrisingha Dey

Abstract Pearl millet is an important C4 cereal crop, mostly grown in marginal areas of arid and semi-arid regions of Africa and Asia. Biotic stresses like fungal, bacterial, and viral diseases adversely affect the crop production. Downy mildew, blast, rust, ergot, etc. are major constraints for pearl millet productivity. Plants have developed various defense processes to address the biotic stresses. Upon pathogen attack, signal transduction pathways activate the stress-related regulatory elements for efficient responses against pathogens. Transcription factors (TFs) are the regulatory proteins that act as molecular switches and regulate the expression of stress-related genes by binding to their cognate cis-acting elements present in respective promoter region. Over the past decade, several TF families such as AP2/ERF, WRKY, NAC, and MYB have been identified, and their role is explored in plant's defense mechanisms. This chapter highlights about current understanding and explores the role of various transcription factors and their involvement in different phytohormonal signaling pathways during defense responses in pearl millet.

Keywords Transcription factors · Phytohormonal signaling · Biotic stress · Regulatory elements

J. Chanwala · D. K. Jha
Division of Plant and Microbial Biotechnology, Institute of Life Sciences, Bhubaneswar, India

Regional Centre for Biotechnology, Faridabad, India

I. S. Sandeep · N. Dey (✉)
Division of Plant and Microbial Biotechnology, Institute of Life Sciences, Bhubaneswar, India
e-mail: ndey@ils.res.in

© The Author(s), under exclusive license to Springer Nature Switzerland AG 2022
S. H. Wani et al. (eds.), *Transcription Factors for Biotic Stress Tolerance in Plants*, https://doi.org/10.1007/978-3-031-12990-2_10

10.1 Introduction

Pennisetum glaucum, also known as pearl millet, belonging to Poaceae family, is the sixth most important cereal crop in the world. Pearl millet occupies approximately 30–40 million hectares in more than 30 countries accounting for a gross production of around 50% of total millets. It is extensively grown as a rain-fed crop in sub-Saharan Africa and Southeast Asia including India, for food and fodder (Jukanti et al. 2016). In Central as well as Western African countries, millets occupy approximately 15.7 million hectares, out of which more than 90% of the area is used for pearl millet cultivation. Pearl millet has a high nutritional value with high protein, low starch content, and high fiber value. It is also a rich source of essential micronutrients such as iron and zinc (Kumar et al. 2016). Being a C4 plant, pearl millet produces biomass efficiently with minimal water requirement (Jaiswal et al. 2018). This crop plants are well adapted to severe abiotic stresses including elevated temperature, high salinity, drought, high soil pH, and Al^{3+} saturation (Dudhate et al. 2018). Pearl millet has the capability to produce adequate yields in harsh climatic conditions, whereas other plants fail to survive, thus making it an attractive choice for understanding mechanisms involved in abiotic stress tolerance (Shinde et al. 2018).

However, pearl millet production is limited by various bacterial, fungal, and viral pathogens. Fungal diseases such as downy mildew, blast, rust, ergot, and smut are major constraints for pearl millet yield. Downy mildew is responsible for maximum yield loss in India and Africa. Rust and blast also affect the leaves that lead to forage loss. The severity of the disease causes drying and shedding of leaves and stunted growth of the plant (Kulkarni et al. 2016). In addition, smut and ergot are floral diseases that affect the pearl millet grain yield. Both the pathogens are soil-borne and infect the plant at the flowering stage through the stigma, where the fungus replaces the millet seeds with its sclerotia, thus making the grain unfit for consumption. Therefore, it is important to understand the molecular mechanisms of plant-pathogen interactions toward developing an efficient strategy for pearl millet crop improvement. Plants have developed various molecular mechanisms to adapt and withstand these biotic stresses. Transcription factors (TFs) are the regulatory proteins that act as molecular switches in controlling stress-responsive gene expression. The role of various transcription factors and their involvement in different phytohormonal signaling pathways during defense responses is an important aspect for development of biotic stress-resistant pearl millet.

10.2 Biotic Constraints

Pearl millet production is affected by several biotic factors such as bacteria, viruses, fungi, parasites, pests, weeds, etc. Mainly biotic stress is caused by fungal diseases such as downy mildew, blast, rust, ergot, and smut. Downy mildew, rust, and blast affect the leaf, characterized by leaf spots contributing to forage loss. The severity of disease causes drying and shedding of leaves and stunted growth of the plant. The smut and ergot are floral diseases in pearl millet that lead to severe loss in grain yield. Both the pathogens causing disease are soil-borne and infect the plant at the flowering stage through the stigma.

10.2.1 Downy Mildew

Downy mildew or "green ear disease" is caused by *Sclerospora graminicola* (Sacc.) Schroet. It is an oomycete that is an obligate parasite to plant. Initial symptoms are developed on the leaf that appear as foliar chlorosis and yellowing of lower leaves. Further, spores are gradually germinated to form 3 to 13 encysted zoospores which are liberated in favorable conditions (Nene and Singh 1976; Nagaraja and Das 2016). The leaf becomes brown and dry due to sexual oospores, then gradually other leaves are affected, plants' growth is stunted, and green ear symptoms of panicles (floral parts are converted into the leafy structure) occur. Annual estimation shows approximately 20–40% crop yield is lost due to downy mildew disease (Shetty et al. 2016).

10.2.2 Blast

Blast commonly refers to grey leafy spots (caused by *Pyricularia grisea*; teleomorph, *Magnaporthe grisea* (Herbert) Barr), the second most severe disease which generally affects leaves and stems. At the initial stage, small grayish lesions surrounded by chlorotic halo appear and then gradual necrosis (concentric ring appearance) of the plant part leading to the drying of young leaves. The premature drying of young leaves leads to a reduction in grain and forage yield. Mean blast disease severity ranges from 10% to 30% (Nayaka et al. 2017).

10.2.3 Rust

Rust characterized by raised reddish-brown spot with yellow halos appears on the surface of the leaf (caused by *Puccinia substriata* var. *indica* Ramachar & Cumm.). Pearl millet acts as the primary host to *P. substriata* species which produces two types of spores urediniospores and later teliospores (have thicker walls) that spread the disease to a distant field. As teliospores fall on the alternative host like brinjal (*Solanum melongena*) that produces spermatia and air-borne aeciospores. Then these aeciospores infect the pearl millet leaves and continue the reproduction cycle (Thakur et al. 2009). The disease symptoms appear as round or elliptical-shaped reddish-orange-hued postulates on the leaf or stem of the pearl millet plant. The premature desiccation of leaves leads to death, thus a loss in forage. The severity of this disease has been observed around various parts of the world such as North America, South America, Asia, and Africa (Nagaraja and Das 2016).

10.2.4 Ergot

Ergot or sugary disease (caused by *Claviceps fusiformis* Lov.) is an important and widespread fungal disease. It is an ascomycete fungus, and symptoms of ergot include mucilaginous creamy-pink fluid that contains conidia in florets (also called honeydew). Gradually, the droplets on the panicles become sclerotia (dark fungal masses) in place of grains in the final stage of the disease. The sclerotia may blend with other seeds during harvest or be left in the soil, which serves as the primary inoculum. In favorable conditions the sclerotia germinate and produce air-borne ascospores, which infect healthy pearl millet floret of next season crops (Nagaraja and Das 2016). This leads to a loss in grain and feed yield. The sclerotia contain alkaloids that have the potential to affect the health of humans and animals. The estimated grain yield losses are as high as 58–70%. The pearl millet ergot has been reported in countries including Africa and Asia (Thakur et al. 2009).

10.2.5 Smut

Smut (caused by *Tolyposporium penicillariae* Bref.) is a major disease that causes grain loss in pearl millet. The grains are replaced by smut sori, generally larger in size. The teleutospores (fertile spores) aggregated to form a black and powdery mass of spores, encapsulated within a film. At an early stage, the color of the sori remains green, but gradually it becomes brownish-black at maturity. Then sori wall burst and the spores are released to the surrounding (Dashora et al. 2008). During the favorable condition, the teleutospores germinate to form promycelium; then chained sporidia is produced. The germination and mating of compatible sporidia lead to the formation of dikaryotic infection hypha, which infects the pearl millet floret by infiltrating the young stigma. The reported grain yield loss is around 30%, and the affected countries are the United States, Western and Central Africa, and India.

10.2.6 Other Diseases

Several bacterial and viral diseases such as bacterial leaf streak, leaf stripe, leaf spot, *Maize dwarf mosaic virus*, and *Maize streak virus* have been reported in pearl millet. However, damage caused by these diseases is minimal and has negligible economic damage. They become important under specific climatic conditions.

10.3 The Role of Plant Transcription Factors in Response to Biotic Stresses

Biotic stresses severely affect crop yield and are major threats to global food security. Plants have evolved with various defense processes to address the biotic stresses, and during pathogen attack, signal transduction pathways activate the stress-related regulatory elements for efficient responses against pathogens. Transcription factors (TFs) play a key role in conversion of stress signal perception to stress-responsive gene expression (Fig. 10.1). TFs regulate expression of stress-related genes by binding to their cognate cis-acting elements present in respective promoter region. Over the past decade, several TF families such as AP2/ERF, WRKY, NAC, and MYB have been identified, and their role is explored in plants' defense mechanism (Table 10.1).

10.3.1 WRKY TFs

The WRKY TFs are identified in several crop plants, and their functions are explored in various plant growth, development processes, and stress responses. WRKY proteins contain a WRKY domain consisting of 60 amino acids. It is composed of a conserved motif "WRKYGQK" followed by a zinc finger-like motif. The WRKY TFs bind to W-box (with core motif TTGACC/T) of downstream target gene promoters to manage their transcription and play a vital role in controlling various developmental and physiological processes as well as in biotic/abiotic stress responses. Based on conservation of WRKY domain and zinc finger motif, WRKY TF family members are distributed into three major groups. Group I members are conserved with two WRKY domains, whereas only one WRKY domain is present in Group II and Group III members. C2H2-type zinc finger motif is conserved in Group I and Group II members, while C2HC-type zinc finger motif is present in Group III members. Further, Group II members are divided into five subgroups based on their phylogenetic relationship analysis. The WRKY family members are known to play a crucial role in regulation of various biotic and abiotic stress responses. WRKY TFs also play an important role in signal transduction pathways under biotic stress (Eulgem et al. 1999). Detailed information is available on various platforms to lay light on the crucial role of the WRKY transcription factor in providing immunity for plants against pathogen attacks.

For example, TIR-NBS-LRR (Toll/interleukin-1 receptor-nucleotide-binding site-leucine-rich repeat) domain containing AtWRKY52 interacts with RPS4 to develop resistance against bacterial pathogen *Pseudomonas syringae* and fungal pathogen *Colletotrichum higginsianum* (Narusaka et al. 2009). AtWRKY8

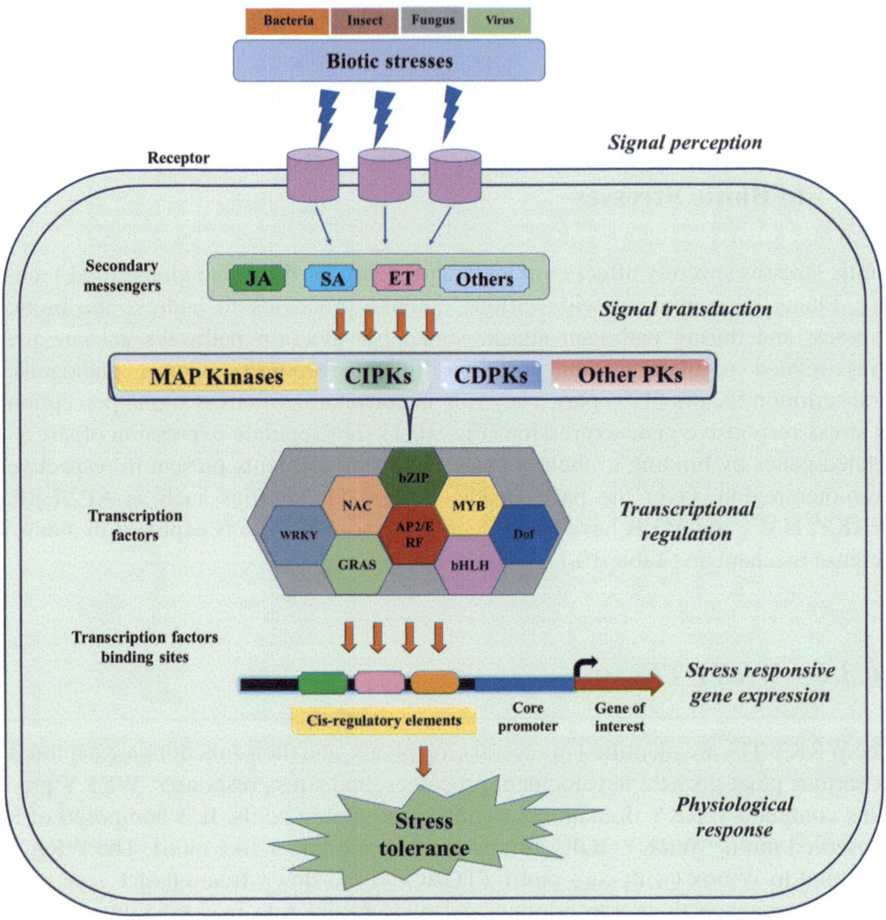

Fig. 10.1 Schematic representation of signaling pathway and TFs as key elements in transcriptional regulatory network in defense responses during biotic stress

mediates basal defense against *Pseudomonas syringae* and *Botrytis cinerea* (Chen et al. 2010). AtWRKY57 negatively regulates plant immunity against *Botrytis cinerea* by transcriptionally regulating the expression of JAZs (repressor of JA signaling pathway) (Jiang and Yu 2016). Similarly, overexpression of OsWRKY13 activates SA-induced defense signaling, suppresses JA signaling, and develops resistance to fungal blast and bacterial blight (Qiu et al. 2007). In rice, broad-spectrum resistance against *Magnaporthe oryzae* is conferred by the physical interaction of OsWRKY45 with a CC-NB-LRR (coiled coil-nucleotide-binding site-leucine-rich repeat) protein Pb1 (Panicleblast1) (Inoue et al. 2013). CaWRKY27 develops resistance against *Ralstonia solanacearum* infection in tobacco (Dang et al. 2014). In grapes, VvWRKY1 overexpression develops stress resistance against downy mildew (Marchive et al. 2013). In pearl millet, WRKYs have been identified and explored

Table 10.1 TFs involved in biotic stress response in different plants

TF	Pathogen	Plant
ATWRKY48	*Pseudomonas syringae*	*Arabidopsis*
AtWRKY57	*Botrytis cinerea*	*Arabidopsis*
ATWRKY75	*Pseudomonas syringae*	*Arabidopsis*
ATWRKY75	*Pseudomonas syringae*	*Arabidopsis*
AtWRKY8	*Pseudomonas syringae*	*Arabidopsis*
BoWRKY6	*Hyaloperonospora parasitica*	Broccoli
CaWKY6	*R. solanacearum*	*Capsicum*
CaWRKY27	*Ralstonia solanacearum*	Pepper
CaWRKY58	*Botrytis cinerea*	Pepper
CsWRKY50	*Pseudoperonospora cubensis*	Cucumber
FaWRKY1	*Pseudomonas syringae*	Strawberry
GhWRKY15	*Colletotrichum gossypii*	Cotton
GhWRKY15	*Phytophthora parasitica*	Cotton
GhWRKY44	*Ralstonia solanacearum*	Cotton
GhWRKY44	*Rhizoctonia solani*	Cotton
MdWRKY1	*Ralstonia solanacearum*	Apple
NtWRKY50	*Ralstonia solanacearum*	Tobacco
OsWRKY13	*Xanthomonas oryzae* and *M. oryzae*	Rice
OsWRKY22	*Magnaporthe oryzae*	Rice
OsWRKY28	*Xanthomonas oryzae*	Rice
OsWRKY6	*Xanthomonas oryzae*	Rice
OsWRKY62	*Xanthomonas oryzae*	Rice
OsWRKY71	*Xanthomonas oryzae*	Rice
OsWRKY76	*Xanthomonas oryzae*	Rice
PtrWRKY73	*Pseudomonas syringae*	Black cottonwood
PtrWRKY89	*Botrytis cinerea*	Black cottonwood
SiWRKY70	*Meloidogyne javanica*	Tomato
SiWRKY70	*Macrosiphum euphorbiae*	Tomato
SiWWRKY39	*Pseudomonas syringae*	Tomato
StWRKY1	*Phytophthora infants*	Potato
VvWRKY1	*Pseudoperonospora cubensis*	Grapes
VvWRKY1	*Plasmopara viticola*	Grapes
ANAC019	*P. syringae*	*Arabidopsis*
ANAC055	*Botrytis cinerea*	*Arabidopsis*
ANAC072	*P. syringae*	*Arabidopsis*
ATAF1, PR1	*Botrytis cinerea*	*Arabidopsis*
Ataf2	*Fusarium oxysporum*	*Arabidopsis*
CaNAC1	*Xanthomonas axonopodis* pv. *glycines*	Pepper
HvNAC6	*Blumeria graminis*	Barley
ONAC122	*Magnaporthe grisea*	Rice
OsNAC19	*Diseases resistance*	Rice
OsNAC4	*Magnaporthe grisea*	Rice

(continued)

Table 10.1 (continued)

TF	Pathogen	Plant
OsNAC6	*Magnaporthe grisea*	Rice
OsNAC66	*Magnaporthe oryzae*	Rice
SiSRN1	*Botrytis cinerea*	Tomato
SlNAC1	*Pseudomonas syringae*	Tomato
TaNAC4	*Puccinia striiformis*	Wheat
TaNAC69	*Puccinia triticina*	Wheat
TaNAC8	*Puccinia striiformis* f. sp. *tritici*	Wheat
VvNAC1	*Botrytis cinerea*	Grapes
GbNAC1	*Verticillium dahliae*	Cotton
GhATAF1	*Verticillium dahliae*	Cotton
GhATAF1	*Botrytis cinerea*	Cotton
GmbZIP19	*Pseudomonas syringae*	Soybean
GmbZIP19	*Sclerotinia sclerotiorum*	Soybean
OBF PROTEIN	SA	*Arabidopsis*
PPI1	Pathogen infection	*Capsicum chinense*
RT42C09	Pathogen infection	*Theobroma cacao*
RT57A09	Pathogen infection	*Theobroma cacao*
rTGA2.1, rTGA2.2, rTGA2.3	SA	Rice
TGA MEMBERS	SA	*Arabidopsis*
VvbZIP23	ABA, ET, JA, and SA	Grapes
AtMYB96	*P. syringae*	*Arabidopsis*
AtMYB44	*P. syringae*	*Arabidopsis*
AtMYB30	*P. syringae*, *Xanthomonas campestris*	*Arabidopsis*
SpMYB	*F. oxysporum*, *B. cinerea*	Currant tomato
SpMYB	*A. alternata*	Currant tomato
TiMYB2R-1	*Gaeumannomyces graminis*	Wheatgrass
TaRIM1	*Rhizoctonia cerealis*	Wheat
TaPIMP1	*B. sorokiniana*	Wheat
TaPIMP1	*R. solanacearum*	Wheat
OsMYB4	Virus infection	Rice
SlERF1	*Rhizopus nigricans*	Tomato
SlERF1	*Botrytis cinerea*, *Xanthomonas campestris*	Tomato
SlERF2	*Botrytis cinerea*	Tomato
SlERF3	*Ralstonia solanacearum*	Tomato
SlERF84	*P. syringae* pv. *DC3000*	Tomato
StERF3	*Phytophthora infestans*	Potato
GmERF3	*Ralstonia solanacearum*, *Alternaria alternata*	Soybean
GmERF3	*Tobacco mosaic virus*	Soybean
GmERF5/GmERF113	*Phytophthora sojae*	Soybean

(continued)

Table 10.1 (continued)

TF	Pathogen	Plant
SlERF5	*R. solanacearum*	Tomato
NtERF5	*Tobacco mosaic virus*	Tobacco
AtERF1	*Botrytis cinerea*	*Arabidopsis*
AtERF2, AtERF4	*Fusarium oxysporum*	*Arabidopsis*
AtERF014	*Pseudomonas syringae*	*Arabidopsis*
AtERF014	*B. cinerea*	*Arabidopsis*
CaPF1	*Bacillus thuringiensis*	*Capsicum*
OsEREBP1	*Xanthomonas oryzae*	Rice
TabHLH060	*Pseudomonas syringae*	Wheat
GmPIB1	*Phytophthora sojae*	Soybean

their involvement in abiotic stress. Kulkarni et al. (2016) have found differentially expressed WRKYs (pm_c21517, pm_c16801, pm_c8821) upon downy mildew infection (Kulkarni et al. 2016). Still, studies required on WRKY members of pearl millet to decipher their role in signal transduction pathways linked to biotic stress response.

10.3.2 NAC TFs

NAC TF family is one of the major families that are involved in various processes related to plant growth, development, and stress responses. All NAC TFs share a conserved NAC domain [no apical meristem (NAM), *Arabidopsis thaliana* transcription activation factor (ATAF1/2), and cup-shaped cotyledon proteins (CUC2)] (Aida et al. 1997). The NAC proteins are conserved with a DNA-binding domain of 150–160 amino acids at N-terminal region, which is further subdivided into five subdomains (A–E). Besides, the NAC domain also plays a role in nuclear localization and homodimer/heterodimer formation with other NAC proteins. NAC family proteins play a vital role in various developmental and abiotic/biotic stress responses. Various studies have established the role of NAC members in plant defense in response to various pathogens. NAC proteins regulate the defense response both by activating and suppressing the expression of PR genes. Accumulating evidence also shows that ATAF, a subfamily of NAC TF, is a key regulator of plant responses under pathogen attack. In *Arabidopsis*, ATAF repressed the expression level of PR genes that results in susceptibility to *F. oxysporum* infection (Delessert et al. 2005). Similarly, it was observed that overexpression of GhATAF1 in cotton resulted in susceptibility to *V. dahliae* and *B. cinerea*. Additionally, it was observed that GhATAF1 was involved in phytohormonal-mediated signaling networks (He et al. 2016). In contrast, Wang et al. (2009) reported that overexpression of ATAF significantly reduced virus accumulation and enhanced the expression of defense genes (Wang et al. 2009). Furthermore, Du et al. (2014) reported that two homologous

NAC TFs (JA2 and JA2L) in tomato play a major role in pathogen-induced stomatal closure and reopening via different mechanisms. SA is an important signaling molecule that has a crucial role in plant immune responses against fungal pathogens (Du et al. 2014). Also, Wang et al. (2015) reported that TaNAC1 acts as a negative regulator of plant defense and is involved in SA and JA cross-talk (Wang et al. 2015). Though pearl millet is considered an ideal crop for ensuring future food security, the productivity is severely limited by various biotic factors. A very limited information is available on NAC TF's role in biotic stress. Kulkarni et al. (2016) have reported differential expression of NAC (INNBLQX01EOCCS) under downy mildew infection (Kulkarni et al. 2016). Thus, taking into consideration their important role in various defense responses, the NAC TFs can be explored for functional profiling and genetic improvement of crop plants for disease resistance in pearl millet.

10.3.3 Myb TFs

Myb transcription factors are a major group of diverse proteins found in the eukaryotes primarily associated with DNA binding, transcriptional and post-transcriptional regulations, and protein-protein interaction. The first identified MYB gene "colored1 (C1)" was involved in the anthocyanin synthesis in maize seeds (Paz-Ares et al. 1987). The characteristics of Myb TF are the presence of Myb DNA-binding domains which comprises one to four repeats of nearly 52 amino acids. This variation in the number of repeats is the basis of Myb TF classification, 1R-MYB (one repeat), R2R3-MYB (two repeats), 3R-MYB (three repeats), and 4R-MYB (four repeats). In plants, R2R3-MYB proteins are the most abundant, which have N-terminal DNA-binding domain (MYB domain) and regulatory domains at the C-terminal region (Dubos et al. 2010). This group of MYB proteins is associated with many important functions in plants such as hormonal signaling, secondary metabolism, meristem development, and cell cycle control (Martin and Paz-Ares 1997; Jin and Martin 1999). Many MYB genes conferring resistance toward biotic and abiotic stresses have been reported in different plant species.

MYB genes have significant involvement in the plant defense system against biotic stresses. Overexpression of some R2R3-MYB TFs provides JA- and SA-mediated defense against bacteria, fungus, and viruses by activating the PR genes. For instance, overexpression of AtMYB96 provided pathogen resistance in *Arabidopsis* through cross-talk between ABA and SA (Seo and Park 2010). In chrysanthemum, CmMYB15 increases the lignin accumulation, which combats the aphid proliferation (An et al. 2019). Similarly, AtMYB102 develops tolerance against insect herbivore *Pieris rapae* (De Vos et al. 2006). Overexpression of AtMYB44 shows elevated PR genes' level and gives better resistance toward *P. syringae* (Zou et al. 2013). In sorghum, MYB TF yellow seed 1 (Y1) is required for the biosynthesis of 3-deoxyanthocyanidin phytoalexins, an essential component against *Colletotrichum sublineolum* resistance; similar results were found in transgenic maize having Y1 MYB TF (Ibraheem et al. 2010). Similarly, R2R3-MYB TF TaPIMP1 overexpression in wheat gives resistance against fungal pathogen *Bipolaris sorokiniana* (Zhang et al. 2012). HvMYB6 provided basal and MLA-mediated immunity to barley against *Blumeria graminis* (Chang et al. 2013). VdMYB1 enhances the expression of defense gene stilbene synthase gene2 (VdSTS2) in response to the fungal attack (*Erysiphe necator*) in grapevine (Yu et al. 2019). In pearl millet, Kulkarni et al. (2016) found significant expression of MYB transcripts (INNBLQX02ID2QT, pm_c25208) upon downy mildew infection (Kulkarni et al. 2016). However functional analysis and their linked signal transduction pathways are not studied yet.

10.3.4 AP2/ERF TFs

Ethylene-responsive element-binding factor (ERF) is a subfamily of AP2/ERF superfamily of TFs and involved in regulating biotic stresses. This superfamily of TFs has a crucial role in the different plant processes such as development, growth, and stress responses. AP2/ERF members are distributed into five groups based on the numbers of AP2 domain (57–60 as DNA-binding domain) and sequence similarity (Okamuro et al. 1997; Sakuma et al. 2002). Single AP2/ERF domain is present in the ERF subfamily, which forms three β-sheets and one α-sheet. GCC-box and DRE/CRT are cis-regulatory domains to which the ERF subfamily essentially binds (Allen et al. 1998).

ERF TFs are linked with both biotic and abiotic stress responses, but they are mainly active in the biotic stress responses. Most ERF TFs bind to the GCC-box, which is abundantly found on the promoter region of many pathogenesis-related (PR) genes. Overexpression of ERF genes in tobacco and *Arabidopsis* has upregulated the expression of PR genes that leads to conferring immunity against pathogens. For instance, overexpression of AtERF5 in tobacco gave resistance against *Alternaria brassicicola* (fungus) and *Pseudomonas syringae* (bacteria) (Son et al. 2011). Similarly, overexpression of NtERF5 provided tolerance toward the *Tobacco mosaic virus* (Fischer and Dröge-Laser 2004). ERF protein acts as both positive and negative regulators, thus maintaining the expression level of target genes as well as other TFs. In transgenic soybean plants, overexpression of GmERF113 elevated the level of PR1 and PR10-1 and resistance against *P. sojae*. A barley ERF TF HvRAF, overexpressed in *Arabidopsis*, protected the plant from *R. solanacearum* infection as well as salinity stress (Jung et al. 2007). Transgenic wheat overexpressing TaPIEP1 showed higher resistance against the fungus *Bipolaris sorokiniana* (Dong et al. 2010). ERF TFs relay the defense responses through phytohormone signaling. For example, Zhang et al. (2007) suggested that in wheat, TaERF3 was responsible for the early response against *Blumeria graminis* through SA signaling and late defense response against *F. graminearum* and *R. cerealis* through ethylene/jasmonic acid signaling pathways (Zhang et al. 2007). Collectively we can say that ERF TFs are essential for triggering the plant immune system against pathogens.

10.3.5 bZIP TFs

Basic leucine zipper domain (bZIP) transcription factors modulate various processes in plants related to biotic or abiotic stress response, plant growth, and development. bZIP TFs are key players to regulate the genes linked to PAMP-triggered immunity, effector-triggered immunity, and hormonal signaling networks. However, a few studies have shown the participation of bZIP TFs in biotic stress responses. bZIP TFs are composed of 40–80 amino acid long bZIP domain with two conserved motifs: a basic region at N-terminal region, which specifically binds to target genes for transcriptional regulation, and a leucine zipper dimerization motif at C-terminal region. bZIP TFs interact with motif that has ACGT core like A-box (TACGTA), C-box (GACGTC), and G-box (CACGTG). G/HBF-1 binds to G-box and H-box motifs that have a positive link with pathogen elicitors in glycine max. Several bZIP TFs like bZIP2, bZIP23, CabZIP1, CabZIP2, CabZIP63, and TGA5 play a key role in biotic stress response (Jakoby et al. 2002; Alves et al. 2013).

10.4 Regulatory Role of Transcription Factors Associated with Biotic Stress

High-throughput sequencing technologies have enabled researchers to sequence large and complex genomes of crop plants quickly. As a result, genome-wide studies have gathered pace, which led to identifying several TFs involved in biotic stress responses. Several databases such as PlantTFDB, TAIR, and ORYZA BASE have been developed, which provide comprehensive information on TFs identified in several species. To date, several TF family genes have been identified in many sequenced species at genome-wide scale, such as *Arabidopsis*, rice, maize, foxtail millet, potato, banana, tobacco, tomato, and cassava. The significant role of WRKY, NAC, MYB, and bZIP TFs in plant defense mechanisms is well established. In rice, overexpression of various NAC genes (OsNAC6, OsNAC122, OSNAC131, and OsNAC111) led to increased resistance against *M. oryzae* (Yuan et al. 2019). Recent studies also indicate that TFs are involved in more than one stress response. For instance, overexpression of OsNAC6 resulted in increased tolerance to blast disease as well as dehydration and salt stress, thus indicating overlapping expression patterns (Sun et al. 2015). TaRIM1 (R2R3-MYB) overexpression enhanced resistance upon *Rhizoctonia cerealis* infection by regulating various defense responsive genes (Shan et al. 2016). Numerous studies have shown that overexpression of several WRKY family members resulted in enhanced tolerance against various pathogens (Baillo et al. 2019). But there are only a few reports on identifying abiotic/biotic stress-related transcription factors (WRKY, MYB, NAC, GRAS) in pearl millet. Most of the transcriptomic profiling studies in pearl millet are focused on abiotic stress. Identified TF family members in pearl millet can be explored for their involvement in biotic stress.

10.5 Transcription Factor: A Molecular Player in the Hormonal Cross-Talk Under Biotic Stress

During biotic stress in plants, various signaling cascades are activated. Defense-related phytohormones or signal molecules such as jasmonic acid (JA), salicylic acid (SA), and ethylene imine (EI) play a vital role in regulating defense-related mechanism against various pathogens. Defense response against necrotrophic pathogens is triggered by JA and ET, whereas SA triggers the response against biotrophic and hemi-biotrophic pathogens (Loake and Grant 2007; Bari and Jones 2009; Glick 2012; Wasternack and Hause 2013). Signaling pathways of these phytohormones interact at a certain point, and hormone-responsive transcription factors play a crucial role in mediating this cross-talk. Various interactions have been reported where SA and JA responded antagonistically to regulate the biotic stress. For instance, WRKY70 is an important mediator of this antagonistic cross-talk. SA-responsive PR genes were constitutively expressed in transgenic plants overexpressing WRKY70, whereas JA-responsive PDF1.2 gene was repressed in same transgenic plants (Li et al. 2004). In *Arabidopsis*, WRKY11 and WRKY17 were also found to have a crucial role in antagonistic cross-talk of SA/JA. WRKY 11 and WRKY 17 double mutant lines showed accumulation of SA-responsive genes, while the level of JA-responsive genes was decreased (Journot-Catalino et al. 2006). WRKY62 was induced by the synergistic interaction between SA and JA pathways (Mao et al. 2007).

JA and ET pathways are mostly found to have a positive interaction in response to pathogen invasion. ERF1 induced by these two pathways synergistically resulted in elevated expression of PR4, b-HI, and PDF1.2 (Lorenzo et al. 2003, 2004). In *Arabidopsis*, JA and ET acted cooperatively to provide resistance against *Verticillium longisporum* through NPR1 (Johnson et al. 2003). In some cases, JA and ET pathways act in an antagonistic manner. In *Arabidopsis*, AtMYC2 and ERF1 mediated this response where JA and ABA activate AtMYC2 in response to a wound, while both ET and JA activate ERF1 to combat the pathogen attack. In pearl millet, no such studies have been reported under biotic stress; however, identified TFs like WRKY, GRAS, and NAC are found to have involvement in phytohormone signaling for mediating the abiotic stress responses (Chanwala et al. 2020; Dudhate et al. 2021; Jha et al. 2021).

10.6 Conclusion and Future Perspectives

Biotic stresses such as downy mildew, smut, ergot, and blast adversely affect the growth and production of pearl millet. However, plants have developed various mechanisms to adjust or respond against different pathogen attacks. TFs are regulated by signaling cascades upon biotic stress in plants. These signal transduction pathways are interconnected for developing efficient plant responses under stress conditions. In recent years, several TFs such as WRKY, NAC, MYB, bZIP, bHLH, and AP2/ERF have been identified and extensively studied for delineating their functional involvement in response to various biotic and abiotic stresses in major crop plants like rice, maize, and wheat. But there is minimal information available on TFs and their role under stress in pearl millet.

This book chapter summarizes the structure and functional role of different TFs related to biotic stress and their current information and understanding in pearl millet. The whole genome sequence of pearl millet was published recently, which has enabled researchers to identify various TF gene family members and explore their role in stress responses. To date, only WRKY, NAC, HSP, and GRAS TF family members have been identified in pearl millet. Still, information of other major families like AP2/ERF, MYB, bZIP, and bHLH is due. In the years ahead, the identification and functional analysis of stress-responsive/inducing TFs and their linked regulatory network under various biotic stresses should be evaluated, which could play a significant role in genetic improvement of crop plants, especially in improving stress tolerance and efficient defense responses in pearl millet.

References

Aida M, Ishida T, Fukaki H, Fujisawa H, Tasaka M (1997) Genes involved in organ separation in Arabidopsis: an analysis of the cup-shaped cotyledon mutant. Plant Cell 9:841–857

Allen MD, Yamasaki K, Ohme-Takagi M, Tateno M, Suzuki M (1998) A novel mode of DNA recognition by a beta-sheet revealed by the solution structure of the GCC-box binding domain in complex with DNA. EMBO J 17:5484–5496

Alves MS, Dadalto SP, Gonçalves AB, De Souza GB, Barros VA, Fietto LG (2013) Plant bZIP transcription factors responsive to pathogens: a review. Int J Mol Sci 14:7815–7828

An C, Sheng L, Du X, Wang Y, Zhang Y, Song A, Jiang J, Guan Z, Fang W, Chen F, Chen S (2019) Overexpression of CmMYB15 provides chrysanthemum resistance to aphids by regulating the biosynthesis of lignin. Horticult Res 6:84

Baillo EH, Kimotho RN, Zhang Z, Xu P (2019) Transcription factors associated with abiotic and biotic stress tolerance and their potential for crops improvement. Genes 10:771

Bari R, Jones JDG (2009) Role of plant hormones in plant defence responses. Plant Mol Biol 69:473–488

Chang C, Yu D, Jiao J, Jing S, Schulze-Lefert P, Shen Q-H (2013) Barley MLA immune receptors directly interfere with antagonistically acting transcription factors to initiate disease resistance signaling. Plant Cell 25:1158–1173

Chanwala J, Satpati S, Dixit A, Parida A, Giri MK, Dey N (2020) Genome-wide identification and expression analysis of WRKY transcription factors in pearl millet (Pennisetum glaucum) under dehydration and salinity stress. BMC Genomics 21:231

Chen L, Zhang L, Yu D (2010) Wounding-induced WRKY8 is involved in basal defense in Arabidopsis. Mol Plant-Microbe Interact 23:558–565

Dang F, Wang Y, She J, Lei Y, Liu Z, Eulgem T, Lai Y, Lin J, Yu L, Lei D, Guan D, Li X, Yuan Q, He S (2014) Overexpression of CaWRKY27, a subgroup IIe WRKY transcription factor of Capsicum annuum, positively regulates tobacco resistance to Ralstonia solanacearum infection. Physiol Plant 150:397–411

Dashora K, Kumar A, Bhansali RR (2008) Smut disease of pearl millet: biology and control. Adv Appl Microbiol 36:238–240

De Vos M, Denekamp M, Dicke M, Vuylsteke M, Van Loon LC, Smeekens SCM, Pieterse C (2006) The Arabidopsis thaliana transcription factor AtMYB102 functions in defense against the insect herbivore Pieris rapae. Plant Signal Behav 1:305–311

Delessert C, Kazan K, Wilson IW, Straeten DVD, Manners J, Dennis ES, Dolferus R (2005) The transcription factor ATAF2 represses the expression of pathogenesis-related genes in Arabidopsis. Plant J 43:745–757

Dong N, Liu X, Lu Y, Du L, Xu H, Liu H, Xin Z, Zhang Z (2010) Overexpression of TaPIEP1, a pathogen-induced ERF gene of wheat, confers host-enhanced resistance to fungal pathogen Bipolaris sorokiniana. Funct Integr Genomics 10:215–226

Du M, Zhai Q, Deng L, Li S, Li H, Yan L, Huang Z, Wang B, Jiang H, Huang T, Li C-B, Wei J, Kang L, Li J, Li C (2014) Closely related NAC transcription factors of tomato differentially regulate stomatal closure and reopening during pathogen attack. Plant Cell 26:3167–3184

Dubos C, Stracke R, Grotewold E, Weisshaar B, Martin C, Lepiniec L (2010) MYB transcription factors in Arabidopsis. Trends Plant Sci. Oct;15(10):573-81. https://doi.org/10.1016/j.tplants.2010.06.005. Epub 2010 Jul 30. PMID: 20674465.

Dudhate A, Shinde H, Tsugama D, Liu S, Takano T (2018) Transcriptomic analysis reveals the differentially expressed genes and pathways involved in drought tolerance in pearl millet [Pennisetum glaucum (L.) R. Br]. Plos One 13:e0195908

Dudhate A, Shinde H, Yu P, Tsugama D, Gupta SK, Liu S, Takano T (2021) Comprehensive analysis of NAC transcription factor family uncovers drought and salinity stress response in pearl millet (Pennisetum glaucum). BMC Genomics 22:70

Eulgem T, Rushton PJ, Schmelzer E, Hahlbrock K, Somssich IE (1999) Early nuclear events in plant defence signalling: rapid gene activation by WRKY transcription factors. EMBO J 18:4689–4699

Fischer U, Dröge-Laser W (2004) Overexpression of NtERF5, a new member of the tobacco ethylene response transcription factor family enhances resistance to tobacco mosaic virus. Mol Plant-Microbe Interact 17:1162–1171

Glick BR (2012) Plant growth-promoting bacteria: mechanisms and applications. Scientifica (Cairo) 2012:963401

He X, Zhu L, Xu L, Guo W, Zhang X (2016) GhATAF1, a NAC transcription factor, confers abiotic and biotic stress responses by regulating phytohormonal signaling networks. Plant Cell Rep 35:2167–2179

Ibraheem F, Gaffoor I, Chopra S (2010) Flavonoid phytoalexin-dependent resistance to anthracnose leaf blight requires a functional yellow seed1 in Sorghum bicolor. Genetics 184:915–926. https://doi.org/10.1534/genetics.109.111831

Inoue H, Hayashi N, Matsushita A, Xinqiong L, Nakayama A, Sugano S, Jiang C-J, Takatsuji H (2013) Blast resistance of CC-NB-LRR protein Pb1 is mediated by WRKY45 through protein–protein interaction. Proc Natl Acad Sci 110:9577

Jaiswal S, Antala TJ, Mandavia MK, Chopra M, Jasrotia RS, Tomar RS, Kheni J, Angadi UB, Iquebal MA, Golakia BA, Rai A, Kumar D (2018) Transcriptomic signature of drought response in pearl millet (Pennisetum glaucum (L.)) and development of web-genomic resources. Sci Rep 8:3382

Jakoby M, Weisshaar B, Dröge-Laser W, Vicente-Carbajosa J, Tiedemann J, Kroj T, Parcy F (2002) bZIP transcription factors in Arabidopsis. Trends Plant Sci 7:106–111

Jha DK, Chanwala J, Sandeep IS, Dey N (2021) Comprehensive identification and expression analysis of <i>GRAS</i> gene family under abiotic stress and phytohormone treatments in pearl millet. Funct Plant Biol 48:1039–1052

Jiang Y, Yu D (2016) The WRKY57 transcription factor affects the expression of jasmonate ZIM-domain genes transcriptionally to compromise Botrytis cinerea resistance. Plant Physiol 171:2771–2782

Jin H, Martin C (1999) Multifunctionality and diversity within the plant MYB-gene family. Plant Mol Biol 41:577–585

Johnson C, Boden E, Arias J (2003) Salicylic acid and NPR1 induce the recruitment of trans-activating TGA factors to a defense gene promoter in Arabidopsis. Plant Cell 15:1846–1858

Journot-Catalino N, Somssich IE, Roby D, Kroj T (2006) The transcription factors WRKY11 and WRKY17 act as negative regulators of basal resistance in Arabidopsis thaliana. Plant Cell 18:3289–3302

Jukanti AK, Gowda CLL, Rai KN, Manga VK, Bhatt RK (2016) Crops that feed the world 11. Pearl Millet (Pennisetum glaucum L.): an important source of food security, nutrition and health in the arid and semi-arid tropics. Food Secur 8:307–329

Jung J, Won SY, Suh SC, Kim H, Wing R, Jeong Y, Hwang I, Kim M (2007) The barley ERF-type transcription factor HvRAF confers enhanced pathogen resistance and salt tolerance in Arabidopsis. Planta 225:575–588

Kulkarni KS, Zala HN, Bosamia TC, Shukla YM, Kumar S, Fougat RS, Patel MS, Narayanan S, Joshi CG (2016) De novo transcriptome sequencing to dissect candidate genes associated with Pearl Millet-Downy Mildew (Sclerospora graminicola Sacc) interaction. Front Plant Sci 7:847

Kumar S, Hash CT, Thirunavukkarasu N, Singh G, Rajaram V, Rathore A, Senapathy S, Mahendrakar MD, Yadav RS, Srivastava RK (2016) Mapping quantitative trait loci controlling high iron and zinc content in self and open pollinated grains of pearl millet [Pennisetum glaucum (L.) R. Br.]. Front Plant Sci 7:1636

Li J, Brader G, Palva ET (2004) The WRKY70 transcription factor: a node of convergence for jasmonate-mediated and salicylate-mediated signals in plant defense[W]. Plant Cell 16:319–331

Loake G, Grant M (2007) Salicylic acid in plant defence—the players and protagonists. Curr Opin Plant Biol 10:466–472

Lorenzo O, Piqueras R, Sánchez-Serrano JJ, Solano R (2003) ETHYLENE RESPONSE FACTOR1 integrates signals from ethylene and jasmonate pathways in plant defense. Plant Cell 15:165–178

Lorenzo O, Chico JM, Sánchez-Serrano JJ, Solano R (2004) JASMONATE-INSENSITIVE1 encodes a MYC transcription factor essential to discriminate between different jasmonate-regulated defense responses in Arabidopsis. Plant Cell 16:1938–1950

Mao P, Duan M, Wei C, Li Y (2007) WRKY62 transcription factor acts downstream of cytosolic NPR1 and negatively regulates jasmonate-responsive gene expression. Plant Cell Physiol 48:833–842

Marchive C, Léon C, Kappel C, Coutos-Thévenot P, Corio-Costet M-F, Delrot S, Lauvergeat V (2013) Over-expression of VvWRKY1 in grapevines induces expression of jasmonic acid pathway-related genes and confers higher tolerance to the downy mildew. PLoS One 8:e54185

Martin C, Paz-Ares J (1997) MYB transcription factors in plants. Trends Genet 13:67–73

Nagaraja A, Das IK (2016) Chapter 3 - Disease resistance in pearl millet and small millets. In: Das IK, Padmaja PG (eds) Biotic stress resistance in millets. Academic Press, Hoboken, pp 69–104

Narusaka M, Shirasu K, Noutoshi Y, Kubo Y, Shiraishi T, Iwabuchi M, Narusaka Y (2009) RRS1 and RPS4 provide a dual resistance-gene system against fungal and bacterial pathogens. Plant J 60:218–226

Nayaka CS, Srivastava RK, Udayashankar AC, Lavanya SN, Prakash G, Bishnoi HR, Kadvani DL, Singh OV, Niranjana SR, Prakash HS, Satyavathi CT (2017) Magnaporthe blast of pearl millet in India-present status and future prospects. All India co-ordinated research project on pearl millet, p 51

Nene YL, Singh SD (1976) Downy mildew and ergot of pearl millet. PANS 22:366–385

Okamuro JK, Caster B, Villarroel R, Van Montagu M, Jofuku KD (1997) The AP2 domain of APETALA2 defines a large new family of DNA binding proteins in Arabidopsis. Proc Natl Acad Sci U S A 94:7076–7081

Paz-Ares J, Ghosal D, Wienand U, Peterson PA, Saedler H (1987) The regulatory c1 locus of Zea mays encodes a protein with homology to myb proto-oncogene products and with structural similarities to transcriptional activators. EMBO J 6:3553–3558

Qiu D, Xiao J, Ding X, Xiong M, Cai M, Cao Y, Li X, Xu C, Wang S (2007) OsWRKY13 mediates rice disease resistance by regulating defense-related genes in salicylate- and jasmonate-dependent signaling. Mol Plant-Microbe Interact 20:492–499

Sakuma Y, Liu Q, Dubouzet JG, Abe H, Shinozaki K, Yamaguchi-Shinozaki K (2002) DNA-binding specificity of the ERF/AP2 domain of Arabidopsis DREBs, transcription factors involved in dehydration- and cold-inducible gene expression. Biochem Biophys Res Commun 290:998–1009

Seo PJ, Park C-M (2010) MYB96-mediated abscisic acid signals induce pathogen resistance response by promoting salicylic acid biosynthesis in Arabidopsis. New Phytol 186:471–483

Shan T, Rong W, Xu H, Du L, Liu X, Zhang Z (2016) The wheat R2R3-MYB transcription factor TaRIM1 participates in resistance response against the pathogen Rhizoctonia cerealis infection through regulating defense genes. Sci Rep 6:28777

Shetty HS, Raj SN, Kini KR, Bishnoi HR, Sharma R, Rajpurohit BS, Yadav OP (2016) Downy mildew of pearl millet and its management. All India coordinated research project on pearl millet (ICAR), Jodhpur

Shinde H, Tanaka K, Dudhate A, Tsugama D, Mine Y, Kamiya T, Gupta S, Liu S, Takano T (2018) Comparative de novo transcriptomic profiling of the salinity stress responsiveness in contrasting pearl millet lines. Environ Exp Bot 155:619–627

Son GH, Wan J, Kim HJ, Nguyen XC, Chung WS, Hong JC, Stacey G (2011) Ethylene-responsive element-binding factor 5, ERF5, is involved in chitin-induced innate immunity response. Mol Plant-Microbe Interact 25:48–60

Sun L, Huang L, Hong Y, Zhang H, Song F, Li D (2015) Comprehensive analysis suggests overlapping expression of rice ONAC transcription factors in abiotic and biotic stress responses. Int J Mol Sci 16:4306–4326

Thakur RP, Sharma R, Rai KN, Gupta SK, Rao VP (2009) Screening techniques and resistance sources for foliar blast in pearl millet. J SAT Agric Res 7:1–5

Veena M, Melvin P, Prabhu SA, Shailasree S, Shetty HS, Kini KR (2016) Molecular cloning of a coiled-coil-nucleotide-binding-site-leucine-rich repeat gene from pearl millet and its expression pattern in response to the downy mildew pathogen. Mol Biol Rep 43:117–128

Wang X, Basnayake BM, Zhang H, Li G, Li W, Virk N, Mengiste T, Song F (2009) The Arabidopsis ATAF1, a NAC transcription factor, is a negative regulator of defense responses against necrotrophic fungal and bacterial pathogens. Mol Plant-Microbe Interact 22:1227–1238

Wang F, Lin R, Feng J, Chen W, Qiu D, Xu S (2015) TaNAC1 acts as a negative regulator of stripe rust resistance in wheat, enhances susceptibility to Pseudomonas syringae, and promotes lateral root development in transgenic Arabidopsis thaliana. Front Plant Sci 6:108

Wasternack C, Hause B (2013) Jasmonates: biosynthesis, perception, signal transduction and action in plant stress response, growth and development. An update to the 2007 review in Annals of Botany. Ann Bot 111:1021–1058

Yu Y, Guo D, Li G, Yang Y, Zhang G, Li S, Liang Z (2019) The grapevine R2R3-type MYB transcription factor VdMYB1 positively regulates defense responses by activating the stilbene synthase gene 2 (VdSTS2). BMC Plant Biol 19:478

Yuan X, Wang H, Cai J, Bi Y, Li D, Song F (2019) Rice NAC transcription factor ONAC066 functions as a positive regulator of drought and oxidative stress response. BMC Plant Biol 19:278

Zhang Z, Yao W, Dong N, Liang H, Liu H, Huang R (2007) A novel ERF transcription activator in wheat and its induction kinetics after pathogen and hormone treatments. J Exp Bot 58(11):2993–3003. https://doi.org/10.1093/jxb/erm151

Zhang L, Zhao G, Xia C, Jia J, Liu X, Kong X (2012) A wheat R2R3-MYB gene, TaMYB30-B, improves drought stress tolerance in transgenic Arabidopsis. J Exp Bot 63:5873–5885

Zou B, Jia Z, Tian S, Wang X, Gou Z, Lü B, Dong H (2013) AtMYB44 positively modulates disease resistance to Pseudomonas syringae through the salicylic acid signalling pathway in Arabidopsis. Funct Plant Biol 40:304–313

Chapter 11
The Role of Transcription Factors in Response to Biotic Stresses in Tomato

Lopamudra Jena, Subhasmita Sahu, Pranaya Pradhan, Kailash Chandra Samal, Jyoti Prakash Sahoo, Laxmipreeya Behera, and Siddhartha Kumar Palai

Abstract Tomato plants continuously confront a host of biotic stressors in the field that harm their life cycle. Due to various stressors, a tightly controlled and highly dynamic regulatory network may be used to re-program the transcriptome, where transcription factors might serve as activators or repressors. Proteins that are responsible for the essential role in improving agricultural yields in aquatic regions and in places where the severity of pathogens is extremely strong to survive continuous climate change are known as the transcription factors. WRKY family, MYB family, NAC family, bZIP family, ERF family, ARF family, and HSF family are the well-known transcription factors, which are directly linked with plant abiotic and biotic stress responses. These families of transcription factors emphasize the potential for increasing yield, increasing stress tolerance, and increasing the efficacy of solanaceous crops cultivated in arid and semi-arid environments. This chapter summarizes the basic framework of plant immunity, tomato plant resistance to pathogens and transcription factors related to this mechanism, R-gene-mediated resistance defense

L. Jena
Department of Floriculture and Landscape Architecture, Faculty of Horticulture, Bidhan Chandra Krishi Viswavidyalaya, Mohanpur, India

S. Sahu
Department of Floriculture and Landscaping, College of Agriculture, Odisha University of Agriculture and Technology, Bhubaneswar, India

P. Pradhan
Department of Nematology, College of Agriculture, Odisha University of Agriculture and Technology, Bhubaneswar, India

K. C. Samal · J. P. Sahoo (✉) · L. Behera
Department of Agricultural Biotechnology, College of Agriculture, Odisha University of Agriculture and Technology, Bhubaneswar, India

S. K. Palai
AICRP on Floriculture, College of Agriculture, Odisha University of Agriculture and Technology, Bhubaneswar, India

© The Author(s), under exclusive license to Springer Nature Switzerland AG 2022

S. H. Wani et al. (eds.), *Transcription Factors for Biotic Stress Tolerance in Plants*, https://doi.org/10.1007/978-3-031-12990-2_11

pathways in tomatoes, and breeding of tomatoes with tolerance to combined biotic and abiotic stress responses.

Keywords Biotic stress · Breeding · Resistance · Tomato · Transcription factors

11.1 Introduction

Stress in plants is explicated as an unfavorable constrain that influences or prevents the normal metabolic process (Lichtenthaler 1996). These stressors have the potential to have a major impact on crop yield. Despite the fact that emergence of plants has sophisticated, speedy reaction to cope with a variant of stresses, the effectiveness of cultivation has remained significantly hampered. Damage in cells of plants can be caused by reactive oxygen species and temperature fluctuations, leading to wilting of the cell followed by bleaching called as cell recession and chlorophyll deterioration, which eventually leads to plant mortality. Due to high nutritional, therapeutic, and vitamin content, tomato, potato, brinjal, and capsicum are found most popular among solanaceous food crops (Matsukura et al. 2008). Tomato (*Solanum lycopersicum*) plants are dicots, as they have branched stems having a bud (terminal), which grows at the tip. This is because of the fact that, while the growing tip steps back growing, whether or not due to pruning or flowering, lateral buds take charge of and develop into vines which is wholly functioning. Tomato vines are regularly pubescent, *i.e.*, they may be protected with tiny, brief hairs. This results in these hairs being used as roots when the plant touches soil and moisture. Anti-cancer properties are alleged to be lycopene, which gives tomatoes their red color. Tomato output worldwide is expected to reach around 4.6 million hectares per year.

Genomic methods are utilized to improve tomatoes and develop new species, as they are economically important for farmers. Conversely, solanaceous crops are found more biotic stress proned that affect both the quality and the amount of food produced. The most common pathogens targeting this crop group include *tomato spotted wilt virus*, *tomato yellow leaf curl virus*, *Phytophthora capsici*, species of *Fusarium*, and species of *Collectricum* (Yeom et al. 2011). These stresses in combination interrupt the cycle of plant life and interfere with its physiological and survival processes. The interactions between proteins have altered and have been aggregated and denaturized (Farooq et al. 2008). Plants throughout their life cycle developed different repressions for various stresses like environmental as well as living factors. Due to unfavorable environmental conditions, a range of stress response mechanisms has developed, including signal perception and cell-level transduction, which accumulates in a pronouncement of certain subsets of defense genes which activates the overall phenotypical defense reaction (Fraire-Velázquez et al. 2011). Changes in plant-associated microbiome composition and functional activities are essential for plant survival (Hacquard et al. 2017). Cellular stress causes a stress reaction from a plant. The expression of several gene networks has therefore changed. Figure 11.1 implies the activation of defense-mediated R-gene

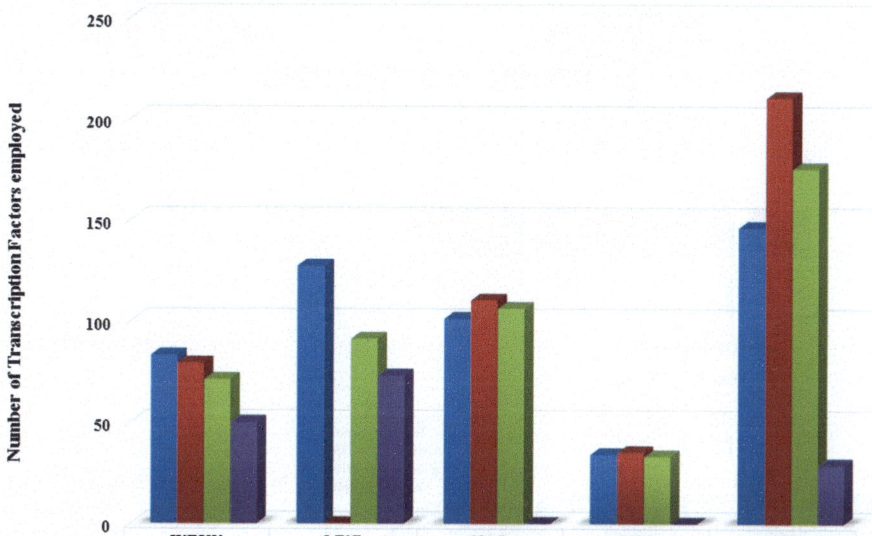

	WRKY	MYB	NAC	DOF	ERF
TOMATO	83	127	101	34	146
POTATO	79	0	110	35	210
HOT PEPPER	71	91	106	33	175
EGG PLANT	50	73	0	0	29

Fig. 11.1 Number of transcriptional factors being employed in major solanaceous food crops

pathways and defense signal transduction pathways, and Fig. 11.2 indicates the key elements that play a principal role of stress response activation within plants.

Transcriptional factors have a crucial function to play in plant development, cell cycles, cellular signals and stress response (Gonzalez 2016). These proteins identify, bind, and regulate cellular operations through gene expression maintenance and regulation in transcription processes to certain DNA sequences, usually in a promoter region (Gupta et al. 2015). DNA-binding transcription elements that understand certain cis-appearing regions govern a strain-triggered turn-round or flip-round of the gene. Transcriptional factors affect the feature of the target gene, in step with several researches (Seo and Choi 2015). So far, the quantity of transcriptional elements being hired in essential solanaceous crops is indexed in Fig. 11.2. The binding and transcriptional activation and repression of DNA are the two most established methods, specifically if the plants are in stress condition (Riechmann et al. 2000). It was believed that, these two domains and others, as a result of the endogenous and external inputs, play a role in activating and/or repressing the transcription processes. Approximately 10% of vegetable genes encode transcription factors (TFs) for specific signaling activities mediated responses (Gonzalez 2016). The study of plant genomes and different molecular studies has diagnosed over 60 transcription elements by using bioinformatics, next-generation sequencing and different methodologies (Dai et al. 2013; Cabello et al. 2008). This chapter covers the fundamentals of plant immunology, tomato plant resistance to pathogens and

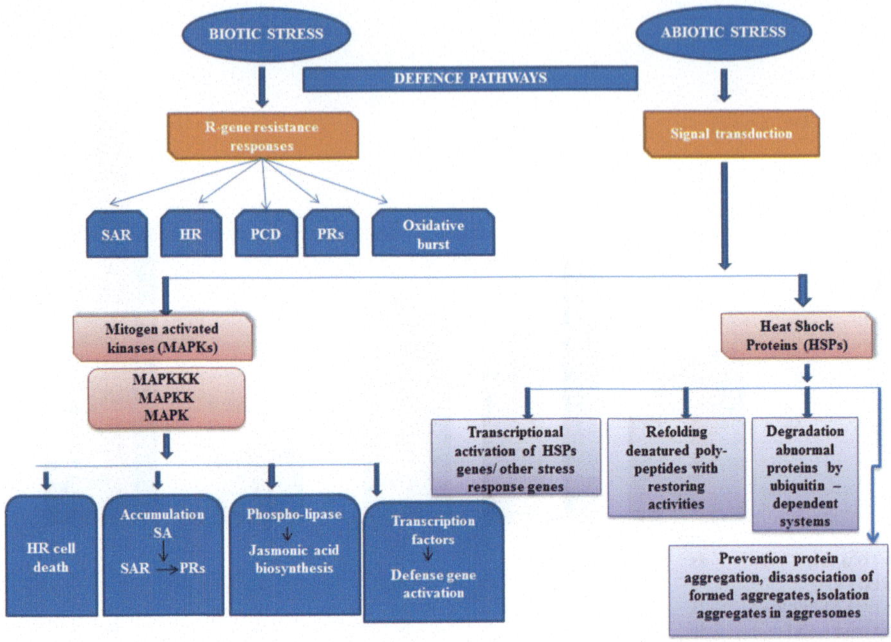

Fig. 11.2 Schematic diagram depicting the important defense pathways present within plants

associated transcription factors, R-gene-mediated resistance defense mechanisms in tomatoes, and tomato breeding for combined abiotic and biotic stress tolerance (Fig. 11.3).

11.2 Plant Immunity Against Biotic and Abiotic Stresses

Plant immunity is primarily based on the activity of inbuilt cell immune receptors, recognizing invasion signals to mount immunity-triggered pattern (PTI) or immune-triggered effector signals (ETI) (Jones and Dangl 2006). Immunity-triggered pattern is fitted with the molecular structures typical of microorganisms and endogenously damaged molecular patterns, namely, microbe- and stress-related molecular pattern (MAMP and DAMP), and cell-localized pattern recognition receiver (PRR), respectively. PRRs include FLS2 and EFR receptor-like (LRR) leucine-rich repetition (LG22) and EF-Tu epitope (LLRR) and RLK CERK1 lysin motif (LysM), which is needed by fungal chitin oligomers and bacterial peptidoglycans to recognize leucine-rich receptor-like kinases and LRRs (Couto and Zipfel 2016). The receptors in the ligand are complexes with the same extracellular domain class kinases of co-receptor or adapter, therefore triggering RLKs and cytoplasmic kinases of protein phosphorylation cascades in the receptor. The sign of PRR

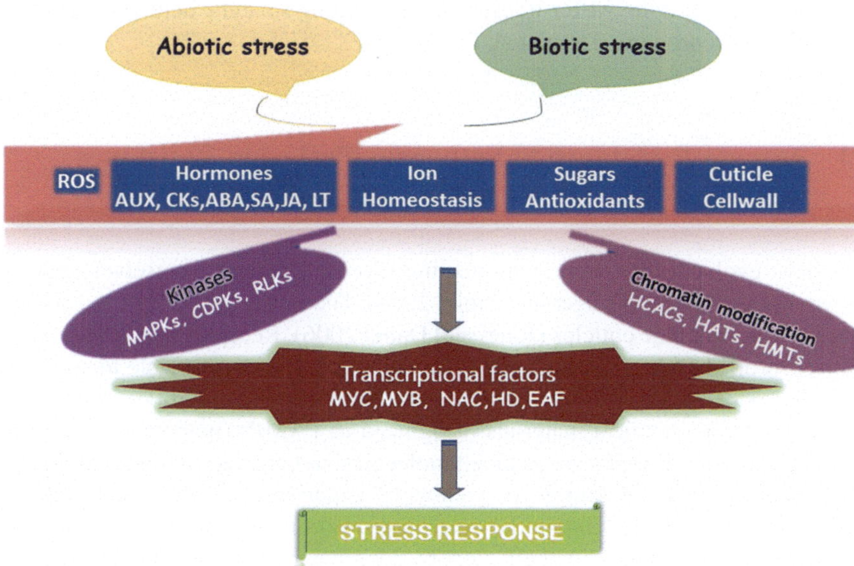

Fig. 11.3 Key factors playing a critical role inside the activation of stress response in plant system

consists of the bursts of cytosolic Ca^{2+}, apoplastic reactive oxygen species (ROS), the explosion of Ca^{2+} protein kinases (CDPKs), the cascading and the programmatic transcription of mitogenic protein kinase (MAPK), and the defensive hormone networks (Yu et al. 2017). Those consequences make contributions together to PTI, preventing contamination of the overwhelming number of microorganisms and proscribing that within the direction of base resistance of adaptive pathogens.

In a charge-limiting step, however, PRR pathways show huge heterogeneity in their susceptibility over biotic or abiotic disturbances (Saijo et al. 2018). Consequently, the continuous use of several PRR pathways can improve the general strength of PTIs closer to pathogenic aggressions and environmental disturbances. Pathogenic microbial had been advanced on the way to subvert PTI thru a mess of effectors which affect the immunity and vulnerability factors within the host (Miwa and Okazaki 2017). So as to overcome this, vegetation advanced an expansion of internal or indoor receptors that pick out microbial effectors directly or not directly, along with the nuclear-binding vicinity and the protein-containing LRR (NLRs). NLR activation consequences in a more perfect protection that is usually known as hypersensitive reaction as the localized cell is dying. In well-known, ETI is extra resistant closer to pathogen-induced disturbances in evaluation with PTI (Cui et al. 2015). Plant resistance or immunity is likewise based on independent non-mobile indicators (Jones and Dangl 2006). Cellular long-distance indicators are created and then disseminated during the plant following the identification of localized pathogen/damage (Toyota et al. 2018). The mechanism is supported via the activation of genes generating pathogenesis-related (PR) proteins, in detail by using balanced transcription law thru the NPR1 transcriptional binding salicylic acid (SA) and the NPR3/NPR4 corepressors (Ding et al. 2018).

11.3 Tomato Resistance to Pathogens/Pathogen Resistance of Tomato

Plants have evolved very superior protecting layers to defend themselves against invasive diseases (Schwessinger and Zipfel 2008). The layers are pre-formed bodily obstacles which include cuticles (Jones and Dangl 2006). Receptor (e.g., PRRs) and RLK (receptor-like kinase) within the plant cellular responses are able to spot the primary layer of pathogen-triggered defenses (Nurnberger et al. 2004) in the preserved pathogenic chemical materials recognized as PAMPs (pathogen-associated molecular patterns). PAMPs are retained molecules and are very tough to alternate to the pathogen; a huge sample of pathogens experiences the claimed PAMP-inducing immunity (PTI). This initial layer of protection can only be overcome by separating viral molecules called effectors by adaptive microorganisms (Dodds and Rathjen 2010). In the second layering of pathogen-driven resistance, effector-driven immunity (ETI), receptors can be recognized as effectors, previously referred to as Avr genes (Lo Presti et al. 2015).

FLS2 (FLAGELLIN-SENSING 2) and its ligand, the flagella epitope flg2, are the number one delineated plant PRR-PAMP. FLS3 is the second flagellin sensor that detects flgII-28 (Hind et al. 2016). For tomatoes, AvrPto, a bacterial pathogen effector of *Pseudomonas syringe*, can be suppressed in pairs FLS2/flg22 and FLS3/flgII-28 by means of way of PTI (Hind et al. 2016). Conversely, the tomato Prf/Pto (R-protein complex), which results in ETI, can be used to distinguish AvrPto (Ntoukakis et al. 2013). PRR and R-protein safety responses encompass the advent of pathogenic proteins, ROS manufacture, and a complicated signaling cascade of hormones which incorporates ethylene (ET), jasmonic acid (JA), and salicylic acid (SA) (Dodds and Rathjen 2010) The R-gene-mediated resistance to bio-transparent pathogens is distinguishable in hypersensitive responses, as the form of programmed cell deceases is contained on the location of a pathogen attempt (Love et al. 2008). In contrast to biotrophic pathogenesis involving immune response, necrotrophic pathogenesis seizes plant immunity through the release of phytotoxins and degenerating enzymes of cell walls to stimulate tissue necrosis of the host before colonization (Laluk and Mengiste 2010; Robert-Seilaniantz et al. 2011).

11.4 Biotic Stress-Related WRKYs Available for Tomato

Jones and Dangl (2006) suggest two tiers of inductive plant defenses where WRKYs perform positive or negative regulatory activities (Sarris et al. 2015). If PAMP and pattern recognition receptors are recognized, PAMP-induced immunity is begun. In this case, the recipient of PAMPs recognizes one another. Effector proteins that suppress PTI responses can be expressed by pathogens adapted to their environment. Pathogenic factors are retrieved by PR proteins' (R) secondary defense (ETI). PR proteins usually feature a repeat rich in nucleotide-binding leucine (NB-LRR). As they generate ROS and activate MAP, PTI and ETI induce resistance reciprocations both locally and systemically (Dodds and Rathjen 2010). The conventional immune hormones are ET, JA, and SA. WRKYs in ETI and PTI include all regulative stages (Bakshi and Oelmuller 2014). WRKYs set off or suppress PTI and ETI mostly when they use PAMPs or effector proteins immediately. When the flg22 (a MAMP), the HvWRKY1, and the HvWRKY2 functions as PTI repressors were active in barley (*Hordeum vulgare*), flg22 was activated.

A special hyperlink among R-protein MLA10 and HvWRKY1/2, the fungal effector AVRA10, also led to HvWRKY1/2 by deactivating its repressive activity in *Arabidopsis* (Xu et al. 2006). The OsWRKY62 gene in rice has bad PTI and ETI rules, supplied with the aid of the Xa21 gene (Peng et al. 2008). The consequences display that participants of this subfamily may also have a retained poor regulatory function in plant protection. These WRKYs are participants of the WRKY II-a subfamily. However, when OsWRKY71 was exaggerated in rice, the plant's strength to *Xanthomonas oryzae* pv. oryzae was raised in rice (Liu et al. 2007). The second research shows MAPKs can also regulate WRKYs (Ishihama and Yoshioka 2012). They might bind the W-Box in the RBOHB promoter and produce ROS bursts in *Nicotiana benthamiana* (Adachi et al. 2015). Interacting with AtWRKY33 have been MPK4 and MKS1 (Andreasson et al. 2005). AtWRKY33 became launched from the trimeric complex and attached to the PAD3 promoting area, boosting the synthesis of antimicrobial camalexin after *P. syringae* or MAMP flg22 has become injected (Ishihama and Yoshioka 2012). WRKYs have a third function in tracking hormonal signposting pathways. AtWRKY18 and AtWRKY70 overexpressions are introduced on defense-associated genes and PR1 expressions induced with SA (Li et al. 2004). The enhanced sensitivity of the AtWRKY33 mutant to *Botrytis cinerea* ends up located by using an SA-mediated-JA pathway blockage (Birkenbihl et al. 2012). WRKYs may provide plant resistance or immunity with the useful resource of regulating quick ribonucleic acids, histone methylation, and retrograde conversation, among organelles (Phukan et al. 2016). Research of WRKY tomatoes was carried out, either via overexpression or silenced, in order to assess their involvement in plant protection.

Several tomato-specific WRKYs play a part as positive regulators to check biotic factor despondence in plants. The AtWRKY33, SlWRKY31, and SlWRKY33 homologs might restore tolerance of the mutant to *B. cinerea* (Zheng et al. 2006). Moreover, the hemi-biotrophic resistances to disease-causing organism, viz., *Phytophthora nicotianae* and *Phytophthora infestans*, in tobacco and tomato were proven by the *Solanum pimpinellifolium* allele SlWRKY33 (named as SpWRKY1 by Li et al. 2015b, c). SlWRKY39, in different circumstances, has been found to have increased resistance to fungus *P. syringes* (Huang et al. 2012) and reported in lines of tomato-overexpressing SlWRKY39 (Sun et al. 2015). The overreaction of SlWRKY45, in another correspondence of AtWRKY40, has improved the

Table 11.1 WRKY genes utilized in some crucial solanaceous vegetable crop plants

Crop	Gene	Functions	References
Potato	StWRKY1	*Phytophthora infants* tolerance	Shahzad et al. (2016)
Potato	StWRKY8	Late blight of potato resistance	Yogendra et al. (2016)
Tomato	SlWRKY45	Resistance to various nematodes	Chinnapandi et al. (2017)
Pepper	CaWRKY30	Pathogen stress response expression	Kang et al. (2016)
Pepper	CaWRKY27	Regulation of *Ralstonia solanacearum* infestation	Dang et al. (2014)
Pepper	CaWRKY58	*Botrytis cinerea* tolerance	Kang et al. (2016)
Tomato	SlWRKY39	*Pseudomonas syringae* pv. tomato DC 30000 resistance	Sun et al. (2015)

sensitivity of root-knot nematodes *Meloidogyne javanica* (Chinnapandi et al. 2017). The SlWRKY72 or SlWRKY74 (SlWRKY72a or SlWRKY72b in Bhattarai et al. 2010) root-knot nematodes (RKNs) (*M. javanica*) and potato aphids from family Aphididae (*Macrosiphum euphorbiae*) favored PTI and Mi-1 ETI-mediated resistance (Bhattarai et al. 2010). SlWRKY80 (SlWRKY70 in Atamian et al. 2012) was also desired to induce micro-resistance in response to potato aphids and nematodes.

The pathogens were shown to change expression after pathogen infection of SlWRKY23 (AtWRKY 23 homolog), SlWRKY46 (AtWRKY40 homolog), SlWRKY53/54 (AtWRKY23), and SlWRKY80 and SmWRKY81 (AtWRKY38 and AtWRKY62 homolog) (Rezzonico et al. 2017). They are used as negative plant protection regulators by using their homologs in *Arabidopsis sp.*, AtWRKY38, AtWRKY48, and AtWRKY62, if challenged against *P. syringae* (Xing et al. 2008). VqWRKY52, AtWRKY53, and SlWRKY53/54 homolog furnished resistance against *Golovinomyces cichoracearum* and *P. syringae*, but better susceptibility toward *B. cinerea* was connected to the multiplied expression of SA pathway gene and hastened dying to cell (Wang et al. 2017). WRKY genes of the solanaceous crop plant must be further analyzed in order to assess their involvement in improving resistance or sensitivity to certain diseases. The WRKY genes utilized in some crucial solanaceous crop plants are indexed including its function in Table 11.1.

11.5 R-Gene-Mediated Defense Pathways Toward Resistance

Stress-responsive pathways of plants involve R-protein setup and lead to hypersensitivity (HR) and plant immunity and are connected with H_2O_2 buildup in pathogen attacks and programmed cell death (PCD), which quickly drive cells to die close to the infection site (Bolwell 1999). The S and R lines have also been compared to an absolutely new pathogen, fungal *S. sclerotiorum*, following inoculation (Cessna et al. 2000). 3,3′-DAB histochemical stains can identify the severity of oxidative burst or buildup of H_2O_2 (Alvarez et al. 1998). DAB stains were prominently reported in leaves by virulent whiteflies, 60 days after inoculation and then when signs of diseases were noticeable. SAR develops in tissues some distance from the authentic confused region marked with the aid of a growth in gene expression coded for proteins related to pathogenesis (PR). Various stressors, including injury, chemistry, hormones, and UV, can control the production of PR proteins (Van Loon and Van Strien 1999).

The transcription factor called as NAC is one of the major factors for the biotic and abiotic stress situations in plants. The NAC TFs in *Petunia*, a solanaceous crop, were identified two decades ago and first characterized (Souer et al. 1996). Many researchers have since been carried out to combat stress tolerance, it may be biotic or abiotic and growth support (Nuruzzaman et al. 2013). Therefore, it can be said that for defense and for stress activation, most genes were linked (Nakashima et al. 2012). If the plant is assaulted by the pathogen, at least three signal mediators ET, JA, and SA combined and harmonized to produce the defense system of plant (Glazebrook 2001). In the wide range of signals to resistance to local and systemic diseases, tomatoes NAC (SLNAC1, SlSRN1, and SlNAC35), brinjal (SmNAC), potato (StNAC4, 5, 18, 48, 81), and hot pepper (CaNAC1) are most striking which are expressed in Fig. 11.4.

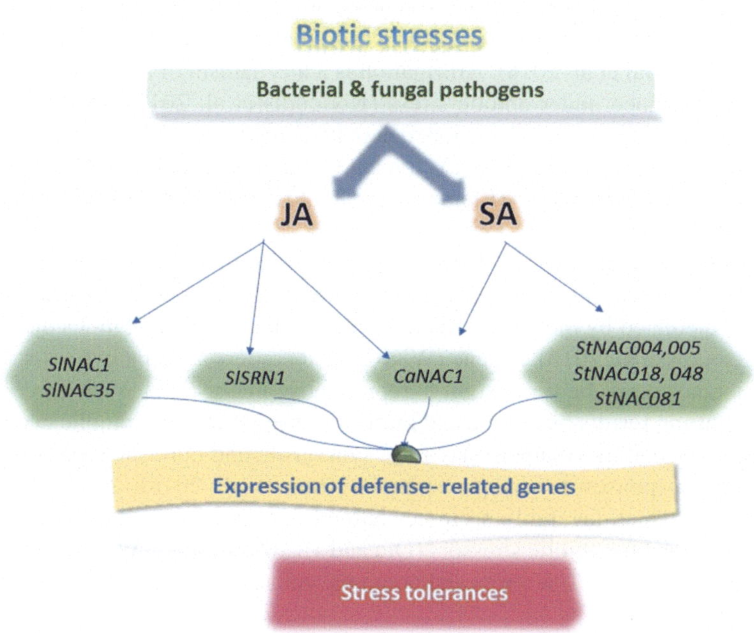

Fig. 11.4 NAC factors in the signal pathway to express defense genes in plants under stress conditions

11.6 WRKYs in Cross-Talk Between Abiotic and Biotic Stress Resistances in Tomato

The various WRKYs described above are active against each biotic and abiotic stress factor in a crosstalk of plant reactions. Previous review showed that AtWRKY33 and two of its tomato counterparts SlWRKY31 and SlWRKY33 are plant protectors for some diseases (Li et al. 2015b). Similarly, the induction of drought or salt stress by SlWRKY31 and SlWRKY33 was determined (Huang et al. 2012). HvWRKY1 and its tomato counterparts SlWRKY39 and SlWRKY45 actively participate in the stimulation to infections and many pathogens and environmental stressors. HvWRKY38 (named in Mare et al. 2004) became an additional diagnosis (Chinnapandi et al. 2017). It should be noted that, WRKYs were studied at the time for their reciprocity with respect to single stress. In tomato plant, SlWRKY23 has been silenced in the interconnectivity of reactive pathways (Kissoudis 2016).

Those plants have proven a greater resistance toward the dusty *Oidium neolycopersici*. But the resistance becomes reconciled with the pressure of the salinity stress. This case in reality establishes a characteristic of WRKY transcription elements inside the interplay between biotic and abiotic stress reactions and demonstrates that the reciprocity of individual stress cannot be additive in the event that the plants react to combinatorial stress. Tomatoes are home to greater than 200 pathogenic species, which includes a number of wild tomato families, regulated with the aid of R genes (Bai et al. 2018). Affirmation is a mechanism of reducing or increasing stress resistance under abiotic stress (Kissoudis et al. 2017). As an example, thermal stress reconciled Mi-1-mediated nematode resistance (Marques de Carvalho et al. 2015). There are four WRKY tomatoes diagnosed as microwave resistant: SlWRKY72 to SlWRKY74 (Bhattarai et al. 2010) and SlWRKY80 (Atamian et al. 2012).

The query is whether or not WRKYs are concerned in Mi-1-mediated resistance instability under warmth stress or, greater precisely, participate in the instability of resistance mediated by means of affected genes in diverse plant R molecular strategies (Kissoudis et al. 2016). A gene (WKRY) that confers a distinctive resistance or stress tolerance could be very useful to reproductive interest, but WRKY genes also can have putting results on abiotic and biotic stress tolerance, as complicated hyperlinks among signaling channels may have synergistic and opposite results on managing plant responses to diverse stressors (Bai et al. 2018). As an example, OsWRKY45, which favorably promotes ailment resistance on a huge scale also as a deterrent model for abiotic stressors (Qiu and Yu 2009), and OsWRKY75 enhance plant tolerance to bloodless stress and boom the vulnerability of rice blast (Yokotani et al. 2013).

Several transcription factors have likewise been stated within the law of reactions to abiotic or biotic stressors, particularly TSRF1 (Zhang et al. 2007), DEAR1 (DREB (Dreb), and EAR (Ethylene), and reaction amphiphilic reaction motif protein 1, respectively (Tsutsui et al. 2009). The plant response regulation to many stressors is based on tightly managed and rather dynamic organized channels wherein WRKYs can function as promoters or suppressors (Phukan et al. 2016). There are five SlWRKY genes in tomatoes that are almost homologous to the three AtWRKY genes and which reply to abiotic and biotic stresses. More records are essential to install vicinity if the additional SlWRKY clusters and clusters still have an area. The feature of unexamined SlWRKY genes in other plant species changed into examined thru their opposite numbers. It has to be stated, however, that small alterations within the area of DNA binding also can have a massive impact on binding specificity, and the sequences may be pretty the same; they however have distinct traits (Du et al. 2014). For instance, the cautiously linked tomato homologs SlWKRY3 and SlWRKY4 are designed to have interaction with the DNA of container won a motif, RKYGQK and WRKYGQK (Aamir et al. 2017).

11.7 ERF Genes in Tomato and Their Response Under Stress Conditions

The modern-day genes that come from only a limited wild species together with *S. peruvianum*, *S. pimpinellifolium*, *S. pennellii*, *S. chilense*, and *S. habrochaites* are capable of genetically restricting approximately 20 sicknesses. In most instances the single dominant genes are monogenic resistant to the variations produced. In maximum cases, all the R genes that dominate tomatoes, which have already been cloned (aside from Gen Ty-1), can be divided into categories, the plasma receptors, which encompass RLK (encoded through the I-3 gene), and receptor-like proteins (RLP, encoded with the resource of the Cf gene and thru the Ve-1 gene) (NBS-LRR). The genome of tomato is consisting of more than 350 genes with NB-LRR-associated domains (Tomato Genome Consortium 2012). Andolfo et al. (2014) have also explained the RenSeq approach in both the Heinz 1706 tomato and *S. pimpinellifolium* LA1589 for the full-length gene of the NB-LRR. The allelic positions and chromosome variants of these NBS-LRR genes will recognize particular alleles that co-segregate with a "quick mapping."

The gene Elf4e is affected by an eIF4E gene mutation. In plants, eIF4E and eIF4G isoforms are important for the translation of *Potyvirus* replication and infestation (Robaglia and Caranta 2006). The Pelo gene (messenger RNA surveillance factor Pelota's tomato homolog) is linked to the host protein biosynthesis. This prevents the proliferation of the *Tomato yellow leaf curl virus*, causing resistance (Lapidot et al. 2015). The genes Mlo, eIF4E/eIF4G, and Pelo are recognized as genes that encode proteins that are misused for their own convenience by a pathogen in the infection process (Pavan et al. 2010).

In addition to the key tomato genes discussed above, there are a lot of quantitative resistance loci (QRLs; Poland et al. 2009). ERF genes and their response beneath pressure conditions in key Solanaceae crops are indexed in Table 11.2. Tomatoes contain three genes which might be recessive, cloned, and immune to

Table 11.2 ERF genes in major solanaceous crops and their response under stress conditions

Name of genes	Crop	Functions	References
SlERF1	Tomato	Resistance to *Xanthomonas campestris*	Pan et al. (2013)
SlERF3	Tomato	*Ralstonia solanacearum* resistance	Pan et al. (2010)
SlERF4	Tomato	Reduction in production of ethylene	Kim et al. (2013)
SlERF84	Tomato	Immunity reductant against *Pseudomonas syringae pv.* DC3000	Li et al. (2018)
StERF3	Potato	Inhibition of resistance to *Phytophthora infestans*	Tian et al. (2015)
NtERF5	Tobacco	*Tobacco mosaic virus* resistance	Fischer and Droge-Laser (2004)
StERF71, StERF47, StERF67, StERF70	Tobacco	Resistance to *Phytophthora infestans*	Charfeddine et al. (2015)
TERF1	Tobacco	Regulate ROS (H_2O_2) during seedling development	Zhang et al. (2016)
SlERF2	Tomato	MeJA-mediated defense with enhancement of resistance against *Botrytis cinerea*	Kim et al. (2013)

powdery mildew and many viruses. E1–2 is a mutant of a tomato MLO ornithologist, SlMLO1, for resistance to powdery mildew (Bai et al. 2008). QTL can be controlled by PTI-relative and PTI-defensive signaling genes and morphological characteristics regulating genes and coding components of chemical warfare (Roux et al. 2014). Two active defensive layers can be created when pathogens are invaded besides a physical layer. The receptors (RLP, R-protein, and S-protein) of tomato linked with pathogenic acumen have been thoroughly investigated (Fig. 11.5). Table 11.3 also lists other significant transcriptional factors that are properly indicated in key solanaceous crops.

11.8 Breeding Tomato with Tolerance to Combined Biotic and Abiotic Stresses

The review of tomato gene and signal cross-references for tomato responses to the above mentioned salinity and disease stressors shows that the development of breeding techniques to attain tomato resilience with combined stress combinations is extremely hard. The evaluation of tomato gene and signal pass-references for tomato retaliation to the above mentioned salinity and disorder stressors indicates that the development of breeding strategies to acquire tomato resilience with mixed stress combinations is extremely tough. The elements stated below should be taken into account.

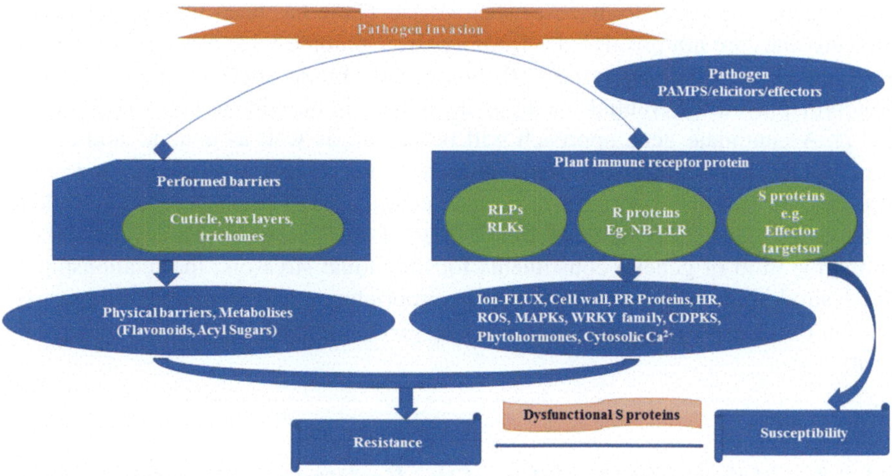

Fig. 11.5 Various layers of plant defense pathways upon pathogen invasion

Table 11.3 Other important transcriptional factors duly expressed in major solanaceous crop plants

Name of genes	Functions	Crop	References
TDDF1	Regulator sequence	Tomato	Ewas et al. (2017)
CaDof17	Resistance to various biotic stresses	Pepper	Kang et al. (2016)
CaDof10, 11	Resistance against *Phytophthora capsici*	Pepper	Kang et al. (2016)
SlMYB14, SlMYB28, SlMYB65, SlMYB66	Jasmonic acid and salicylic acid	Tomato	Li et al. (2016)
SlNAC1	Defense mechanism to check *Pseudomonas* infection	Tomato	Ma et al. (2013)
SlNAC35	Bacterial resistance	Tomato	Wang et al. (2016)
StNAC4, StNAC5, StNAC18, StNAC48, StNAC81	Resistance to *Phytophthora infestans* infection	Potato	Collinge and Boller (2001)
SlSRN1	Activates the response factor against *Botrytis cinerea*	Tomato	Liu et al. (2014)
SmNAC	Bacterial wilt resistance	Brinjal	Chen et al. (2016)
SlbZIP06, SlbZIP32, SlbZIP46, SlbZIP12, SlbZIP6	Regulates SA, JA, and ACC	Tomato	Li et al. (2015a)

11.8.1 Exploring Genes Controlling Biotic and Abiotic Stress in Tomato

The optimal way to pyramiding the favorable alleles of these genes in a single tomato cultivar is that few genes contribute to the tolerance to both abiotic and biotic stressors. It is important to undertake genetic research under combined stress circumstances in order to discover such genes. There has been little research, and the outcomes are not positive. In *Arabidopsis*, a genome-wide association analysis revealed QTLs which underlie both biotic and abiotic stressors, most QTLs of which displayed differential reactions to biotic and abiotic stresses (Thoen et al. 2017). A candidate gene approach will be useful, as well as genetic studies that perceive the shared variables that contribute to the tolerance of abiotic and biotic stresses. The following genomic techniques can be used to discover genes, and allelic versions that gene, have biotic or abiotic tolerant characteristics (Thoen et al. 2017). In the pyramid of genetic constituents for individual stressors, the relationship of the desired allies with the elements of the proprietary stress as well as the approximation to the gene characteristic should be given special consideration (Kissoudis et al. 2016).

11.8.2 Maintaining Stability of the Resistance of Tomato Against Biotic and Abiotic Stress

Genes that alter numerous pathogenic merchandises have been added to cultivated tomatoes in tomatoes from wild species. The question is whether resistance under abiotic settings is strong. Complete resistance to PM doesn't become altered by way of salt stress, but the incomplete resistance given through the Ol-1 gene and *S. habrochaites* LYC4 QTLs has been impaired (Kissoudis et al. 2017). These consequences show that the identification of resistance genes needs to be discovered and that their purposeful modes need to be understood approximately thru the goal crops. In salt and hormonal imbalance, the Ol-4 gene exhibited resilience to PM, which shows that the resistance of the NBS LRR gene might be resistant. Ol-4 is a Mi-1 tomato homolog, the encoder of the protein of NBS-LRR (Seifi et al. 2014). Heat stresses can triumph over the nematode resistance of the Mi-1 gene (Marques de Carvalho et al. 2015). Therefore, the steadiness cannot be generalized of an individual resistance gene. Warning must be made, consequently, to draw inferences at the sturdiness of the R-gene resistance while affected on the ground considering the fact that, it may be brought about by means of a new pathogen race or combined infection of pathogen (Kissoudis et al. 2016).

11.8.3 Gene Pyramiding for Biotic and Abiotic Stress Resistance in Tomato

In combining salt and pathogen stress, the role of the ion flux alterations is critical. The only element of salt stress is disrupted ion homeostasis that interferes early signals in the defense of pathogen, like changes in streams of ion caused by high levels of Na+ and Cl (Yoshioka et al. 2006). Cell mortality can also occur with K+

ion leakage (Demidchik et al. 2014). Calcium waves like early ion patterns are markers of reactions which is stress specific (Stephan and Schroeder 2014). Changes in ion shift also are one of the first occurrences when pathogenic reactions are identified. A number of ion channels, e.g., CNGC (Clough et al. 2000), permeable good enough K+ channels (Demidchik et al. 2014), as well as Na+ and Cl emitters (NHX and CLC), were based on the idea of the salinity model and to sign protection induction responses and cellular loss of life (Guo et al. 2014).

CNGC, for instance, has been proven to be one of the most important conduction pathways of calcium. Several CNGC households are concerned in salt tolerance in *Arabidopsis* (e.g., AtCNGC10 in Guo et al. 2008) and disease resistance (e.g., AtCNGC2 in Clough et al. 2000). AtCNGC2 silencing tomatoes and orthological potatoes ended in resistance to *Phytophthora* infestations. Moreover, the activities of the NHX1 Na+ -H+ antiporters have demonstrated their involvement inside the manager of vacuolar pH and cell oxidation for maximum induction of plant protection (Chen et al. 2014). Cross-talking through signaling pathways (JA, ET, and ABA) precipitated the instability of PM resistance at a sure degree of saline stress in studies (Kissoudis et al. 2017). The signaling potential of ABA or expanded cleaning of ROS can negatively affect SA and cell loss of life (De Pinto et al. 2012). Thoen et al. (2017) checked out gene polymorphism and discovered SNPs that had an awesome effect on biotic stress and a horrible impact on abiotic responses.

Minimizing adverse interactions inside the pyramid of abiotic and biotic stress tolerance genes is consequently the cornerstone and may be balanced using the aggregate of gene alleles. It can additionally be wonderful if changes to an element that consists of RLK or transcription factors were to cause blended stress tolerance. As an example, overexpression of ERF1 wheat caused tolerance to *Rhizoctonia* (Zhu et al. 2014). In the inter-law of abiotic and biotic stress signals, RLKs have emerged as essential regulatory hubs, particularly on the subject of ABA indicators (Paparella et al. 2014). ABA signaling results on senescence capability, and protective damping signs may be best for the use of RLK. Moreover, cytokine modulation can help fight senescence caused by a blended strain and boom shielding responses (Jiang et al. 2013).

Resistance to biotrophic and necrotrophic diseases is well known to include several signaling routes (Pieterse et al. 2012). In addition, abiotic stress can lead to a re-programming of plants which might influence resistance differently according to pathogenic lifestyle (Bostock et al. 2014). Senescence is often encouraged by cytokines, auxins, and ABAs (Haffner et al. 2015) and is delayed by cytokinetics. In order to attain combined stress resilience through hormone-balancer, it is important to comprehend the intricacy of signaling pathways. In order to achieve this, implementing high-performance phenotype (Furbank and Tester 2011) and high-resolution environmental recording processes (Campbell et al. 2015) for stress control purposes can provide clear genotype associations across the environment (Nagano et al. 2012). Wild species of accessions can be the best source for allelic diversity in stress tolerance because, in marginal environments, they often thrive and propagate (Ortiz et al. 2015). The development of molecular markers and advanced genetic engineering technology facilitates the development of natural variation (Bolger et al. 2017). The knowledge combined from the variation of allele and its outcome by means of phenotypic allele variation can be the element of departure for progressed allele manufacturing thru specified gene editing (Rinaldo and Ayliffe 2015).

11.9 Conclusion

Nearly about all stimuli trigger a generally overlapping collection of signaling pathways at the outset, although some are frequently redundant or damaging and hence disadvantaged to maximize the total defense response. The plants will have to make early decisions regarding their priority pathways and also feedback control on their activation intensity, kinetics, and length depending on their efficacy to allow adaptation to particular stress. In addition, a knowledge gap between biotic and abiotic stress signaling researches continues to exist, notwithstanding significant advances in recent decades. In contrast, environmental variables can modify plant-microbe relationships dynamically along a continuum, from suitable to mutual or parasite. Particularly, in unfavorable situations, mutual and parasitic relationships are noteworthy to the host (Hacquard et al. 2017). Plant immunity is based on and fine-tuned by the integration of signals into biotic and abiotic indicators reflecting tight connections of pathogenic infection or beneficial microbial interactions with abiotic stress. Experimental environments with native growth circumstances (Song et al. 2018) and the use of naturally occurring stress resistance in wild plant species include effective techniques (Zhang et al. 2018). Biotic and abiotic stress research for tomato will have to be integrated in order to clarify how tomato plants are adapted toward changing environmental conditions.

Conflicts of Interest The authors declare no conflict of interest.

References

Aamir M, Singh VK, Meena M, Upadhyay RS, Gupta VK, Singh S (2017) Structural and functional insights into WRKY3 and WRKY4 transcription factors to unravel the WRKY–DNA (W-Box) complex interaction in tomato (Solanum Lycopersicum L.). A computational approach. Front Plant Sci 8:819

Adachi H, Nakano T, Miyagawa N, Ishihama N, Yoshioka M, Katou Y, Yoshioka H (2015) WRKY transcription factors phosphorylated by MAPK regulate a plant immune NADPH oxidase in Nicotiana benthamiana. Plant Cell 27(9):2645–2663

Alvarez ME, Pennell RI, Meijer PJ, Ishikawa A, Dixon RA, Lamb C (1998) Reactive oxygen intermediates mediate a systemic signal network in the establishment of plant immunity. Cell 92(6):773–784

Andolfo G, Jupe F, Witek K, Etherington GJ, Ercolano MR, Jones JD (2014) Defining the full tomato NB-LRR resistance gene repertoire using genomic and cDNA RefSeq. BMC Plant Biol 14(1):1–12

Andreasson E, Jenkins T, Brodersen P, Thorgrimsen S, Petersen NH, Zhu S et al (2005) The MAP kinase substrate MKS1 is a regulator of plant defense responses. EMBO J 24(14):2579–2589

Atamian HS, Eulgem T, Kaloshian I (2012) SlWRKY70 is required for Mi-1-mediated resistance to aphids and nematodes in tomato. Planta 235(2):299–309

Bai Y, Pavan S, Zheng Z, Zappel NF, Reinstädler A, Lotti C et al (2008) Naturally occurring broad-spectrum powdery mildew resistance in a Central American tomato accession is caused by loss of Mlo function. Mol Plant-Microbe Interact 21(1):30–39

Bai Y, Kissoudis C, Yan Z, Visser RG, van der Linden G (2018) Plant behaviour under combined stress: tomato responses to combined salinity and pathogen stress. Plant J 93(4):781–793

Bakshi M, Oelmüller R (2014) WRKY transcription factors: Jack of many trades in plants. Plant Signal Behav 9(2):e27700

Bhattarai KK, Atamian HS, Kaloshian I, Eulgem T (2010) WRKY72-type transcription factors contribute to basal immunity in tomato and Arabidopsis as well as gene-for-gene resistance mediated by the tomato R gene Mi-1. Plant J 63(2):229–240

Birkenbihl RP, Diezel C, Somssich IE (2012) Arabidopsis WRKY33 is a key transcriptional regulator of hormonal and metabolic responses toward Botrytis cinerea infection. Plant Physiol 159(1):266–285

Bolger M, Schwacke R, Gundlach H, Schmutzer T, Chen J, Arend D, Usadel B (2017) From plant genomes to phenotypes. J Biotechnol 261:46–52

Bolwell GP (1999) Role of active oxygen species and NO in plant defense responses. Curr Opin Plant Biol 2(4):287–294

Bostock RM, Pye MF, Roubtsova TV (2014) Predisposition in plant disease: exploiting the nexus in abiotic and biotic stress perception and response. Annu Rev Phytopathol 52:517–549

Cabello JV, Arce L, Chan RL (2008) Patents on plant transcription factors. Recent Pat Biotechnol 2:209–217

Campbell MT, Knecht AC, Berger B, Brien CJ, Wang D, Walia H (2015) Integrating image-based phenomics and association analysis to dissect the genetic architecture of temporal salinity responses in rice. Plant Physiol 168(4):1476–1489

Cessna SG, Sears VE, Dickman MB, Low PS (2000) Oxalic acid, a pathogenicity factor for Sclerotinia sclerotiorum, suppresses the oxidative burst of the host plant. Plant Cell 12(11):2191–2199

Charfeddine M, Saïdi MN, Charfeddine S, Hammami A, GargouriBouzid R (2015) Genome-wide analysis and expression profiling of the ERF transcription factor family in potato (Solanumtuberosum L.). Mol Biotechnol 57(4):348–358

Chen X, Bao H, Guo J, Jia W, Tai F, Nie L, Li Y (2014) Na+/H+ exchanger 1 participates in tobacco disease defence against Phytophthora parasitica var. nicotianae by affecting vacuolar pH and priming the antioxidative system. J Exp Bot 65(20):6107–6122

Chen N, Wu S, Fu J, Cao B, Lei J, Chen C, Jiang J (2016) Overexpression of the eggplant (Solanum melongena) NAC family transcription factor SmNAC suppresses resistance to bacterial wilt. Sci Rep 6:31568

Chinnapandi B, Bucki P, Braun Miyara S (2017) SlWRKY45, nematode-responsive tomato WRKY gene, enhances susceptibility to the root-knot nematode; M. javanica infection. Plant Signal Behav 12(12):e1356530

Clough SJ, Fengler KA, Yu IC, Lippok B, Smith RK, Bent AF (2000) The Arabidopsis dnd1 "defense, no death" gene encodes a mutated cyclic nucleotide-gated ion channel. Proc Natl Acad Sci 97(16):9323–9328

Collinge M, Boller T (2001) Differential induction of two potato genes, Stprx2 and StNAC, in response to infection by Phytophthora infestans and to wounding. Plant Mol Biol 46(5):521–529

Couto D, Zipfel C (2016) Regulation of pattern recognition receptor signalling in plants. Nat Rev Immunol 16(9):537

Cui H, Tsuda K, Parker JE (2015) Effector-triggered immunity: from pathogen perception to robust defense. Annu Rev Plant Biol 66:487–511

Dai X, Sinharoy S, Udvardi M, Zhao PX (2013) PlantTFcat: an online plant transcription factor and transcriptional regulator categorization and analysis tool. BMC Bioinform 14(1):1–6
Dang F, Wang Y, She J, Lei Y, Liu Z, Eulgem T, He S (2014) Overexpression of CaWRKY27, a subgroup IIe WRKY transcription factor of Capsicum annuum, positively regulates tobacco resistance to Ralstonia solanacearum infection. Physiol Plant 150(3):397–411
De Pinto MC, Locato V, De Gara L (2012) Redox regulation in plant programmed cell death. Plant Cell Environ 35(2):234–244
Demidchik V, Straltsova D, Medvedev SS, Pozhvanov GA, Sokolik A, Yurin V (2014) Stress-induced electrolyte leakage: the role of K+-permeable channels and involvement in programmed cell death and metabolic adjustment. J Exp Bot 65(5):1259–1270
Ding Y, Sun T, Ao K, Peng Y, Zhang Y, Li X, Zhang Y (2018) Opposite roles of salicylic acid receptors NPR1 and NPR3/NPR4 in transcriptional regulation of plant immunity. Cell 173(6):1454–1467
Dodds PN, Rathjen JP (2010) Plant immunity: towards an integrated view of plant-pathogen interactions. Nat Rev Genet 11(8):539–548
Du M, Zhai Q, Deng L, Li S, Li H, Yan L, Li C (2014) Closely related NAC transcription factors of tomato differentially regulate stomatal closure and reopening during pathogen attack. Plant Cell 26(7):3167–3184
Ewas M, Khames E, Ziaf K, Shahzad R, Nishawy E, Ali F, Luo J (2017) The tomato DOF daily fluctuations 1, TDDF1 acts as flowering accelerator and protector against various stresses. Sci Rep 7(1):1–16
Farooq M, Basra SMA, Wahid A, Cheema ZA, Cheema MA, Khaliq A (2008) Physiological role of exogenously applied glycinebetaine to improve drought tolerance in fine grain aromatic rice (Oryza sativa L.). J Agron Crop Sci 194(5):325–333
Fischer U, Dröge-Laser W (2004) Overexpression of NtERF5, a new member of the tobacco ethylene response transcription factor family enhances resistance to tobacco mosaic virus. Mol Plant-Microbe Interact 17(10):1162–1171
Fraire-Velázquez S, Rodríguez-Guerra R, Sánchez-Calderón L (2011) Abiotic and biotic stress response crosstalk in plants. In: Abiotic stress response in plants—physiological, biochemical and genetic perspectives. InTech, London, pp 3–26
Furbank RT, Tester M (2011) Phenomics–technologies to relieve the phenotyping bottleneck. Trends Plant Sci 16(12):635–644
Glazebrook J (2001) Genes controlling expression of defense responses in Arabidopsis-2001 status. Curr Opin Plant Biol 4:301–308
Gonzalez DH (2016) Introduction to transcription factor structure and function. In: Plant transcription factors. Elsevier, Amsterdam, pp 3–11
Guo KM, Babourina O, Christopher DA, Borsics T, Rengel Z (2008) The cyclic nucleotide-gated channel, AtCNGC10, influences salt tolerance in Arabidopsis. Physiol Plant 134(3):499–507
Guo W, Zuo Z, Cheng X, Sun J, Li H, Li L, Qiu JL (2014) The chloride channel family gene CLCd negatively regulates pathogen-associated molecular pattern (PAMP)-triggered immunity in Arabidopsis. J Exp Bot 65(4):1205–1215
Gupta S, Malviya N, Kushwaha H, Nasim J, Bisht NC, Singh VK, Yadav D (2015) Insights into structural and functional diversity of Dof (DNA binding with one finger) transcription factor. Planta 241(3):549–562
Hacquard S, Spaepen S, Garrido-Oter R, Schulze-Lefert P (2017) Interplay between innate immunity and the plant microbiota. Annu Rev Phytopathol 55:565–589
Häffner E, Konietzki S, Diederichsen E (2015) Keeping control: the role of senescence and development in plant pathogenesis and defense. Plan Theory 4(3):449–488
Hind SR, Strickler SR, Boyle PC, Dunham DM, Bao Z, O'Doherty IM, Martin GB (2016) Tomato receptor FLAGELLIN-SENSING 3 binds flgII-28 and activates the plant immune system. Nat Plants 2(9):1–8
Huang S, Gao Y, Liu J, Peng X, Niu X, Fei Z, Liu Y (2012) Genome-wide analysis of WRKY transcription factors in Solanum lycopersicum. Mol Gen Genomics 287(6):495–513

Ishihama N, Yoshioka H (2012) Post-translational regulation of WRKY transcription factors in plant immunity. Curr Opin Plant Biol 15(4):431–437
Jiang CJ, Shimono M, Sugano S, Kojima M, Liu X, Inoue H, Takatsuji H (2013) Cytokinins act synergistically with salicylic acid to activate defense gene expression in rice. Mol Plant-Microbe Interact 26(3):287–296
Jones JD, Dangl JL (2006) The plant immune system. Nature 444(7117):323–329
Kang WH, Kim S, Lee HA, Choi D, Yeom SI (2016) Genome-wide analysis of Dof transcription factors reveals functional characteristics during development and response to biotic stresses in pepper. Sci Rep 6(1):1–12
Kim JG, Stork W, Mudgett MB (2013) Xanthomonas type III effector XopD desumoylates tomato transcription factor SlERF4 to suppress ethylene responses and promote pathogen growth. Cell Host Microbe 13(2):143–154
Kissoudis C (2016) Genetics and regulation of combined abiotic and biotic stress tolerance in tomato (Doctoral dissertation, Wageningen University)
Kissoudis C, Sunarti S, Van De Wiel C, Visser RG, van der Linden CG, Bai Y (2016) Responses to combined abiotic and biotic stress in tomato are governed by stress intensity and resistance mechanism. J Exp Bot 67(17):5119–5132
Kissoudis C, Seifi A, Yan Z, Islam ATM, van der Schoot H, van de Wiel C et al (2017) Ethylene and abscisic acid signaling pathways differentially influence tomato resistance to combined powdery mildew and salt stress. Front Plant Sci 7:2009
Laluk K, Mengiste T (2010) Necrotroph attacks on plants: wanton destruction or covert extortion? Arabidopsis Book/Am Soc Plant Biol 8:e0136
Lapidot M, Karniel U, Gelbart D, Fogel D, Evenor D, Kutsher Y, Levin I (2015) A novel route controlling begomovirus resistance by the messenger RNA surveillance factor pelota. PLoS Genet 11(10):e1005538
Li J, Brader G, Palva ET (2004) The WRKY70 transcription factor: a node of convergence for jasmonate-mediated and salicylate-mediated signals in plant defense. Plant Cell 16(2):319–331
Li D, Fu F, Zhang H, Song F (2015a) Genome-wide systematic characterization of the bZIP transcriptional factor family in tomato (Solanum Lycopersicum L.). BMC Genomics 16(1):1–18
Li JB, Luan YS, Liu Z (2015b) SpWRKY1 mediates resistance to Phytophthora infestans and tolerance to salt and drought stress by modulating reactive oxygen species homeostasis and expression of defense-related genes in tomato. Plant Cell Tissue Organ Cult 123(1):67–81
Li JB, Luan YS, Liu Z (2015c) Over expression of SpWRKY1 promotes resistance to Phytophthora nicotianae and tolerance to salt and drought stress in transgenic tobacco. Physiol Plant 155(3):248–266
Li Z, Peng R, Tian Y, Han H, Xu J, Yao Q (2016) Genome-wide identification and analysis of the MYB transcription factor superfamily in Solanumlycopersicum. Plant Cell Physiol 57(8):1657–1677
Li Z, Tian Y, Xu J, Fu X, Gao J, Wang BO, Yao Q (2018) A tomato ERF transcription factor, SlERF84, confers enhanced tolerance to drought and salt stress but negatively regulates immunity against Pseudomonas syringaepv. Tomato DC3000. Plant Physiol Biochem 132:683–695
Lichtenthaler HK (1996) Vegetation stress: an introduction to the stress concept in plants. J Plant Physiol 148(1–2):4–14
Liu X, Bai X, Wang X, Chu C (2007) OsWRKY71, a rice transcription factor, is involved in rice defense response. J Plant Physiol 164(8):969–979
Liu B, Ouyang Z, Zhang Y, Li X, Hong Y, Huang L, Song F (2014) Tomato NAC transcription factor SlSRN1 positively regulates defense response against biotic stress but negatively regulates abiotic stress response. PLoS One 9(7):e102067
Lo Presti L, Lanver D, Schweizer G, Tanaka S, Liang L, Tollot M, Kahmann R (2015) Fungal effectors and plant susceptibility. Annu Rev Plant Biol 66:513–545
Love AJ, Milner JJ, Sadanandom A (2008) Timing is everything: regulatory overlap in plant cell death. Trends Plant Sci 13(11):589–595

Ma NN, Zuo YQ, Liang XQ, Yin B, Wang GD, Meng QW (2013) The multiple stress-responsive transcription factor SlNAC1 improves the chilling tolerance of tomato. Physiol Plant 149(4):474–486

Mare C, Mazzucotelli E, Crosatti C, Francia E, Cattivelli L (2004) Hv-WRKY38: a new transcription factor involved in cold-and drought-response in barley. Plant Mol Biol 55(3):399–416

Marques de Carvalho L, Benda ND, Vaughan MM, Cabrera AR, Hung K, Cox T (2015) Mi-1-mediated nematode resistance in tomatoes is broken by short-term heat stress but recovers over time. J Nematol 47:133–140

Matsukura C, Aoki K, Fukuda N, Mizoguchi T, Asamizu E, Saito T, Ezura H (2008) Comprehensive resources for tomato functional genomics based on the miniature model tomato Micro-Tom. Curr Genomics 9(7):436–443

Miwa H, Okazaki S (2017) How effectors promote beneficial interactions. Curr Opin Plant Biol 38:148–154

Nagano AJ, Sato Y, Mihara M, Antonio BA, Motoyama R, Itoh H, Izawa T (2012) Deciphering and prediction of transcriptome dynamics under fluctuating field conditions. Cell 151(6):1358–1369

Nakashima K, Takasaki H, Mizoi J, Shinozaki K, Yamaguchi-Shinozaki K (2012) NAC transcription factors in plant abiotic stress responses. BBA 1819:97–103

Ntoukakis V, Balmuth AL, Mucyn TS, Gutierrez JR, Jones AM, Rathjen JP (2013) The tomato Prf complex is a molecular trap for bacterial effectors based on Pto transphosphorylation. PLoS Pathog 9(1):e1003123

Nürnberger T, Brunner F, Kemmerling B, Piater L (2004) Innate immunity in plants and animals: striking similarities and obvious differences. Immunol Rev 198(1):249–266

Nuruzzaman M, Sharoni AM, Kikuchi S (2013) Roles of NAC transcription factors in the regulation of biotic and abiotic stress responses in plants. Front Microbiol 4:248

Ortiz R, Redden R, Yadav SS, Maxted N, Dulloo ME, Guarino L, Smith P (2015) The importance of crop wild relatives, diversity, and genetic potential for adaptation to abiotic stress-prone environments. In: Crop wild relatives and climate change. Wiley, Hoboken, pp 80–87

Pan IC, Lin CW, Su RC, Cheng CP, Lin CS, Chan MT (2010) Ectopic expression of an EAR motif deletion mutant of SlERF3 enhances tolerance to salt stress and Ralstonia solanacearum in tomato. Planta 232(5):1075–1086

Pan XQ, Fu DQ, Zhu BZ, Lu CW, Luo YB (2013) Overexpression of the ethylene response factor SlERF1 gene enhances resistance of tomato fruit to Rhizopus nigricans. Postharvest Biol Technol 75:28–36

Paparella C, Savatin DV, Marti L, De Lorenzo G, Ferrari S (2014) The Arabidopsis LYSIN MOTIF-CONTAINING RECEPTOR-LIKE KINASE3 regulates the cross talk between immunity and abscisic acid responses. Plant Physiol 165(1):262–276

Pavan S, Jacobsen E, Visser RG, Bai Y (2010) Loss of susceptibility as a novel breeding strategy for durable and broad-spectrum resistance. Mol Breed 25(1):1–12

Peng Y, Bartley LE, Chen X, Dardick C, Chern M, Ruan R, Ronald PC (2008) OsWRKY62 is a negative regulator of basal and Xa21-mediated defense against Xanthomonas oryzae pv. oryzae in rice. Mol Plant 1(3):446–458

Phukan UJ, Jeena GS, Shukla RK (2016) WRKY transcription factors: molecular regulation and stress responses in plants. Front Plant Sci 7:760

Pieterse CM, Van der Does D, Zamioudis C, Leon-Reyes A, Van Wees SC (2012) Hormonal modulation of plant immunity. Annu Rev Cell Dev Biol 28:489–521

Poland JA, Balint-Kurti PJ, Wisser RJ, Pratt RC, Nelson RJ (2009) Shades of gray: the world of quantitative disease resistance. Trends Plant Sci 14(1):21–29

Qiu Y, Yu D (2009) Over-expression of the stress-induced OsWRKY45 enhances disease resistance and drought tolerance in Arabidopsis. Environ Exp Bot 65(1):35–47

Rezzonico F, Rupp O, Fahrentrapp J (2017) Pathogen recognition is compatible plant-microbe interactions. Sci Rep 7(1):1–12

Riechmann JL, Heard J, Martin G, Reuber L, Jiang CZ, Keddie J, Yu GL (2000) Arabidopsis transcription factors: genome-wide comparative analysis among eukaryotes. Science 290(5499):2105–2110

Rinaldo AR, Ayliffe M (2015) Gene targeting and editing in crop plants: a new era of precision opportunities. Mol Breed 35(1):1–15

Robaglia C, Caranta C (2006) Translation initiation factors: a weak link in plant RNA virus infection. Trends Plant Sci 11(1):40–45

Robert-Seilaniantz A, Grant M, Jones JD (2011) Hormone crosstalk in plant disease and defense: more than just jasmonate-salicylate antagonism. Annu Rev Phytopathol 49:317–343

Roux F, Voisin D, Badet T, Balagué C, Barlet X, Huard-Chauveau C, Raffaele S (2014) Resistance to phytopathogens e tuttiquanti: placing plant quantitative disease resistance on the map. Mol Plant Pathol 15(5):427–432

Saijo Y, Loo EPI, Yasuda S (2018) Pattern recognition receptors and signaling in plant-microbe interactions. Plant J 93(4):592–613

Sarris PF, Duxbury Z, Huh SU, Ma Y, Segonzac C, Sklenar J, Jones JD (2015) A plant immune receptor detects pathogen effectors that target WRKY transcription factors. Cell 161(5):1089–1100

Schwessinger B, Zipfel C (2008) News from the frontline: recent insights into PAMP-triggered immunity in plants. Curr Opin Plant Biol 11(4):389–395

Seifi A, Gao D, Zheng Z, Pavan S, Faino L, Visser RF, Wolters AM, Bai Y (2014) Genetics and molecular mechanisms of resistance to powdery mildews in tomato (Solanum Lycopersicum) and its wild relatives. Eur J Plant Pathol 138:641–665

Seo E, Choi D (2015) Functional studies of transcription factors involved in plant defenses in the genomics era. Brief Funct Genomics 14(4):260–267

Shahzad R, Harlina PW, Cong-hua X, Ewas M, Nishawy E, Zhenyuan P, Foly MM (2016) Overexpression of potato transcription factor (StWRKY1) conferred resistance to Phytophthora infestans and improved tolerance to water stress. Plant Omics 9(2):149–158

Song YH, Kubota A, Kwon MS, Covington MF, Lee N, Tan ER, Imaizumi T (2018) Molecular basis of flowering under natural long-day conditions in Arabidopsis. Nat Plants 4(10):824–835

Souer E, van Houwelingen A, Kloos D, Mol J, Koes R (1996) The no apical meristem gene of Petunia is required for pattern formation in embryos and flowers and is expressed at meristem and primordia boundaries. Cell 85:159–170

Stephan AB, Schroeder JI (2014) Plant salt stress status is transmitted systemically via propagating calcium waves. Proc Natl Acad Sci USA 111:6126–6127

Sun XC, Gao YF, Li HR, Yang SZ, Liu YS (2015) Over-expression of SlWRKY39 leads to enhanced resistance to multiple stress factors in tomato. J Plant Biol 58(1):52–60

Thoen MP, Davila Olivas NH, Kloth KJ, Coolen S, Huang PP, Aarts MG et al (2017) Genetic architecture of plant stress resistance: multi-trait genome-wide association mapping. New Phytol 213(3):1346–1362

Tian Z, He Q, Wang H, Liu Y, Zhang Y, Shao F, Xie C (2015) The potato ERF transcription factor StERF3 negatively regulates resistance to Phytophthora infestans and salt tolerance in potato. Plant Cell Physiol 56(5):992–1005

Tomato Genome Consortium (2012) The tomato genome sequence provides insights into fleshy fruit evolution. Nature 485(7400):635

Toyota M, Spencer D, Sawai-Toyota S, Jiaqi W, Zhang T, Koo AJ, Gilroy S (2018) Glutamate triggers long-distance, calcium-based plant defense signaling. Science 361(6407):1112–1115

Tsutsui T, Kato W, Asada Y, Sako K, Sato T, Sonoda Y, Yamaguchi J (2009) DEAR1, a transcriptional repressor of DREB protein that mediates plant defense and freezing stress responses in Arabidopsis. J Plant Res 122(6):633–643

Van Loon LC, Van Strien EA (1999) The families of pathogenesis-related proteins, their activities, and comparative analysis of PR-1 type proteins. Physiol Mol Plant Pathol 55(2):85–97

Wang G, Zhang S, Ma X, Wang Y, Kong F, Meng Q (2016) A stress-associated NAC transcription factor (SlNAC35) from tomato plays a positive role in biotic and abiotic stresses. Physiol Plant 158(1):45–64

Wang X, Guo R, Tu M, Wang D, Guo C, Wan R, Wang X (2017) Ectopic expression of the wild grape WRKY transcription factor VqWRKY52 in Arabidopsis thaliana enhances resistance to the biotrophic pathogen powdery mildew but not to the necrotrophic pathogen Botrytis cinerea. Front Plant Sci 8:97

Xing DH, Lai ZB, Zheng ZY, Vinod KM, Fan BF, Chen ZX (2008) Stress-and pathogen-induced Arabidopsis WRKY48 is a transcriptional activator that represses plant basal defense. Mol Plant 1(3):459–470

Xu X, Chen C, Fan B, Chen Z (2006) Physical and functional interactions between pathogen-induced Arabidopsis WRKY18, WRKY40, and WRKY60 transcription factors. Plant Cell 18(5):1310–1326

Yeom SI, Baek HK, Oh SK, Kang WH, Lee SJ, Lee JM et al (2011) Use of a secretion trap screen in pepper following Phytophthora capsici infection reveals novel functions of secreted plant proteins in modulating cell death. Mol Plant-Microbe Interact 24(6):671–684

Yogendra KN, Dhokane D, Kushalappa AC, Sarmiento F, Rodriguez E, Mosquera T (2016) StWRKY8 transcription factor regulates benzylisoquinoline alkaloid pathway in potato conferring resistance to late blight. Plant Sci 256:208–216

Yokotani N, Sato Y, Tanabe S, Chujo T, Shimizu T, Okada K, Nishizawa Y (2013) WRKY76 is a rice transcriptional repressor playing opposite roles in blast disease resistance and cold stress tolerance. J Exp Bot 64(16):5085–5097

Yoshioka K, Moeder W, Kang HG, Kachroo P, Masmoudi K, Berkowitz G, Klessig DF (2006) The chimeric Arabidopsis CYCLIC NUCLEOTIDE-GATED ION CHANNEL11/12 activates multiple pathogen resistance responses. Plant Cell 18(3):747–763

Yu X, Feng B, He P, Shan L (2017) From chaos to harmony: responses and signaling upon microbial pattern recognition. Annu Rev Phytopathol 55:109–137

Zhang H, Li W, Chen J, Yang Y, Zhang Z, Zhang H, Huang R (2007) Transcriptional activator TSRF1 reversely regulates pathogen resistance and osmotic stress tolerance in tobacco. Plant Mol Biol 63(1):63–71

Zhang H, Li A, Zhang Z, Huang Z, Lu P, Zhang D, Huang R (2016) Ethylene response factor TERF1, regulated by ETHYLENE-INSENSITIVE3-like factors, functions in reactive oxygen species (ROS) scavenging in tobacco (Nicotianatabacum L.). Sci Rep 6(1):1–10

Zhang H, Li Y, Zhu JK (2018) Developing naturally stress-resistant crops for a sustainable agriculture. Nat Plants 4(12):989–996

Zheng Z, Qamar SA, Chen Z, Mengiste T (2006) Arabidopsis WRKY33 transcription factor is required for resistance to necrotrophic fungal pathogens. Plant J 48(4):592–605

Zhu X, Qi L, Liu X, Cai S, Xu H, Huang R, Zhang Z (2014) The wheat ethylene response factor transcription factor pathogen-induced ERF1 mediates host responses to both the necrotrophic pathogen Rhizoctonia cerealis and freezing stresses. Plant Physiol 164(3):1499–1514

Chapter 12
The Role of Transcription Factors in Response to Biotic Stresses in Potato (*Solanum tuberosum* L.)

Namo Dubey and Kunal Singh

Abstract Potato is an important horticultural crop worldwide, not only as a food but also as an emerging model plant. Understanding the gene pool regulation is essential for the future of potato crop to counter diseases caused by multitude of pathogens. These plant-pathogen interactions entail complex molecular pathways of tolerance, resistance, susceptibility, and responsiveness happening inside a plant cell, singularly and in whole tissue, collectively. Transcription factors are one such well known and researched molecular component of plant cell that regulate the target transcript behaviour by expression activation or inactivation. As a result, TFs play diverse functions at different phases of pathogen attack and interact in multiple metabolic pathways to define more accurately response patterns. A catalogue of potato TFs implicated in biotic stress was gathered in this chapter. We are hopeful that information provided would be significant for potato breeding efforts aiming for biotic stress tolerance in order to boost overall yield and productivity.

Keywords Potato · Transcription factors · Biotic stress · *Phytophthora infestans* · Plant defence

12.1 Introduction

Plants are constantly exposed to biotic and abiotic stresses, which affect their health and total yield. Among biotic factors that affect plant health and integrity, microbes, small invertebrates like insects and aphids and even other plants are responsible. Fungi, viruses, viroids, bacteria, protists, oomycetes, nematodes and mycoplasmas

N. Dubey · K. Singh (✉)
CSIR-Institute of Himalayan Bioresource Technology, Palampur, India

Academy of Scientific and Innovative Research (AcSIR), Ghaziabad, India
e-mail: kunal@ihbt.res.in

© The Author(s), under exclusive license to Springer Nature Switzerland AG 2022

S. H. Wani et al. (eds.), *Transcription Factors for Biotic Stress Tolerance in Plants*, https://doi.org/10.1007/978-3-031-12990-2_12

are group of microbes that adversely affect plants (Dubey and Singh 2018; Zhang et al. 2021). Due to adverse effect caused by these microbes, distinct symptoms developed in plants are called as plant disease. These disease and pests further impact plants and agriculture and are broadly considered as biotic stress. Understanding how agriculturally important crop plants fight against these biotic stress-causing factors is essential for developing long-term management techniques and methodology (Chacon-Cerdas et al. 2020). One such crop plant is potato (*Solanum tuberosum* L.) that is widely grown worldwide and faces a myriad of biotic stress-causing factors at native and newly introduced places. Potatoes have a pretty broad distribution pattern, ensuring availability, homogeneity and consistency (Devaux et al. 2020), and currently grown as a commercial crop in 149 countries (Kroschel et al. 2020). Among fungi and oomycetes, derived diseases known to cause major harm to potato plants are late blight of potato caused by *Phytophthora infestans*, early blight disease caused by *Alternaria solani*, wart disease caused by *Synchytrium endobioticum* and black rot caused by *Fusarium solani* (Adolf et al. 2020). Among the bacterial diseases affecting potato health are bacterial wilt caused by *Ralstonia* spp., zebra chip by *Candidatus Liberibacter solanacearum*, common scab by *Streptomyces* spp., blackleg by *Pectobacterium* spp. and powdery scab by *Spongospora subterranea*. There are 49 bug species identified as potato pests from temperate, subtropical and tropical zones, impacting yield with almost 16% of global crop losses (Kroschel et al. 2020). In addition, 2 viroid and 40 viruses are reported to infect potatoes.

To develop efficient varieties tolerant to biotic stresses, resistant genetic germplasm is both desirable and essential. Being a cultivated crop, potato has been bred in the past primarily for yield and productivity with disease concern being secondary. This approach led to great famine of Ireland caused primarily due to destruction of potato crop by *Phytophthora infestans* (Goss et al. 2014). In the last century, many varieties have been developed using diverse pre-existing *S. tuberosum* germplasm for disease resistance along with application of wild varieties and wild species of potato. One major concern that arises in such breeding programme is difficulty of selection of desirable traits that come with introgression of many non-desirable genotypes and traits. To mitigate this issue, marker-assisted selection (MAS) is one preferred methodology. But the question remains about how to choose markers for disease-related genotype. Identification and involvement of master genes like resistance genes or major transcriptional regulator can be used for such MAS-assisted breeding. Transcription factors fall into category of transcriptional regulator with many of them having a role as master regulator in different biotic and abiotic stress signalling (Hussain et al. 2018; Shinozaki et al. 2015).

Transcription factors (TFs) are proteins that bind to cis-acting sites in promoter regions of the target gene to regulate their transcript expression. These transcription factors can function as repressors or activators, and how well they interact with co-repressor molecules and other cofactors to modify target promoter expression is frequently the determining factor. According to one estimate by Hong (2016), around 2000 genes in plant genomes encode diverse transcription factors (TFs). The PlnTF database (http://plntfdb.bio.uni-potsdam.de/v3.0/) lists 63 TF families and 22 classes of additional plant regulatory proteins. According to the iTAK database (http://itak.feilab.net/cgi-bin/itak/dbfamily.cgi?plant=4113), TFs are classified into 67 families, 24 of which are transcriptional regulators (Zheng et al. 2016). The regulation of biotic stress response in potatoes has been linked to NAC (NAM, no apical meristem; ATAF1, two *Arabidopsis thaliana*-activating factors 1 and 2; CUC, cup-shaped cotyledon), ARF (auxin-responsive factor), AP2/ERF/DREB (apetala-2/ethylene-responsive factor/dehydration-responsive element-binding protein), WRYK (WRKYGQK), TCP (teosinte branched 1; CYC in cycloidea; PCF, proliferating cell factors 1 and 2), ZFP (zinc finger protein), BELL (BEL-like) and bZIP (basic leucine zipper) families (Chacon-Cerdas et al. 2020).

The goal of this study was to collect data on transcription factors implicated in potato biotic stress responses up to this point. Though many genes has been discovered or linked having a role in potato-microbe interaction, the literature available on the role of transcription factors in potato biotic stress are very limited and still in infancy. Yet, the overall information collected and analysed here will be useful as a starting point as knowledge-based regarding TFs and potato defence mechanism.

12.2 Transcription Factors' Common Way of Action in Biotic Stress

The mechanisms by which transcription factors impact the expression pattern of genes linked to biotic stimuli vary depending on the TF family. The general methods by which TFs exercise their central role in the defence reaction are cell wall reinforcement, hypersensitive response regulation, PR gene expression control, SA pathway gene regulation and antioxidant enzyme activity activation (Fig. 12.1). In this context, the most widely documented method is PR gene regulation, which is linked to *F. solani*, *P. infestans* and plant defence priming chemicals and elicitors.

12.3 Transcription Factors Involvement Against Fungal and Oomycete Pathogens in Potato

As previously discussed, fungus and oomycetes are the most destructive diseases of potato crop. Devastating effect of *P. infestans* was such in the 1840s that it led to a famine. Along with legacy issue, the continuing harmful and impactful effect of late blight disease is the reason that a major portion of published literature on biotic stresses is tilted towards *S. tuberosum*-*P. infestans* interaction. Therefore, TFs involved against *P. infestans* attack are more studied. One such transcription factor family is ARF. Analysis of RNA-Seq data for members of ARF family revealed that none of the member got upregulated under *P. infestans* attack, but expression of four genes *AtARF6a*, *StARF6b*, *StARF17* and *StARF19a* prominently got induced under

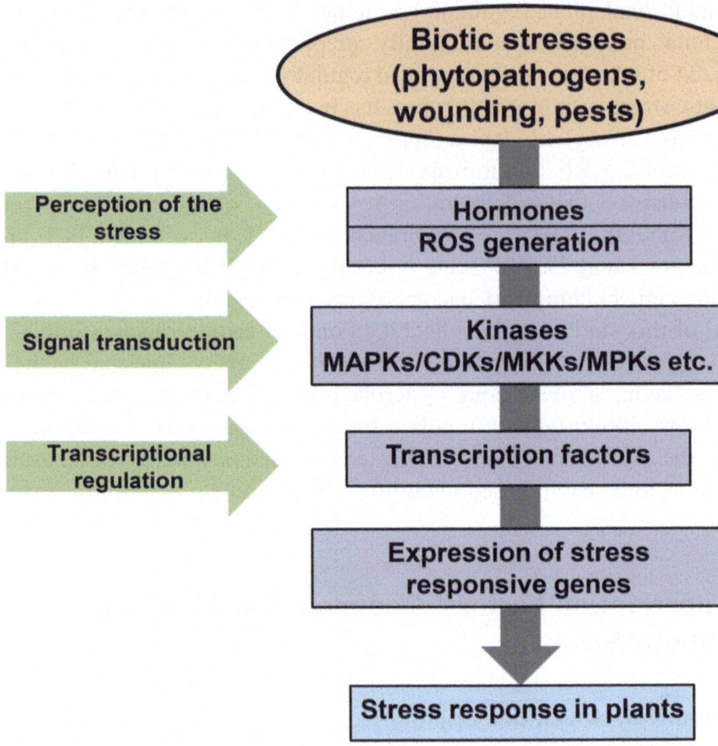

Fig. 12.1 A diagrammatic representation of the role of transcription factors in potato against biotic stress signalling involving the component of signalling cascade under stress by pathogen and pests

benzothiadiazole S-methyl ester (BTH) treatment. The downregulation of most of the ARF members under DL-β-amino-n-butyric acid (BABA) treatment and *P. infestans* attack indicates the probable negative regulation implied by ARF TF family in resistance priming by former and necrotrophic attack of pathogen (Song et al. 2019). BTH is a salicylic acid (SA) analogue that has been implicated as a priming agent of plant defence and systemic acquired resistance (Gorlach et al. 2006). Induced expression under BTH indirectly implies that these four TFs get regulated via SA-mediated defence pathway. All the four are also shown to be differentially expressed under ethylene, abscisic acid (ABA) and IAA treatments. Another set of TFs shown to be involved under *P. infestans* attack are StNAC43, StMYB8 and StERF3. Yogendra et al. (2017) showed that under *P. infestans* attack, ethylene response factor – StERF3 – may bind with an ethylene response element (ERE) at the promoter of *StNAC43*. After getting expressed, StNAC43 protein binds at the promoter region of *StMYB8*, activating it in response. Afterwards, StMYB8 protein provides resistance to *P. infestans* via synthesis of many defence-related downstream genes like flavanone 3-hydroxylase (F3H), putrescine hydroxycinnamoyl transferase (PHT), hydroxycinnamoyl transferase (HCT) and chalcone synthase activity (CHS). All these genes are involved in secondary cell wall formation that is considered an effective step towards late blight resistance.

One of the most widely studied transcription factors during biotic stress to plants are the members of WRKY family. They function as co-repressors or repressors as well as activators of key pathways such as terpene, alkaloid and other metabolite production and have been linked to the activation of a variety of immune response cascades (Schluttenhofer and Yuan 2015). In potato, empirical evidence suggest a major role of StWRKY1, StWRKY8 and StWRKY33 under *P. infestans* infection. In an experiment where potato was primed with priming agent potassium phosphite, the plant was able to restrict *P. infestans* growth, and *AtWRKY1* was getting induced (Machinandiarena et al. 2012). It was suggested that interaction of potassium phosphite with StWRKY1 protects via interacting with SA. In gene silenced plants with virus-induced gene silencing approach for *AtWRKY1*, the resistance to *P. infestans* was highly compromised in resistant potato genotype – F06025 (Yogendra et al. 2015). Later on, plants with *AtWRKY1* over-expression clearly showed the role of the protein in potato defence as transgenic lines were resistant to *P. infestans* (Shahzad et al. 2016). *StWRKY1* was found getting upregulated under HA, SA, ABA and ethylene while inducing expression of many pathogenesis-related (PR) genes in transgenics including PR2, PR3 and PR9. These results are in corroboration with the role of WRKY1 in barley and *Arabidopsis* where it functions as repressor and activator of basal defence, ETI (effector-triggered immunity) and MTI (microbial-triggered immunity). In barley, WRKY1/WRKY2 suppresses the immune response, but it is suppressed when it interacts with the MLA (activated by AVRa) that triggers defence responses (Rushton et al. 2010). Transcriptome assessment of Chinese potato cultivar Qingshu 9 which shows excellent resistance against late blight disease revealed the high transcript induction of AP2/ERF-domain protein-PGSC0003DMG400016812, WRKY5-PGSC0003DMG400028469 and MYB-PGSC0003DMG400024572, while downregulation of MYBRL3-PGSC0003DMG400028098 under *P. infestans* attack (He et al. 2021). *StWRKY5* acts through BABA and provides resistance to transgenic potatoes against late blight disease (Yang et al. 2018). StWRKY8 has also been found to control the benzylisoquinoline alkaloid pathway, which results in defence to *P. infestans* (Yogendra et al. 2017), confirming the link between StWRKY8 and the promoters of the NCS, COR-2 and TyDC genes, which are all implicated in this metabolic process.

Members of ERF, bZIP and ZFP TFs have also been reported having a role in response to late blight disease. Though reports of ERF in biotic stress in potato are few, Tian et al. (2015) reported that StERF3 negatively regulates the late blight disease occurrence. SA and ABA also found to regulate the transcript expression of StERF3 that contain EAR motif (ethylene-responsive element-binding factor-associated amphiphilic repression). In a genome-wide study of ERFs in potatoes, a number of ERF TFs containing EAR motifs were identified (Charfeddine et al. 2015) and implicated the *StERF41*, *StERF67* and *StERF71* transcript accumulation during *P. infestans* infection. *StERF3* and *StERF6* by Bouaziz et al. (2015) and *StERF4* and *StERF5* by Massa et al. (2011) have also been reported to be expressed and induced under *P. infestans* exposure. Salicylic acid inhibits the StbZIP61-StNPR3L interaction in potato defence trials against *P. infestans*, while StbZIP61 promotes SA production by stimulating *StICS1* expression. However, StNPR3L protein diminishes *StbZIP61* activity, and the StbZIP61-StNPR3L relationship is inhibited by salicylic acid (Zhou et al. 2018). StTCP23 (class I TCP) analogous to *Arabidopsis* was reported for involvement in the regulation of plant defence in potato (Bao et al. 2019). Transcription factors also play a role in plant defence against fungal pathogens. For example, *StERF94* transcript gets induced under *Fusarium solani* infection, and transgenic lines over-expressing the gene provide resistance to the pathogen (Charfeddine et al. 2019). StERF94 probably provides resistance by activating multiple antioxidant enzymes, viz. glutathione peroxidase, superoxide dismutase and catalase, and inducing PR genes like PR2, PR3 and PR9. *StDREB1*, a member of AP2/ERF-DREB TF family, has also been shown to be induced under *F. solani* exposure (Charfeddine et al. 2019).

12.4 Response of Potato Transcription Factors Against Virus and Bacterial Pathogen

There are only few reports to demonstrate the transcription factors' significance in the resistance of potato against virus and bacterial pathogens. In a few situations, such as interactions with *Erwinia carotovora* subsp. *atroseptica*, StWRKY1 was shown to be enhanced in transcript expression. The findings suggested that the pectate lyase genes (pelB and pelD) caused a protective response in *E. carotovora* that was connected to StWRKY1 (Dellagi et al. 2000). Study on *Candidatus Liberibacter solanacearum*, a phloem-limited, gram-negative bacteria; the causal agent of zebra chip disease in potato; and interaction with "Atlantic", a susceptible variety, and "Waneta", a zebra chip-resistant variety, showed that most of the transcription factor assessed were downregulated in Waneta including *MYB185*, *WRKY30* and *ERF5*. Under susceptible interaction MYB185 showed downregulation, while ERF5 showed upregulation (Levy et al. 2017). The expression analysis suggests that under pathogenic bacterial interaction resistance is led by transcriptional repressors with negative regulations. The AP2 was only exception in the study that got induced even in incompatible interaction. Assessment of wild-type potato, *Solanum commersonii*, after treatment with bacterial pathogen (*Pectobacterium carotovorum*) and viral pathogen (*Potato virus Y*) (Villano et al. 2020), revealed the induced expression of *ScWRKY016* and *ScWRKY023* under bacterial and *ScWRKY023* and *ScWRKY055* under viral treatment. Recently, *StERF3* was also found to be induced under *Potato virus Y* exposure under StPIP1 priming which acts as a defence priming agent in potato (Combest et al. 2021).

12.5 The Role of Transcription Factor Against Wound and Insect

In line with the assessment of many TFs in potato acting as repressor, potato lines over-expressing *StGR1106* showed downregulation of *StWRKY22* while showing susceptibility to both *Globodera rostochiensis* (a nematode) and *Verticillium dahliae* and repression of *StWRKY53* under influence of *V. dahliae* (https://core.ac.uk/download/pdf/29234801.pdf). It has been shown that in potato, the BEL1 or BELL transcription factor is primarily responsible for wound and insect protection. According to Bürglin (1997), the BELL (BEL1) transcription factor family is one of many within the TALE superclass. It is encoded by 1 of 14 families of homeobox genes, which are responsible for encoding the homeodomain (Mukherjee et al. 2009; Kerstetter et al. 1994). The homeodomain is a 60-amino-acid DNA-binding domain that has been found to be highly conserved throughout evolution. A helix-spin-helix motif formed by the second and third helices, as described by Chatterjee et al. (2007), interacts with DNA sequences and is involved in the regulation of gene expression. On the other hand, the second and third helices, which interact with DNA sequences and regulate gene expression, form a helix-spin-helix motif, which interacts with DNA sequences and regulates gene expression (Desplan et al. 1988). Homeobox genes belonging to the BELL family have been implicated in a wide range of hormone response pathways and are associated with plant health and development. Over-expression of the BELL transcription factor has been linked to tuber growth, insect herbivory in leaves, wounding response to mechanical stress and a variety of photoperiod-dependent responses in potatoes. The BELL genes in potato have been discovered to the best of our knowledge (Sharma et al. 2014), with StBEL5 being the gene that has been most commonly mentioned in connection with biotic response (Chatterjee et al. 2007). The transcriptional and post-transcriptional regulation of BELL TF expression and activity is a difficult process to understand. There have been numerous accounts of light stimuli having the ability to regulate behaviour. Exactly as predicted, the GATA, GT1 and AT1 motifs are among the light-sensitive promoter elements detected in the StBEL5 promoter region, with the AT1 motif being the most prominent.

Previous reports have identified the Q-type C2H2 ZFP (StZFP2) as a wound-responsive transcription factor. Q-type C2H2 ZFPs have two additional motifs: B-box, which contains a nuclear localization signal (NLS), and L-box, which is rich in leucine residues, which are required for transcriptional regulation. The expression profiles of the selected ZFPs revealed that they were involved in the defence against *Manduca sexta* infection (Lawrence and Novak 2018). Other transcription factors, such as StMEV47 (a zinc finger protein related to ZPT2–13), were shown to be upregulated in the presence of the Colorado potato beetle (CPB). Involvement of WRKY transcription factor has also been reported against wounding in potato. For instance, two transcription factors such as *ScWRKY023* and *ScWRKY045* were found to be expressed in *S. commersonii*, but no change in the expression was achieved in wild type. Wound also induced ScWRKY023 and ScWRKY045 in wild-resistant genotype *Solanum commersonii* (Villano et al. 2020). Collinge and Boller investigated NAC transcription factors in potato for the first time in 2001, discovering that a potato NAC gene, *StNAC*, was promptly and substantially activated after injury, but under *P. infestans* infection, its transcript was discovered only after 48 h. In this study, *StNAC005*, *StNAC004* and *StNAC019* were found to be upregulated in the presence of wound (Singh et al. 2013). The regulatory mechanisms of the transcription factor (TF) families are still substantially unknown, particularly when it comes to biotic stress reactions. Exploration with models for the programming of these transcription factors (TFs) in response to abiotic stimuli like drought and hormonal interactions can serve as a starting point for unravelling these complicated forms of interactions, among other factors.

12.6 The Role of Jasmonic Acid, Salicylic Acid and Other Priming Agents on Potato TFs

Phytohormones such as jasmonic acid (JA) and salicylic acid (SA) are involved in the regulation of TF expression and play a critical role in plant defence (Robert-Seilaniantz et al. 2011). JA is a phytohormone, classified as a jasmonate and expressed mostly as JA and methyl jasmonate (MeJA) variants. JA has reported frequently in plant immunity as a signalling molecule in response to necrotrophic pathogens, wounds and herbivore attack (Bishop et al. 2015). JA/MeJA works either alone or in combination with ethylene to activate the transcription of various systemic wound response proteins (SWRPs) and proteinase inhibitor (PI) genes. When it reaches the target tissues, the systemin protein causes an oxidative cascade to be activated, MAP kinase to be stimulated and ions to move across the membrane, all of which enhance the production of jasmonate (JA). Later, JA can be coupled with amino acids, such as isoleucine (JA-Ile), which is essential for JA-mediated signalling because it gets detected and recognised by mediator proteins. COI1 (coronatine-insensitive 1) is a jasmonate receptor that forms complexes with SCFCOI1 (Skp1/Cullin1/F-box protein-COI1). The accumulation of JA leads to the synthesis of transcription factors that are vital in the defence response against plant diseases and pests, such as the over-expression of the *StZFP2* gene, which has been linked to herbivore resistance in this species, according to Lawrence et al. (2014). JA has also been linked for induced expression of *StERF5* and *StERF94*. StZFP2 has shown changes in expression with changes in salicylic and jasmonic acid concentrations in late blight infections in potato leaves (Lawrence et al. 2019); however the regulatory mechanisms are still unknown. Salicylic acid (SA) is a phenolic chemical that is part of a larger family of compounds known as the salicylates, which also include substances such as methyl salicylate, saligenin and the glycosides of these compounds. Phytochemically, it can be produced by plants via two separate routes: phenylpropanoic acid derivation and isochorismic acid derivation. Through the hypersensitive reaction (HR) and systemic acquired resistance (SAR), it is associated with inhibition of ethylene production and defence against biotrophic and hemibiotrophic diseases. Although it serves as a local defence signal, it also triggers a signal amplification loop with reactive oxygen species (ROS), which pushes HR to the infection site. A variety of transcription factors in potatoes have been shown to be activated by exogenous administration of SA, including StZFP1 (Tian et al. 2010), numerous StERFs (*StERF3*, *StERF4*, *StERF5* and *StERF6*) (Bouaziz et al. 2015), StNAC genes (*StNAC4* and *StNAC5*) (Singh et al. 2013) and StbZIP61 (Zhou et al. 2018). In a wide expression analysis for 22 WRKY genes, four genes, viz. *StWRKY1*, *STWRKY34*, *STWRKY39* and *STWRKY72*, also showed transcript induction after treatment with SA (Zhang et al. 2017). Gene expression studies by Villano et al. (2020) reveal induced expression of *StWRKY042*, *StWRKY075*, *StWRKY078* and *StWRKY080* under treatment of defence priming agents BABA and BTH. It has also been postulated that IAA, ABA and ET perform a variety of roles in regulatory cross-talk, especially between ethylene and auxin signalling but also by ABA in some circumstances. Many transcription factors were tested for their transcript induction under these hormones along with their behaviour in the presence of pathogen and pests to understand the mechanism of their regulation in plant cell. A model has been presented in Fig. 12.2 summarizing the role of various hormones on transcript regulation of few transcription factors with reported role in plant-pathogen interaction. For example, *StARF6a* is activated by ABA and ethylene, but *StARF6b* is activated by IAA along with ABA and ethylene. Meanwhile, in the presence of high levels of IAA and ABA, *StARF19a* is downregulated. *StERF94* gene is found to be upregulated in the presence of JA and ethylene application, while *StZFP2* and *StDREB1* get stimulated in the presence of JA only. Treatment of ethylene also induces *StERF47*, *StERF67* and *StERF71*.

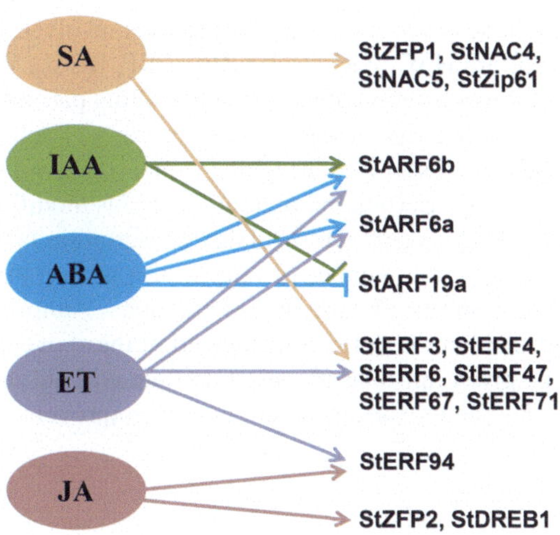

Fig. 12.2 An overview of hormone regulation of few transcription factors reported to induce/involve in the response to biotic stress in potatoes. Positive regulation is represented by arrows, whereas negative regulation is represented by blocked arrows. Majority of the transcription factors get regulated themselves under complex interplay of multiple hormones

12.7 Post-Translational Regulation of Transcription Factors Under Potato Biotic Stress

Certain regulatory mechanisms for the TF discovered in potatoes have been uncovered as a result of the synthesis of these TFs. In the structure of several ARFs, there were possible miRNA-binding sites, indicating that miRNAs are involved in the control of these proteins' function (Zhang et al. 2006). According to the study, MiR167, which was discovered in potato leaves, has been demonstrated to have significant complementarity with ARF6 and ARF8, indicating that it may be able to function as a proteolytic activity for the mRNA of these ARFs. Therefore, it is plausible that some of the transcription factors associated with biotic stress are likewise controlled by miRNAs (Zhang et al. 2018). Phosphorylation with kinases via the N-end rule pathway and ubiquitin-mediated protein degradation are all known to regulate transcription factors (Licausi et al. 2011). StWRKY1 phosphorylation helps in interaction with StMEK1 (MAPK) to create resistance to *P. infestans* (Yogendra et al. 2015). StBEL5 and POTH1 work together to regulate transcription (Chatterjee et al. 2007). To illustrate an example, in potato, BEL1 mRNAs such as StBEL5 (stolon tip BEL1) are synthesized in the leaves and transported to the tips of the stolon either by phloem, where the long-distance transport of mRNA is activated, as well as additional BEL1 mRNAs (Sharma et al. 2014). When StERF3 interacts with certain other proteins, the TF-regulation function of the protein may be impaired, resulting in StERF3 being re-localized between the cytoplasm and the nucleus (Tian et al. 2015). Three of these proteins (StCYN, StKIN1 and StCIP) have a favourable connection with StERF3 in BY2 tobacco cell lines.

12.8 Conclusion

The nine transcription factor families involved in biotic stress responses in potato crops are NAC, ARF, bZIP, ZFP, ERF, BELL, WRKY, BRED and TCP. They control the expression of hormone signalling pathway genes, the formation of cell walls and the activity of antioxidant enzymes. Each TF family responds and interacts, either as messengers, triggers or final transcript products.

Post-transcriptional regulation entails modifying the protein to inactivate or activate it. This approach regulates transcription factor over-expression and activity by regulating the number of copies of a gene transcribed as well as the components present in its structure and interactions. Pathogen pressure and the environment in which plants develop influence all of the above, pushing co-evolutionary processes. The given information based on data will aid in developing biotic stress tolerance variety under various potato genetic improvement programmes.

Acknowledgement The authors are thankful to the Director of CSIR-IHBT for providing a workplace environment. ND is grateful to the University Grants Commission of India for the financial support in the form of a fellowship. This manuscript's IHBT communication number is 10091.

References

Adolf B, Andrade-Piedra J, Bittara Molina F, Przetakiewicz J, Hausladen H, Kromann P, Lees A, Lindqvist-Kreuze H, Perez W, Secor GA (2020) Fungal, oomycete, and plasmodiophorid diseases of potato. In: The potato crop. Springer, Cham, pp 307–350

Bao S, Zhang Z, Lian Q, Sun Q, Zhang R (2019) Evolution and expression of genes encoding TCP transcription factors in *Solanum tuberosum* reveal the involvement of StTCP23 in plant defence. BMC Genet 20(1):1–15

Bishop G, Sakakibara H, Seo M, Yamaguchi S (2015) Biosynthesis of hormones. In: Biochemistry & molecular biology of plants. Wiley, Chichester, pp 769–833

Bouaziz D, Charfeddine M, Jbir R, Saidi MN, Pirrello J, Charfeddine S, Bouzayen M, Gargouri-Bouzid R (2015) Identification and functional characterization of ten AP2/ERF genes in potato. Plant Cell Tissue Organ Cult 123(1):155–172

Bürglin TR (1997) Analysis of TALE superclass homeobox genes (MEIS, PBC, KNOX, Iroquois, TGIF) reveals a novel domain conserved between plants and animals. Nucleic Acids Res 25(21):4173–4180

Chacón-Cerdas R, Barboza-Barquero L, Albertazzi FJ, Rivera-Mendez W (2020) Transcription factors controlling biotic stress response in potato plants. Physiol Mol Plant Pathol 112:101527

Charfeddine M, Saïdi MN, Charfeddine S, Hammami A, Bouzid RG (2015) Genome-wide analysis and expression profiling of the ERF transcription factor family in potato (*Solanum tuberosum* L.). Mol Biotechnol 57(4):348–358

Charfeddine M, Samet M, Charfeddine S, Bouaziz D, Bouzid RG (2019) Ectopic expression of StERF94 transcription factor in potato plants improved resistance to *fusarium solani* infection. Plant Mol Biol Report 37(5):450–463

Chatterjee M, Banerjee AK, Hannapel DJ (2007) A BELL1-like gene of potato is light activated and wound inducible. Plant Physiol 145(4):1435–1443

Combest MM, Moroz N, Tanaka K, Rogan CJ, Anderson JC, Thura L, Rakotondrafara AM, Goyer A (2021) St PIP1, a PAMP-induced peptide in potato, elicits plant defenses and is associated with disease symptom severity in a compatible interaction with Potato Virus Y. J Exp Bot 72(12):4472–4488

Dellagi A, Heilbronn J, Avrova AO, Montesano M, Palva ET, Stewart HE, Toth IK, Cooke DE, Lyon GD, Birch PR (2000) A potato gene encoding a WRKY-like transcription factor is induced in interactions with *Erwinia carotovora* subsp. atroseptica and *Phytophthora infestans* and is coregulated with class I endochitinase expression. Mol Plant-Microbe Interact 13(10):1092–1101

Desplan C, Theis J, O'Farrell PH (1988) The sequence specificity of homeodomain-DNA interaction. Cell 54(7):1081–1090

Devaux A, Jean-Pierre G, Athanasios P, Peter K, Marcel G, Julius O, Victor S, Guy H (2020) Global food security, contributions from sustainable potato agrifood systems. In: The potato crop. Springer International Publishing, Cham, pp 3–35

Dubey N, Singh K (2018) Role of NBS-LRR proteins in plant defense. In: Molecular aspects of plant-pathogen interaction. Springer, Singapore, pp 115–138

Görlach A, Klappa P, Kietzmann DT (2006) The endoplasmic reticulum: folding, calcium homeostasis, signaling, and redox control. Antioxid Redox Signal 8:1391–1418

Goss EM, Tabima JF, Cooke DE, Restrepo S, Fry WE, Forbes GA, Fieland VJ, Cardenas M, Grünwald NJ (2014) The Irish potato famine pathogen *Phytophthora infestans* originated in central Mexico rather than the Andes. Proc Natl Acad Sci 111(24):8791–8796

He M, Zhou Y, Ye G, Zheng J, Meng Y, Wang J, Shan W (2021) Serial transcriptome analysis reveals genes associated with late blight resistance in potato cultivar Qingshu 9. Agronomy 11(10):1919

Hong JC (2016) General aspects of plant transcription factor families. In: Plant transcription factors. Elsevier Inc., pp 35–56

Hussain HA, Hussain S, Khaliq A, Ashraf U, Anjum SA, Men S, Wang L (2018) Chilling and drought stresses in crop plants: implications, cross talk, and potential management opportunities. Front Plant Sci 9:393

Kerstetter R, Vollbrecht E, Lowe B, Veit B, Yamaguchi J, Hake S (1994) Sequence analysis and expression patterns divide the maize knotted1-like homeobox genes into two classes. Plant Cell 6(12):1877–1887

Kroschel J, Mujica N, Okonya J, Alyokhin A (2020) Insect pests affecting potatoes in tropical, subtropical, and temperate regions. In: The potato crop. Springer International Publishing, Cham, pp 251–306

Lawrence SD, Novak NG (2018) The remarkable plethora of infestation-responsive Q-type C2H2 transcription factors in potato. BMC Res Notes 11(1):1–7

Lawrence SD, Novak NG, Jones RW, Farrar RR Jr, Blackburn MB (2014) Herbivory responsive C2H2 zinc finger transcription factor protein StZFP2 from potato. Plant Physiol Biochem 80:226–233

Lawrence SD, Novak NG, Perez FG, Jones RW (2019) Over expression of the Q-type ZFP StZFP2 in potato increases resistance to potato late blight (*Phytophthora infestans*) infection. J Plant Interact 14(1):129–136

Levy JG, Mendoza A, Miller JC, Tamborindeguy C, Pierson EA (2017) Global gene expression in two potato cultivars in response to *Candidatus Liberibacter* solanacearuminfection. BMC Genomics 18(1):1–6

Licausi F, Kosmacz M, Weits DA, Giuntoli B, Giorgi FM, Voesenek LA, Perata P, Van Dongen JT (2011) Oxygen sensing in plants is mediated by an N-end rule pathway for protein destabilization. Nature 479(7373):419–422

Machinandiarena MF, Lobato MC, Feldman ML, Daleo GR, Andreu AB (2012) Potassium phosphite primes defense responses in potato against *Phytophthora infestans*. J Plant Physiol 169(14):1417–1424

Massa AN, Childs KL, Lin H, Bryan GJ, Giuliano G, Buell CR (2011) The transcriptome of the reference potato genome *Solanum tuberosum* Group Phureja clone DM1-3 516R44. Plos One 6(10):e26801

Mukherjee K, Brocchieri L, Bürglin TR (2009) A comprehensive classification and evolutionary analysis of plant homeobox genes. Mol Biol Evol 26(12):2775–2794

Robert-Seilaniantz A, Grant M, Jones JD (2011) Hormone crosstalk in plant disease and defense: more than just jasmonate-salicylate antagonism. Annu Rev Phytopathol 49:317–343

Rushton PJ, Somssich IE, Ringler P, Shen QJ (2010) WRKY transcription factors. Trends Plant Sci 15:247–258

Schluttenhofer C, Yuan L (2015) Regulation of specialized metabolism by WRKY transcription factors. Plant Physiol 167(2):295–306

Shahzad R, Harlina PW, Cong-hua X, Ewas M, Nishawy E, Zhenyuan P, Foly MM (2016) Overexpression of potato transcription factor (StWRKY1) conferred resistance to *Phytophthora infestans* and improved tolerance to water stress. Plant Omics 9(2):149–158

Sharma P, Lin T, Grandellis C, Yu M, Hannapel DJ (2014) The BEL1-like family of transcription factors in potato. J Exp Bot 65(2):709–723

Shinozaki K, Uemura M, Bailey-Serres J, Bray E, Weretilnyk E (2015) Responses to abiotic stress. In: Buchanan BB, Gruissem W, Jones RL (eds) Biochemistry and molecular biology of plants. Wiley Blackwell, UK, John Wiley & Sons, Chichester

Singh AK, Sharma V, Pal AK, Acharya V, Ahuja PS (2013) Genome-wide organization and expression profiling of the NAC transcription factor family in potato (*Solanum tuberosum* L.). DNA Res 20(4):403–404

Song S, Liaoyang H, Pan Z, Ya X, Naiqin Z, Hongji Z, Ning L (2019) Genome-wide identification, expression profiling and evolutionary analysis of Auxin Response Factor gene family in potato (*Solanum tuberosum* Group Phureja). Sci Rep 9:1–13

Tian ZD, Zhang Y, Liu J, Xie CH (2010) Novel potato C2H2-type zinc finger protein gene, StZFP1, which responds to biotic and abiotic stress, plays a role in salt tolerance. Plant Biol 12(5):689–697

Tian Z, He Q, Wang H, Liu Y, Zhang Y, Shao F, Xie C (2015) The potato ERF transcription factor StERF3 negatively regulates resistance to *Phytophthora infestans* and salt tolerance in potato. Plant Cell Physiol 56(5):992–1005

Villano C, Esposito S, D'Amelia V, Garramone R, Alioto D, Zoina A, Aversano R, Carputo D (2020) WRKY genes family study reveals tissue-specific and stress-responsive TFs in wild potato species. Sci Rep 10(1):1–2

Yang X, Guo X, Yang Y, Ye P, Xiong X, Liu J, Dong D, Li G (2018) Gene profiling in late blight resistance in potato genotype SD20. Int J Mol Sci 19(6):1728

Yogendra KN, Kumar A, Sarkar K, Li Y, Pushpa D, Mosa KA, Duggavathi R, Kushalappa AC (2015) Transcription factor StWRKY1 regulates phenylpropanoid metabolites conferring late blight resistance in potato. J Exp Bot 66(22):7377–7389

Yogendra KN, Dhokane D, Kushalappa AC, Sarmiento F, Rodriguez E, Mosquera T (2017) StWRKY8 transcription factor regulates benzylisoquinoline alkaloid pathway in potato conferring resistance to late blight. Plant Sci 256:208–216

Zhang B, Pan X, Cannon CH, Cobb GP, Anderson TA (2006) Conservation and divergence of plant microRNA genes. Plant J 46(2):243–259

Zhang C, Wang D, Yang C, Kong N, Shi Z, Zhao P, Nan Y, Nie T, Wang R, Ma H, Chen Q (2017) Genome-wide identification of the potato WRKY transcription factor family. PLoS One 12(7):e0181573

Zhang L, Chen M, Shao X (2018) Inhibition of cloud cavitation on a flat hydrofoil through the placement of an obstacle. Ocean Eng 155:1–9

Zhang J, Cook J, Nearing JT, Zhang J, Raudonis R, Glick BR, Langille MG, Cheng Z (2021) Harnessing the plant microbiome to promote the growth of agricultural crops. Microbiol Res 245:126690

Zheng Y, Chen J, Honghe S, Hernán R, Marina AP, Peifen Z, Michael B, Dai X, Martin GB, Giovannoni JJ, Zhao PX (2016) iTAK: a program for genome-wide prediction and classification of plant transcription factors, transcriptional regulators, and protein kinases. Mol Plant 9(12):1667–1670

Zhou XT, Jia LJ, Wang HY, Zhao P, Wang WY, Liu N, Song SW, Wu Y, Su L, Zhang J, Zhong NQ (2018) The potato transcription factor StbZIP61 regulates dynamic biosynthesis of salicylic acid in defense against *Phytophthora infestans* infection. Plant J 95(6):1055–1068

The manufacturer's authorised representative in the EU is Springer Nature Customer Service Centre GmbH, Europaplatz 3, 69115 Heidelberg, Germany. If you have any concerns regarding our products, please contact ProductSafety@springernature.com

Printed and bound by CPI Group (UK) Ltd, Croydon, CR0 4YY
30/09/2026
02225493-0003